CLIMATE STABILIZATION TARGETS

Emissions, Concentrations, and Impacts over Decades to Millennia

Committee on Stabilization Targets for Atmospheric Greenhouse Gas Concentrations

Board on Atmospheric Sciences and Climate

Division on Earth and Life Studies

NATIONAL RESEARCH COUNCIL
OF THE NATIONAL ACADEMIES

THE NATIONAL ACADEMIES PRESS
Washington, D.C.
www.nap.edu

THE NATIONAL ACADEMIES PRESS 500 Fifth Street, N.W. Washington, DC 20001

NOTICE: The project that is the subject of this report was approved by the Governing Board of the National Research Council, whose members are drawn from the councils of the National Academy of Sciences, the National Academy of Engineering, and the Institute of Medicine. The members of the committee responsible for the report were chosen for their special competences and with regard for appropriate balance.

This study was supported by the The Energy Foundation under contract number G-0812-10616 and The Environmental Protection Agency under contract number EP10H001368. Any opinions, findings, and conclusions, or recommendations expressed in this material are those of the author(s) and do not necessarily reflect the views of the sponsors.

International Standard Book Number-13: 978-0-309-15176-4 (Book)
International Standard Book Number-10: 0-309-15176-7 (Book)
International Standard Book Number-13: 978-0-309-15177-1 (PDF)
International Standard Book Number-10: 0-309-15177-5 (PDF)
Library of Congress Control Number: 2010 936129

Additional copies of this report are available from the National Academies Press, 500 Fifth Street, N.W., Lockbox 285, Washington, DC 20055; (800) 624-6242 or (202) 334-3313 (in the Washington metropolitan area); Internet, *http://www.nap.edu.*

National Research Council. 2011. Climate Stabilization Targets: Emissions, Concentrations, and Impacts over Decades to Millennia. Washington, DC: National Academies Press.

Printed in the United States of America

THE NATIONAL ACADEMIES

Advisers to the Nation on Science, Engineering, and Medicine

The **National Academy of Sciences** is a private, nonprofit, self-perpetuating society of distinguished scholars engaged in scientific and engineering research, dedicated to the furtherance of science and technology and to their use for the general welfare. Upon the authority of the charter granted to it by the Congress in 1863, the Academy has a mandate that requires it to advise the federal government on scientific and technical matters. Dr. Ralph J. Cicerone is president of the National Academy of Sciences.

The **National Academy of Engineering** was established in 1964, under the charter of the National Academy of Sciences, as a parallel organization of outstanding engineers. It is autonomous in its administration and in the selection of its members, sharing with the National Academy of Sciences the responsibility for advising the federal government. The National Academy of Engineering also sponsors engineering programs aimed at meeting national needs, encourages education and research, and recognizes the superior achievements of engineers. Dr. Charles M. Vest is president of the National Academy of Engineering.

The **Institute of Medicine** was established in 1970 by the National Academy of Sciences to secure the services of eminent members of appropriate professions in the examination of policy matters pertaining to the health of the public. The Institute acts under the responsibility given to the National Academy of Sciences by its congressional charter to be an adviser to the federal government and, upon its own initiative, to identify issues of medical care, research, and education. Dr. Harvey V. Fineberg is president of the Institute of Medicine.

The **National Research Council** was organized by the National Academy of Sciences in 1916 to associate the broad community of science and technology with the Academy's purposes of furthering knowledge and advising the federal government. Functioning in accordance with general policies determined by the Academy, the Council has become the principal operating agency of both the National Academy of Sciences and the National Academy of Engineering in providing services to the government, the public, and the scientific and engineering communities. The Council is administered jointly by both Academies and the Institute of Medicine. Dr. Ralph J. Cicerone and Dr. Charles M. Vest are chair and vice chair, respectively, of the National Research Council.

www.national-academies.org

COMMITTEE ON STABILIZATION TARGETS FOR ATMOSPHERIC GREENHOUSE GAS CONCENTRATIONS

SUSAN SOLOMON (*Chair*), National Oceanic and Atmospheric Administration, Boulder, CO
DAVID BATTISTI, University of Washington, Seattle, WA
SCOTT DONEY, Woods Hole Oceanographic Institution, Woods Hole, MA
KATHARINE HAYHOE, Texas Tech University, Lubbock, TX
ISAAC M. HELD, Geophysical Fluid Dynamics Laboratory, Princeton, NJ
DENNIS P. LETTENMAIER, University of Washington, Seattle, WA
DAVID LOBELL, Stanford University, Stanford, CA
H. DAMON MATTHEWS, Concordia University, Montreal, Quebec
RAYMOND PIERREHUMBERT, University of Chicago, Chicago, IL
MARILYN RAPHAEL, University of California, Los Angeles, CA
RICHARD RICHELS, Electric Power Research Institute, Inc., Washington, DC
TERRY L. ROOT, Stanford University, Stanford, CA
KONRAD STEFFEN, University of Colorado, Boulder, CO
CLAUDIA TEBALDI, Climate Central, Vancouver, British Columbia
GARY W. YOHE, Wesleyan University, Middletown, CT

NRC Staff

TOBY WARDEN, Study Director
LAUREN BROWN, Research Associate
EDWARD DUNLEA, Senior Program Officer
DAVID REIDMILLER, Christine Mirzayan Science and Technology Fellow
SHELLY FREELAND, Senior Program Assistant
RICARDO PAYNE, Senior Program Assistant
DANIEL BEARRS, Senior Librarian

AMANDA PURCELL, Senior Program Assistant
JANEISE STURDIVANT, Program Assistant
RICARDO PAYNE, Senior Program Assistant
SHUBHA BANSKOTA, Financial Associate
DAVID REIDMILLER, Christine Mirzayan Science and Technology Fellow

Acknowledgments

This report has been reviewed in draft form by individuals chosen for their diverse perspectives and technical expertise, in accordance with procedures approved by the NRC's Report Review Committee. The purpose of this independent review is to provide candid and critical comments that will assist the institution in making its published report as sound as possible and to ensure that the report meets institutional standards for objectivity, evidence, and responsiveness to the study charge. The review comments and draft manuscript remain confidential to protect the integrity of the deliberative process. We wish to thank the following individuals for their participation in their review of this report:

Marcia Baker, University of Washington
Virginia Burkett, U.S. Geological Survey
William Easterling, Pennsylvania State University
Jay Gulledge, Pew Center on Global Climate Change
Prasad Kasibhatla, Duke University
Haroon Khesghi, ExxonMobil Research and Engineering Company
Jeffrey T. Kiehl, National Center for Atmospheric Research/University Corporation for Atmospheric Research
Corinne LeQuere, University of East Anglia
Gerald R. North, Texas A&M University
Matthias Ruth, University of Maryland

Although the reviewers listed above have provided many constructive comments and suggestions, they were not asked to endorse the conclusions or recommendations nor did they see the final draft of the report before its release. The review of this report was overseen by Robert E. Dickinson, The University of Texas at Austin, appointed by the Division on Earth and Life Studies, and George M. Hornberger, Vanderbilt University Institute for

Energy and Environment, appointed by the Report Review Committee, who were responsible for making certain that an independent examination of this report was carried out in accordance with institutional procedures and that all review comments were carefully considered. Responsibility for the final content of this report rests entirely with the authoring committee and the institution.

The committee would like to thank the following individuals who offered direct input to the committee with meeting presentations and personal, phone, or email discussions, including: Todd J. Sanford, National Oceanic and Atmospheric Administration and CIRES, Kirsten Zickfield, Canadian Centre for Climate Modelling and Analysis, Michael Eby, University of Victoria, Jonathan Patz, University of Wisconsin, Dan Cayan, Scripps Institution of Oceanography, Joseph Goffman, Senate Committee on Environment and Public Works, Dan Reifsnyder, U.S. Department of State, Reto Knutti, Institute for Atmospheric and Climate Science, Jeffrey Kiehl, NCAR, Leon Clark, PNNL, Eric Steig, University of Washington, Nigel Arnell, University of Reading, Phil Mote, Oregon Climate Change Research Institute and Oregon Climate Services, Samuel Myers, Harvard University, and Andrew Weaver, Concordia University.

Table of Contents

Section I

Synopsis

Emissions of carbon dioxide from the burning of fossil fuels have ushered in a new epoch where human activities will largely determine the evolution of Earth's climate. Because carbon dioxide in the atmosphere is long lived, it can effectively lock Earth and future generations into a range of impacts, some of which could become very severe. Therefore, emissions reductions choices made today matter in determining impacts experienced not just over the next few decades, but in the coming centuries and millennia. Policy choices can be informed by recent advances in climate science that quantify the relationships between increases in carbon dioxide and global warming, related climate changes, and resulting impacts, such as changes in streamflow, wildfires, crop productivity, extreme hot summers, and sea level rise.

Since the beginning of the industrial revolution, concentrations of greenhouse gases from human activities have risen substantially. Evidence now shows that the increases in these gases very likely (>90 percent chance) account for most of Earth's warming over the past 50 years. Carbon dioxide is the greenhouse gas produced in the largest quantities, accounting for more than half of the current impact on Earth's climate. Its atmospheric concentration has risen about 35 percent since 1750 and is now at about 390 ppmv, the highest level in at least 800,000 years. Depending on emissions rates, carbon dioxide concentrations could double or nearly triple from today's level by the end of the century, greatly amplifying future human impacts on climate.

Society is beginning to make important choices regarding future greenhouse gas emissions. One way to inform these choices is to consider the projected climate changes and impacts that would occur if greenhouse gases in the atmosphere were stabilized at a particular concentration level. The information needed to understand such targets is multifaceted: how do emissions affect global atmospheric concentrations and in turn global warming and its impacts?

This report quantifies, insofar as possible, the outcomes of different stabilization targets for greenhouse gas concentrations using analyses and information drawn from the scientific literature. It does not recommend or justify any particular stabilization target. It does provide important scientific insights about the relationships among emissions, greenhouse gas concentrations, temperatures, and impacts.

CLIMATE CHANGE DUE TO CARBON DIOXIDE WILL PERSIST MANY CENTURIES

Carbon dioxide flows into and out of the ocean and biosphere in the natural breathing of the planet, but the uptake of added human emissions depends on the net change between flows, occurring over decades to millennia. This means that climate changes caused by carbon dioxide are expected to persist for many centuries even if emissions were to be halted at any point in time.

Such extreme persistence is unique to carbon dioxide among major agents that warm the planet. Choices regarding emissions of other warming agents, such as methane, black carbon on ice/snow, and aerosols, can affect global warming over coming decades but have little effect on longer-term warming of Earth over centuries and millennia. Thus, long-term effects are primarily controlled by carbon dioxide (see Figure Syn.1).

The report concludes that the world is entering a new geologic epoch, sometimes called the Anthropocene, in which human activities will largely control the evolution of Earth's environment. Carbon emissions during this century will essentially determine the magnitude of eventual impacts and whether the Anthropocene is a short-term, relatively minor change from the current climate or an extreme deviation that lasts thousands of years. The higher the total, or cumulative, carbon dioxide emitted and the resulting atmospheric concentration, the higher the peak warming that will be experienced and the longer the duration of that warming. Duration is critical; longer warming periods allow more time for key, but slow, components of the Earth system to act as amplifiers of impacts, for example, warming of the deep ocean that releases carbon stored in deep-sea sediments. Warming sustained over thousands of years could lead to even bigger impacts (see Box Syn.1).

IMPACTS CAN BE LINKED TO GLOBAL MEAN TEMPERATURES

To date, climate stabilization goals have been most often discussed in terms of stabilizing *atmospheric concentrations* of carbon dioxide (e.g., 350

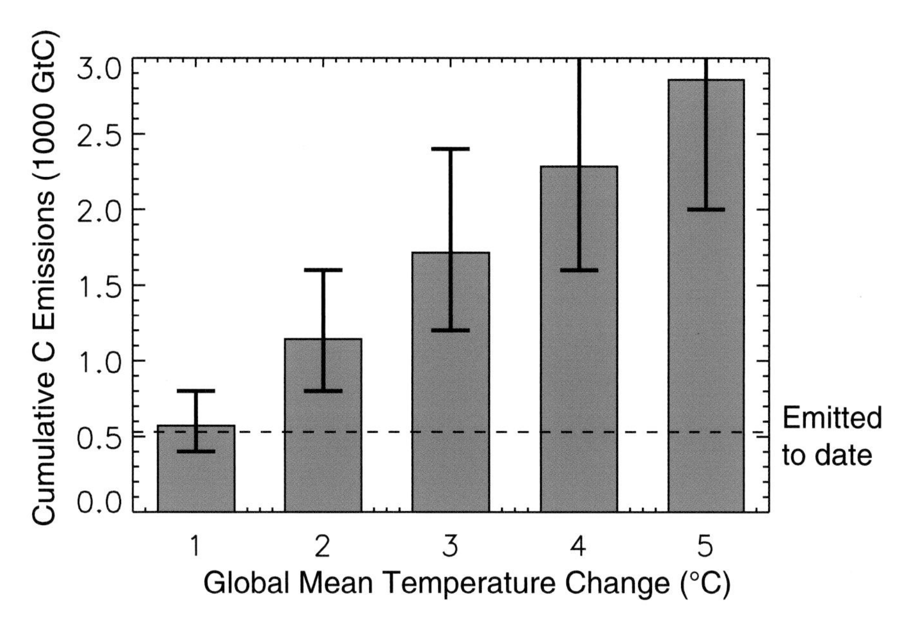

FIGURE Syn.1 Recent studies show that cumulative carbon dioxide emission is a useful metric for linking emissions to impacts. Error bars reflect uncertainty in carbon cycle and climate responses to carbon dioxide emissions due to observational constraints and the range of model results. Cumulative carbon emissions are in teratonnes of carbon (trillion metric tonnes or 1,000 gigatonnes).

ppmv, 450 ppmv, etc.). This report concludes that, for a variety of conceptual and practical reasons, it is more effective to assess climate stabilization goals by using global mean temperature change as the primary metric. Global temperature change can in turn be linked both to concentrations of atmospheric carbon dioxide (Table 1) and to accumulated carbon emissions.

An important reason for using warming as a reference is that scientific research suggests that many key impacts can be quantified for given tem-

BOX SYN-1 SUSTAINED WARMING COULD LEAD TO SEVERE IMPACTS

Widespread coastal flooding would be expected if warming of several degrees is sustained for millennia. Model studies suggest that a cumulative carbon emission of about 1,000 to 3,000 gigatonnes (billion metric tonnes carbon) implies warming levels above about 2°C sustained for millennia. This could lead to eventual sea level rise on the order of 1 to 4 m due to thermal expansion of the oceans and to glacier and small ice cap loss alone. Melting of the Greenland ice sheet could contribute an additional 4 to 7.5 m over many thousands of years.

TABLE 1 Relationship of Atmospheric Concentrations of Carbon Dioxide to Temperature

Stabilization CO_2-Equivalent Concentration (ppmv): Range and Best Estimate	Equilibrium Global Average Warming (°C)
320 ← **340** → 380	1
370 ← **430** → 540	2
440 ← **540** → 760	3
530 ← **670** → 1060	4
620 ← **840** → 1490	5

Note: **Green** and **red** numbers represent low and high ends of ranges, respectively; **black bolded** numbers represent best estimates.

The report calculates the "likely" range (66% chance) of atmospheric concentrations associated with various degrees of warming, consistent with model results[1] and roughly consistent with paleoclimate evidence. There are large uncertainties in **'climate sensitivity'**—the amount of warming expected from different atmospheric concentrations of greenhouse gas—the range is 30% below and 40% above the best estimates.

[1]The estimated "likely" range presented in this report corresponds to the range of model results in the Climate Modelling Intercomparison Project (CMIP3) global climate model archive.

perature increases. This is done by scaling local to global warming and by "coupled linkages" that show how other climate changes, such as alterations in the water cycle, scale with temperature.

There is now increased confidence in how global warming levels of 1°C, 2°C, 3°C etc. (for °F conversion, see Figure Syn.2) would relate to certain future impacts. This report lists some of these effects ***per degree (°C) of global warming*** (see Figure Syn.3), including:

- 5-10 percent changes in precipitation in a number of regions
- 3-10 percent increases in heavy rainfall
- 5-15 percent yield reductions of a number of crops
- 5-10 percent changes in streamflow in many river basins worldwide

- About 15 percent and 25 percent decreases in the extent of annually averaged and September Arctic sea ice, respectively

For warming of 2 to 3°C, summers that are among the warmest recorded or the warmest experienced in people's lifetimes, would become frequent. For warming levels of 1 to 2°C, the area burned by wildfire in parts of western North America is expected to increase by 2 to 4 times for each degree (°C) of global warming.

Many other important impacts of climate change are difficult to quantify for a given change in global average temperature, in part because temperature is not the only driver of change for some impacts; multiple environmental and other human factors come into play. It is clear from scientific studies, however, that a number of projected impacts scale approximately with temperature. Examples include shifts in the range and abundance of some terrestrial and marine species, increased risk of heat-related human health impacts, and loss of infrastructure in the coastal regions and the Arctic.

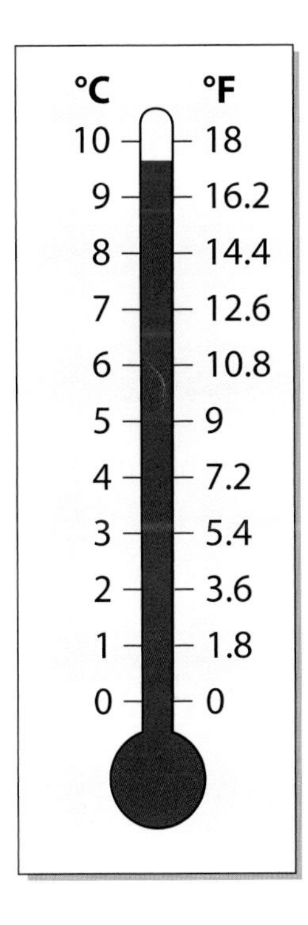

FIGURE Syn.2 Illustration of how temperature change in degrees Celsius (left side of thermometer) relates to temperature change in degrees Fahrenheit (right side of thermometer). For example, a warming of 5°C is equal to a warming of 9°C. In this report estimates of temperature change are in made in degrees Celsius in accordance with international scientific practice.

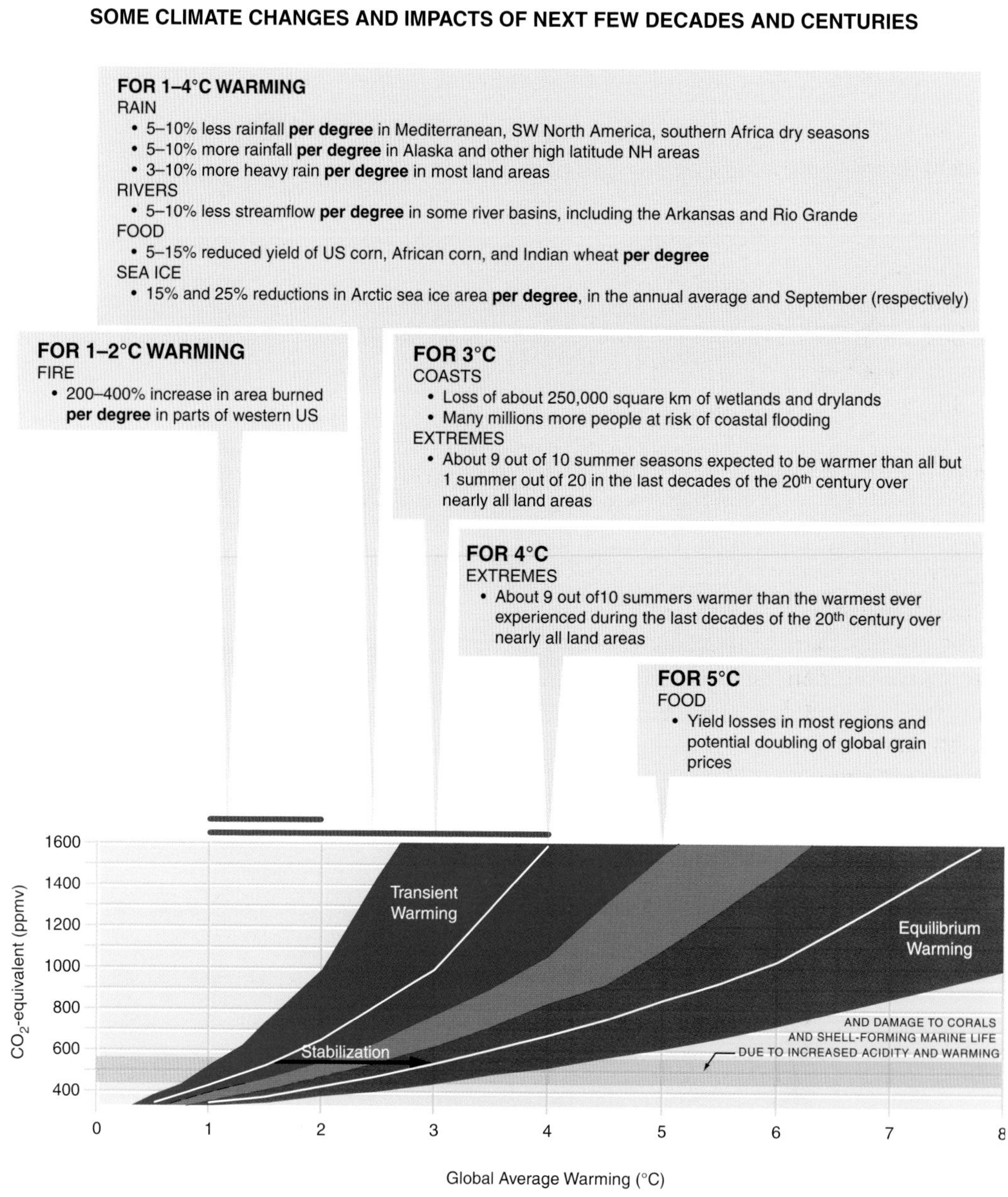

SOME CLIMATE CHANGES AND IMPACTS OF NEXT FEW DECADES AND CENTURIES

FOR 1–4°C WARMING

RAIN
- 5–10% less rainfall **per degree** in Mediterranean, SW North America, southern Africa dry seasons
- 5–10% more rainfall **per degree** in Alaska and other high latitude NH areas
- 3–10% more heavy rain **per degree** in most land areas

RIVERS
- 5–10% less streamflow **per degree** in some river basins, including the Arkansas and Rio Grande

FOOD
- 5–15% reduced yield of US corn, African corn, and Indian wheat **per degree**

SEA ICE
- 15% and 25% reductions in Arctic sea ice area **per degree**, in the annual average and September (respectively)

FOR 1–2°C WARMING

FIRE
- 200–400% increase in area burned **per degree** in parts of western US

FOR 3°C

COASTS
- Loss of about 250,000 square km of wetlands and drylands
- Many millions more people at risk of coastal flooding

EXTREMES
- About 9 out of 10 summer seasons expected to be warmer than all but 1 summer out of 20 in the last decades of the 20th century over nearly all land areas

FOR 4°C

EXTREMES
- About 9 out of10 summers warmer than the warmest ever experienced during the last decades of the 20th century over nearly all land areas

FOR 5°C

FOOD
- Yield losses in most regions and potential doubling of global grain prices

1600
1400
1200
1000
800
600
400
CO_2-equivalent (ppmv)
Transient Warming
Equilibrium Warming
Stabilization
AND DAMAGE TO CORALS AND SHELL-FORMING MARINE LIFE DUE TO INCREASED ACIDITY AND WARMING
0 1 2 3 4 5 6 7 8
Global Average Warming (°C)

FIGURE Syn.3 What impacts can be expected? The report quantifies—per degree of warming—several anticipated effects and impacts of global warming, including changes in streamflow, wildfires, crop productivity, extreme hot summers, and sea level rise. The graphical part of the diagram shows how atmospheric concentrations of carbon dioxide correspond to temperatures—transient, or near-term warming (in blue), is only a fraction of the total warming (the equilibrium warming) expected to occur (in red).

STABILIZATION REQUIRES DEEP EMISSIONS REDUCTIONS

The report demonstrates that stabilizing atmospheric carbon dioxide concentrations will require deep reductions in the amount of carbon dioxide emitted. Because human carbon dioxide emissions exceed removal rates through natural carbon "sinks," keeping emission rates the same will not lead to stabilization of carbon dioxide. Emissions reductions larger than about 80 percent, relative to whatever peak global emissions rate may be reached, are required to approximately stabilize carbon dioxide concentrations for a century or so at any chosen target level (see Figure Syn.3).

But stabilizing atmospheric concentrations does not mean that temperatures will stabilize immediately. Because of time lags inherent in the Earth's climate, warming that occurs in response to a given increase in the concentration of carbon dioxide ("transient climate change") reflects only about half the eventual total warming ("equilibrium climate change") that would occur for stabilization at the same concentration (see Figure Syn.4). For example, if concentrations reached 550 ppmv, transient warming would be about 1.6°C, but holding concentrations at 550 ppmv would mean that warming would continue over the next several centuries, reaching a best estimate of an equilibrium warming of about 3°C.

Estimates of warming are based on models that incorporate "climate sensitivities"—the amount of warming expected at different atmospheric concentrations of carbon dioxide (Table 1). Because there are many factors that shape climate, uncertainty in the climate sensitivity is large; the possibility of greater warming, implying additional risk, cannot be ruled out, and smaller warmings are also possible. In the example given above, choosing a concentration target of 550 ppmv could produce a likely global warming at equilibrium as low as 2.1°C, but warming could be as high as 4.3°C, increasing the severity of impacts. Thus, choices about stabilization targets will depend upon value judgments regarding the degree of acceptable risk.

CONCLUSION

This report provides a scientific evaluation of the implications of various climate stabilization targets. The report concludes that certain levels of warming associated with carbon dioxide emissions could lock Earth and many future generations of humans into very large impacts; similarly, some targets could avoid such changes. It makes clear the importance of 21st century choices regarding long-term climate stabilization.

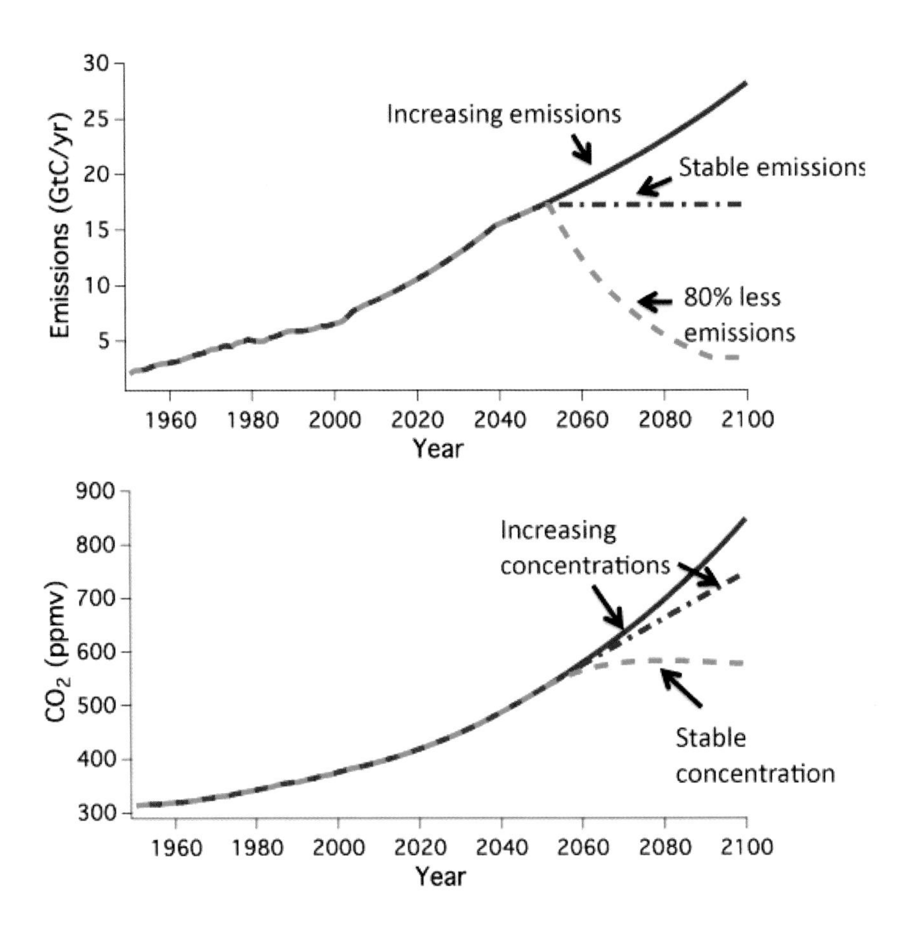

FIGURE Syn.4 Because emissions of carbon dioxide are greater than the sinks that remove it, emissions reductions larger than about 80% (green line-top graph) are required if concentrations are to be stabilized (green line-bottom graph). The lower graph shows how carbon dioxide concentrations would be expected to evolve depending upon emissions for one illustrative case, but this applies for any chosen target.

Summary

Society faces important choices in the coming century regarding future greenhouse gas emissions and the resulting effects on Earth's climate, ecosystems, and people. Atmospheric concentrations of several important greenhouse gases have increased markedly since the start of the 20th century because of human activities, and the increased concentrations of these gases *very likely*[1] account for most of the globally averaged warming of the past 50 years. Carbon dioxide is responsible for more than half of the current impact of human emissions of greenhouse gases on Earth's climate, or radiative forcing,[2] and its influence is projected to grow. Its atmospheric concentration has increased by more than 35 percent since 1750, and is now higher than at any time in at least 800,000 years. Looking to the future, the concentration of carbon dioxide could undergo a further doubling or tripling by the end of the century, greatly amplifying the human impact on climate.

Because of the long atmospheric lifetime of carbon dioxide and the time lags in the climate system (particularly slow processes in the ocean, see Section 3.2), human choices in the near-term have long-term ramifications on Earth's climate not only for the rest of the century but also for the next sev-

---

[1]In this report, uncertainty ranges indicated as *likely* correspond to >66% probability (2 out of 3 chance), while *very likely* is used for >90% (9 out of 10 chance). Assessed uncertainty intervals are not always symmetric about the corresponding best estimate, and include statistical information and expert judgment.

[2]**Radiative forcing (RF)** refers to the radiative flux change evaluated at the tropopause (which has been adjusted for stratospheric changes, see Ramaswamy et al., 2007). Greenhouse gases such as carbon dioxide, methane, and nitrous oxide exert a warming influence on climate, and differ in their radiative forcing of the global climate system due mainly to their different radiative properties and abundances in the atmosphere. Some greenhouse gas changes (e.g., stratospheric ozone depletion) and aerosols produce negative radiative forcing. The net RF is the sum of positive and negative terms, and each term is defined as the change relative to 1750. These warming influences may be expressed as **CO_2-equivalent concentrations**, corresponding to the concentration of CO_2 that would cause the same amount of radiative forcing as a given mixture of CO_2 and other forcing components.

eral millennia. Indeed, some effects of 21st century human choices would contribute to climate change for more than 100,000 years. {2.1, 3.2}[3]

One way of informing these choices is to consider the projected climate changes and impacts that would occur if greenhouse gases increase to particular concentration levels and then stabilize, as highlighted in the Statement of Task (see Appendix A). Alternative futures then can be represented by a broad range of atmospheric concentration "target" levels (hereafter referred to as stabilization targets). The committee was charged to evaluate different stabilization targets with particular emphasis on the avoidance of serious or irreversible impacts on Earth's climate system. This report does not evaluate the plausibility of any stabilization target, nor does it make any recommendations regarding desirable or "safe" targets.

It should be emphasized that choosing among different targets is a policy issue rather than strictly a scientific one, because such choices involve questions of values, e.g., regarding how much risk to people or to nature might be considered too much. Some climate changes could be beneficial for some people or regions, while being damaging to others.

The primary challenge for this study is to quantify, insofar as possible, the outcomes of different stabilization targets using analyses and information drawn from the scientific literature. Expected changes based on broad scientific understanding are discussed, as well as projected values based upon models. Where there is sufficient understanding to be quantitative, numerical values for projected climate change and impacts are provided as a function of stabilization target. A number of important aspects of climate change that are currently understood in a qualitative manner, or for which the time horizon of the response is poorly constrained, are also reviewed. The report represents a brief summary of a vast scientific literature and seeks to be illustrative and representative rather than comprehensive. Special emphasis is placed on climate changes and impacts in North America and the United States.

The report focuses on human forcing of the climate system from carbon dioxide emissions and rising atmospheric concentrations because of the dominant role and unique influences of carbon dioxide on long-term climate change. The role of other anthropogenic greenhouse gases, such as methane, nitrous oxide and halocarbon, and aerosols are also briefly discussed. For many purposes, the total radiative forcing of the suite of anthropogenic greenhouse gases and aerosols can be cast in terms of an

[3]Throughout this summary and the technical overview presented in the next section, numbers in curly brackets refer to sections of the main report where details and references are to be found.

equivalent level of atmospheric carbon dioxide, also known as the CO_2-equivalent concentration.

APPROACH

The goal and implications of stabilizing climate change are most often discussed in terms of stabilizing atmospheric concentrations of CO_2. This report takes a different approach by (1) using global temperature change as the frame of reference and (2) focusing in part on the relationship between accumulated carbon emissions and global mean temperature change.

The motivation for this approach is both practical and conceptual. Available data and modeling suggest that the magnitudes of many key impacts can be quantified for given amounts of global warming through scaling of local to global warming and through coupled linkages to warming (such as alterations in the water cycle that scale with warming). But although published analyses of future climate impacts can be tied to specific warming levels in particular studies, this information often cannot readily be linked to CO_2-equivalent concentrations (because, for example, of lack of information on aerosol forcing used in many future climate impact studies based on emission scenarios).

Moreover, using warming as the frame of reference provides a picture of impacts and their associated uncertainties in a warming world—uncertainties that are distinct from the uncertainties in the relationship of CO_2-equivalent concentrations to warming. Use of warming as a metric of change also permits coverage of the transient climate changes and impacts while concentrations increase, as well as the lock-in to further changes after stabilization. Further, the approach taken here facilitates cataloging ranges of impacts that should be expected for 1°C, 2°C, 3°C, or other levels of warming. The reader can thus consider how much warming s/he considers to be an appropriate target. Information is also provided to translate warming into best estimates of associated CO_2-equivalent target concentrations with these best estimates accompanied by estimated likely uncertainty ranges derived from uncertainty in climate sensitivity.

Furthermore, this report also describes the cumulative carbon framework, a perspective that has recently received considerable attention. Rather than CO_2-equivalent concentration levels, this approach considers the amount of carbon emissions accumulated over time and the implications of different accumulated emissions targets. Models consistently suggest a persistent temperature response to a given level of cumulative carbon emissions. Accumulated carbon emission targets link to impacts through temperature (or warming) and are clearly relevant to policy aimed at controlling emissions and reducing the risk of dangerous impacts. The approaches used here thereby provide additional policy-relevant information that would be lost in an analysis that only related impacts to CO_2-equivalent concentration levels.

KEY FINDINGS

There are three key findings of this report, which correspond to the structure of this summary:

1. **Climate change in the very long term:** Future stabilization targets correspond to altered states of Earth's climate that would be nearly irreversible for many thousands of years, even long after anthropogenic greenhouse gas emissions ceased. The capacity to adapt to slow changes is generally greater than for near-term rapid climate change, but different stabilization levels can lock the Earth and many future generations of humans into large impacts that can occur very slowly over time, such as the melting of the polar ice sheets; similarly, some stabilization levels could prevent such changes.

2. **Climate change in the next few decades and centuries:** Understanding the implications of future stabilization targets requires paying attention to the expected climate change and to the emissions required to achieve stabilization. Because of time lags inherent in Earth's climate, the observed climate changes as greenhouse gas emissions increase reflect only about half of the eventual total warming that would occur for stabilization at the same concentrations. Moreover, emissions reductions larger than about 80% (relative to whatever peak global emission rate may be reached) are required to approximately stabilize carbon dioxide concentrations for a century or so at any chosen target level (e.g., 450 ppmv, 550 ppmv, 650 ppmv, 750 ppmv, etc.).[4] Even greater reductions in emissions would be required to maintain stabilized concentrations in the longer term. It should be emphasized that this finding is not linked to any particular policy choice about time of stabilization or stabilization concentration, but applies broadly, and is due to the fundamental physics of the carbon cycle presented in Chapter 2.

3. **Climate changes, impacts, and choices among stabilization targets:** A number of key climate changes and impacts for the next few decades and

[4]In this report the mixing ratio for any compound, CO_2 for example, is expressed in either ppm (parts per million, i.e., the number of molecules of CO_2 for every million molecules of air) or in ppmv (parts per million volume, the ratio of CO_2 to air calculated in volumes) but used equivalently.

centuries can now be identified and estimated at different levels of warming. Many impacts can be shown to scale with warming (see Figure S.5). Scientific progress has resulted in increased confidence in the understanding how global warming levels of 2°C, 3°C, 4°C, 5°C, etc. (see Figure S.1) affect precipitation patterns, extreme hot seasons, streamflow, sea ice retreat, reduced crop yields, coral bleaching, and sea level rise. This increased confidence provides direct scientific support for evaluating the implications of different stabilization targets. However, other climate changes and impacts are currently understood only in a qualitative manner. Many potential effects on human societies and the natural environment cannot presently be quantified as a function of stabilization target (see Figure S.6). This shortcoming does not imply that these changes and impacts are negligible. Some of these impacts, such as species changing their ranges or behavior, could be very important; indeed, some may dominate future risks due to anthropogenic climate change. Uncertainty in the carbon dioxide emissions and concentrations corresponding to a given temperature target is large, and choices about stabilization targets depend upon judgments regarding the degree

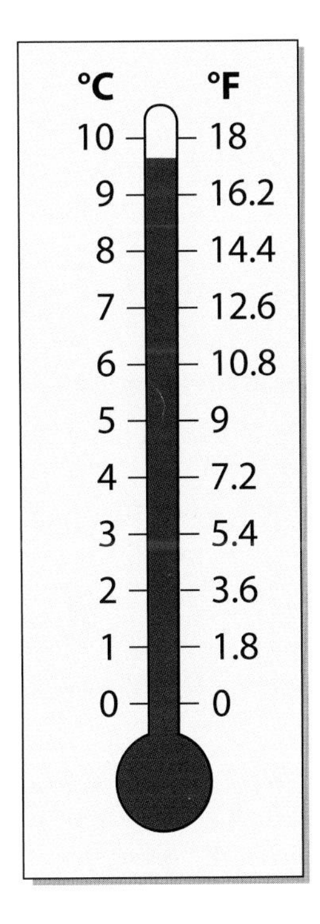

FIGURE S.1 Illustration of how temperature change in degrees Celsius (left side of thermometer) relates to temperature change in degrees Fahrenheit (right side of thermometer). For example, a warming of 5°C is equal to a warming of 9°C. In this report estimates of temperature change are made in degrees Celsius in accordance with international scientific practice.

of acceptable risk associated with both quantifiable and non-quantifiable impacts and changes.

SUPPORTING EVIDENCE

1. Cumulative Carbon Dioxide Emissions and Climate Change over Millennia

Climate changes that occur because of carbon dioxide increases are expected to persist for thousands of years[5] even if emissions were to be halted at any point in time. Recent scientific literature has shown that the contribution to global warming caused by anthropogenic CO_2 can be directly related to the cumulative emissions of carbon dioxide.

For example, our best estimate (see Figure S.2) is that 1,000 gigatonnes of anthropogenic carbon (GtC) (Box S.1) emissions lead to about 1.75°C increase in global average temperature,[6] implying that approximately 1,150 gigatonnes of carbon (or 4,200 Gt CO_2) would lead to a global mean warming of 2°C (the stated aspirational goal of the "group of eight" nations). Based on current understanding, this warming is expected to be nearly irreversible for more than 1,000 years (Figure S.3). Figure S.2 shows best estimates and likely uncertainty ranges for cumulative carbon emissions leading to a range of warming levels, along with cumulative emissions to date (about 500 GtC). Carbon dioxide alone accounted for about 55 percent of the total CO_2-equivalent concentration of the sum of all greenhouse gases in 2005. The contribution of carbon dioxide increases to between 75 and 85 percent of total CO_2-equivalent by the end of this century based on a range of current emission scenarios. Some anthropogenic carbon dioxide is removed by the oceans and biosphere in decades to centuries, but the slow time scales of the long-term uptake of carbon in the ocean means that some is expected to persist in the atmosphere for many thousands of years. This behavior is unique to carbon dioxide among major radiative forcing agents. Choices regarding continued emissions or mitigation of other warming agents such

---

[5]Approaches to "geoengineer" future climate, e.g., to actively remove carbon from the atmosphere or reflect sunlight to space using particulate matter or mirrors are topics of active research. If effective, these may be able to reduce or reverse global warming that would otherwise be effectively irreversible. This study does not evaluate geoengineering options, and statements throughout this report regarding the commitment to climate change over centuries and millennia from near-term emissions should be read as assuming no geoengineering. Reforestation or other methods of sequestration of carbon are also not considered.

[6]The quasi-linear response of temperature to cumulative carbon is discussed in detail in Section 3.4.

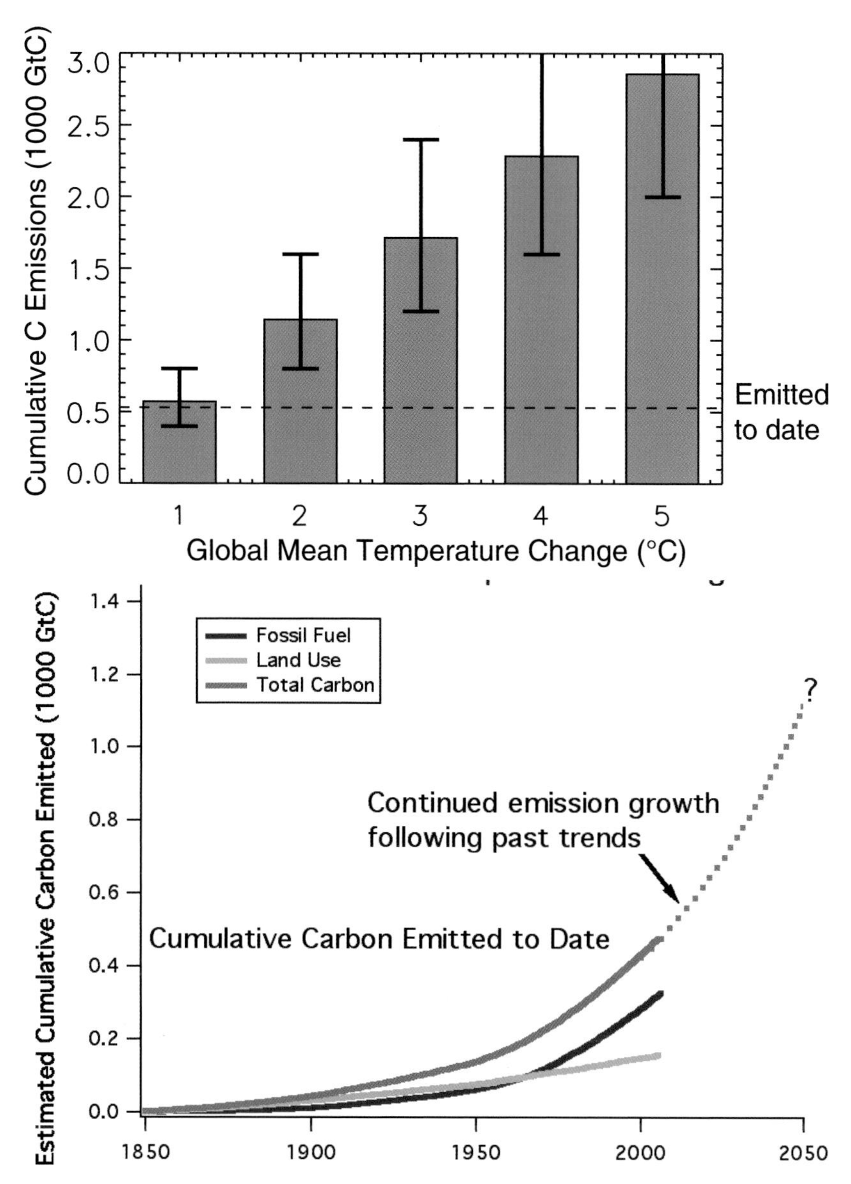

FIGURE S.2 (top) Best estimates and likely range of cumulative carbon emissions that would result in global warming of 1°C, 2°C, 3°C, 4°C, or 5°C (see Figure S.1), based on recent studies that have demonstrated a near linearity in the temperature response to cumulative emissions (see Section 3.4). Error bars reflect uncertainty in carbon cycle and climate responses to CO_2 emissions, based on both observational constraints and the range of climate-carbon cycle model results (see Section 3.4). (bottom) Estimated global cumulative carbon emissions to date from fossil fuel burning and cement production, land use, and total. The figure also shows how much cumulative carbon would be emitted by 2050 if past trends in emission growth rates were to continue in the future, based upon a best fit to the past emission growth curve. {3.4}

BOX S.1 GTC (GIGATON OF CARBON)

One gigaton of carbon is 1 billion tons of carbon, where "carbon" refers literally to the mass of carbon, *not* the mass of a molecule as a whole (i.e., all the atoms), but just the mass of carbon atoms.

Example: Burning 1 gallon of gasoline emits approximately 19.6 lbs of CO_2 (*http://cdiac.ornl.gov/pns/faq.html*), so if you assume a typical American vehicle gets 20 miles per gallon and it travels 15,000 miles per year, then the typical American vehicle emits about 1.8 tons of carbon per year. Stated differently, about 550 million average American vehicles would emit 1 GtC per year.

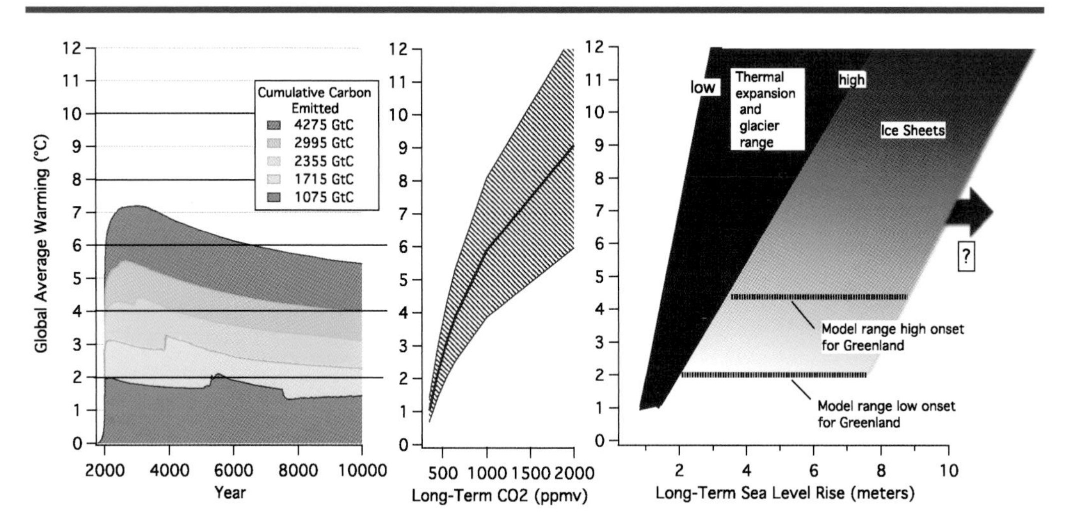

FIGURE S.3 Commitments to global warming over thousands of years, expressed as best estimates depending upon the cumulative anthropogenic carbon emitted (direct human emission plus possible induced feedbacks such as release of carbon from clathrates, see below) by the end of the next few centuries from a model study (left, from the calculations presented in Eby et al., 2009), the corresponding long-term carbon dioxide concentrations, shown as best estimates and likely ranges (middle, from Table 3.1 of this report), and estimated range of corresponding global average sea level rise (right, see Section 6.1; the adopted equilibrium long-term thermal sea level rise is 0.2-0.6 m per degree as noted in Meehl et al., 2007). The "low" and "high" onset values in the right panel reflect differences between available climate models in the global mean temperature at which the Greenland ice sheet will disappear after thousands of years since the accumulation cannot sustain the ice loss by melt in the ablation area and rapid ice flow-related loss along the margins. This depends not only on increased ice loss from warming but also on increased accumulation from greater snowfall in a warmer world, and the balance between these terms differs from model to model. The range across models is taken from Meehl et al. (2007) based on a detailed analysis of the models evaluated in the Intergovernmental Panel on Climate Change (IPCC) report. Additional contributions from rapid ice discharge are possible (see Chapters 4 and 6). The climate sensitivity used to construct the likely ranges shown in the middle panel is discussed in Chapter 3 where it is noted that larger or smaller warmings than the estimated likely value for a given carbon dioxide concentration cannot be ruled out. Bumps in the warming curves in the left panel are because of adjustments in ocean circulation in response to warming in this particular climate model and should be thought of as illustrative only. {3.2, 6.1}

as methane, black carbon on ice/snow, and aerosols can affect the global warming of coming decades but have little effect on the lock-in to longer-term warming of Earth over centuries and millennia; that commitment is primarily controlled by carbon dioxide. {2.1, 2.2, 2.3, 3.4}

Earth is now entering a new geological epoch, sometimes called the Anthropocene, during which the evolution of the planet's environment will be largely controlled by the effects of human activities, notably emissions of carbon dioxide. Actions taken during this century will determine whether the Anthropocene climate anomaly will be a relatively short-term and minor deviation from the Holocene climate, or an extreme deviation extending over many thousands of years.

Higher cumulative carbon emissions result in both a higher peak warming and a longer duration of the warming (see Figure S.4). The duration of the warming is critical, because an extended period of warming provides more time for the components of the Earth system that may respond very slowly (such as the deep oceans and the great ice sheets) to assert themselves, even very long after anthropogenic emissions have ceased. {6.1}

The sea level rise implications of the Anthropocene could lead to major changes in the geography of the Earth over the coming millennia. Model studies suggest that a cumulative carbon emission of about 1,000 to 3,000 GtC would lead to eventual sea level rise due to thermal expansion and glacier and small ice cap loss alone of the order of 1 to 4 meters. Additional contributions from Greenland could contribute as much as a further 4 to 7.5 m on multi-millennial time scales, for a possible total of order 5 to 11.5 meters from thermal expansion, glaciers and small ice caps, and Greenland.

Widespread coastal inundation would be expected if anthropogenic warming of several degrees is sustained for millennia; although these slow changes allow time for adaptation, they are essentially irreversible. The projected fragility of the Greenland ice sheet is in accord with studies suggesting that Greenland was essentially free of ice during the Pliocene era (which was probably about 3°C warmer than pre-industrial times in the mid-Pliocene, about 3-3.3 million years ago). Changes in Antarctica are less clear, in part because both the West and East Antarctic ice sheets must be considered: one study suggests that cumulative carbon emission of about 2,000-5,000 GtC could also contribute up to 5 meters of additional sea level rise from West Antarctic ice sheet loss. Future changes in East Antarctica could offset at least part of West Antarctic changes. While carbon emissions in the 21st

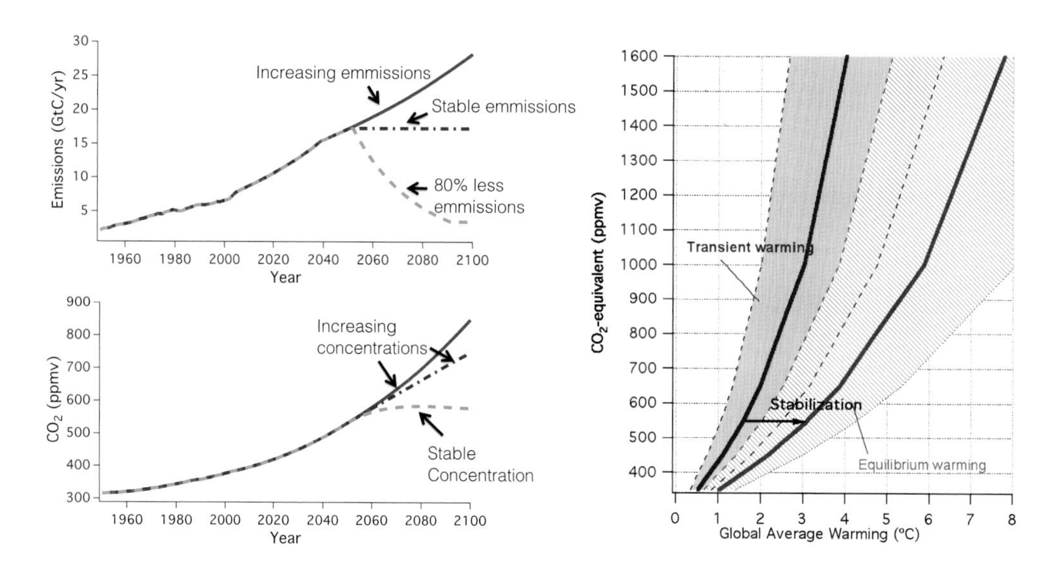

FIGURE S.4 The left panel shows illustrative examples (from calculations using the Bern Earth Model of Intermediate Complexity, see Chapter 2 and Methods) of how carbon dioxide concentrations would be expected to evolve depending upon emissions. Stable emissions (blue lines) do not result in stable concentrations because the source of carbon is much larger than the sink. Emission reductions larger than about 80% are required if concentrations are to be stabilized (green lines). The right panel shows the best estimates and likely ranges of global warming projected for various levels of carbon dioxide concentration in the transient (blue) and equilibrium states, or climate sensitivity (red); see Table 3.1. As carbon dioxide emissions increase, average global warming is projected to follow the blue curve. If concentrations of carbon dioxide were to be stabilized, the global warming is expected to increase from the blue to the red curve, as depicted by the arrow. Note that the equilibrium warming indicated in the figure incorporates only feedbacks from water vapor, clouds, sea ice, or snow changes; the slower acting feedbacks incorporated in Earth System Sensitivity may increase the warming (by about 50% over the values shown according to one study by Lunt et al., 2010) {2.1, 3.2, 3.3}

century are expected to determine the commitments to these eventual future changes, the sea level rise expected to occur in the 21st century is considerably smaller, in the range of 0.5 to 1.0 m. Some semi-empirical models predict sea level rise up to 1.6 m by 2100 for a warming scenario of 3.1°C, a possible upper limit which cannot be excluded. {4.8, 6.1}

Some slow climate components could act as amplifiers that would greatly increase the size and duration of the Anthropocene.

If elevated global temperatures were to persist for a thousand years or more, some studies suggest that the resulting warming of the deep ocean could release deep-sea carbon stored in the form of methane clathrates[7] in marine sediments. Other contributions could come from the substantial reservoir of near-surface organic carbon in soils and permafrost, whose stability is poorly understood. For example, a release rate of a half GtC per year from such sources would add 2,500 Gt of carbon over 5,000 years to the carbon emitted directly by humans. For reference, paleoclimate studies suggest that during the Paleocene-Eocene Thermal Maximum (about 55 million years ago), similar amounts of carbon were released in less than 10,000 years. A number of recent studies show that large methane releases from particular local sites have been observed, but these are too limited to imply that globally significant changes are already occurring or will occur for warming levels in the near term. {6.1}

2. Stabilization and Climate Change of the Next Few Decades and Next Several Centuries

Because the global anthropogenic source of carbon dioxide greatly exceeds the net global sink (through removal mechanisms in the ocean, land, and biosphere), stabilization of carbon dioxide concentrations at any selected target level would require reductions in total emissions of at least 80 percent (relative to any peak emission level).

Unless the source matches the sink, concentrations of carbon dioxide (and resulting warming influences) will continue to rise, much like the water in a bathtub when water is coming in faster than it is going out. Because current carbon dioxide emissions exceed removal rates, stabilization of carbon dioxide *emissions* at current rates will not lead to stabilization of carbon dioxide concentrations (see Figure S.4). A robust consequence of the stock and flow nature of atmospheric carbon and the physics of the carbon cycle is that emissions reductions larger than about 80% (relative to whatever peak emission level occurs) are required to approximately stabilize carbon dioxide concentrations for a century or so and even greater reductions in emissions would be required in the longer term; this applies for any chosen stabilization target.

Observed climate responses in coming decades will be smaller than the longer-term temperature response to any given stabilization level. If carbon dioxide equivalent concentrations were to be stabilized at some point

[7]Methane clathrates, also called methane hydrates, are material in which methane is trapped inside a larger crystalline water chemical structure.

in the future, there would be a lock-in to further warming of comparable magnitude to that already occurring at the time of stabilization.

The instantaneous response of Earth's atmosphere and oceans to increases in greenhouse gases and net radiative forcing represents a transient climate change, which can be linked to "transient climate response."[8] The transient climate response is smaller than the longer-term "climate sensitivity" that includes adjustments by the oceans to the added heat (Table S.1). For example, if carbon dioxide equivalent concentrations (including aerosols and other gases) were to increase from today's best estimate levels of about 390 parts per million by volume (ppmv) to 550 ppmv at rates of growth similar to those occurring today, averaged warming would be expected to increase in a manner that scales with the change in radiative forcing relative to the transient climate response; for 550 ppmv the best estimate total warming since pre-industrial times is about 1.6°C (within a likely uncertainty range of 1.3-2.2°C). In the hypothetical case where concentrations are then immediately stabilized at 550 ppm, further warming would subsequently occur over the next several centuries, reaching a best estimate "climate sensitivity" of about 3°C (likely in the range of 2.1-4.3°C). The horizontal arrow in Figure S.4 depicts such a transition from transient to equilibrium warming. {2.2, 2.4, 3.2, 3.3}

Climate sensitivity remains subject to considerable uncertainty. The estimated "likely" range presented in this report corresponds to the range of model results in the CMIP3 global climate model archive, and it is roughly consistent with paleoclimate evidence. However, the possibility of climate sensitivities substantially higher than this range cannot at present be ruled out. This report should be read with this proviso in mind, as these high sensitivities, if realized, would amplify many of the impacts discussed and associated risk. {3.2, 3.3, 6.1}

3. Climate Changes, Future Impacts, and Choices among Stabilization Targets

Increases in global mean temperature caused by higher anthropogenic greenhouse gas concentrations would be expected to lead to a diverse range

[8]The transient climate response is defined as the warming at the time of doubling of CO_2 concentration (compared to a pre-industrial value of 278 ppm this is about 550 ppm). Scaled by radiative forcing, the same relationship characterizes warming that has occurred during the 20th century as well as further warming that is projected to continue with growing CO_2 concentrations in the 21st century for a broad range of plausible scenarios.

TABLE S.1 Estimated Likely Ranges and Best Estimate Values for Transient and Equilibrium Global Averaged Warming Versus Carbon Dioxide Equivalent Concentrations.

CO_2-Equlvalent Concentration (ppmv)	Best Estimate Transient Warming (°C)	Estimated Likely Range of Transient Warming (°C)	Best Estimate Equilibrium Warming (°C)	Estimated Likely Range of Equilibrium Warming (°C)
350	0.5	0.4-0.7	1	0.7-1.4
450	1.1	0.9-1.5	2.2	1.4-3.0
550	1.6	1.3-2.1	3.1	2.1-4.3
650	2	1.6-2.7	3.9	2.6-5.4
1000	3	2.4-4.0	5.9	3.9-8.1
2000	4.7	3.7-6.2	9.1	6.0-12.5

of changes in potentially damaging climate-related parameters and impacts, affecting many aspects of human society and the natural environment.

The magnitude of some near-term (next few decades and centuries) climate changes and impacts can be estimated for specific levels of global mean temperature change experienced, illustrating how stabilization at different levels of greenhouse gas forcing would be expected to alter our world (see Figure S.5). Approximate estimates of these effects, per degree C of global warming, include:

- 5-10% changes in precipitation in a number of regions
- 3-10% increases in heavy rainfall[9]
- 5-15% yield reductions of a number of crops[10]
- 5-10% changes in streamflow in many river basins worldwide, including several in the U.S.
- about 15% and 25% decreases, in the extent of annually averaged and September Arctic sea ice, respectively

In addition, effects at particular levels of warming include:

- Increases in the number of exceptionally warm summers (i.e., 9 out of 10 boreal summers that are "exceptionally warm" in nearly all land areas for about 3°C of global warming, and every summer "exceptionally warm" in nearly all land areas for about 4°C, where an "exceptionally warm" summer is defined as one that is warmer than all but about 1 of the 20 summers in the last decades of the 20th century).

[9]Heaviest 15% of daily falls

[10]Unless adaptation measures not presently in hand become available

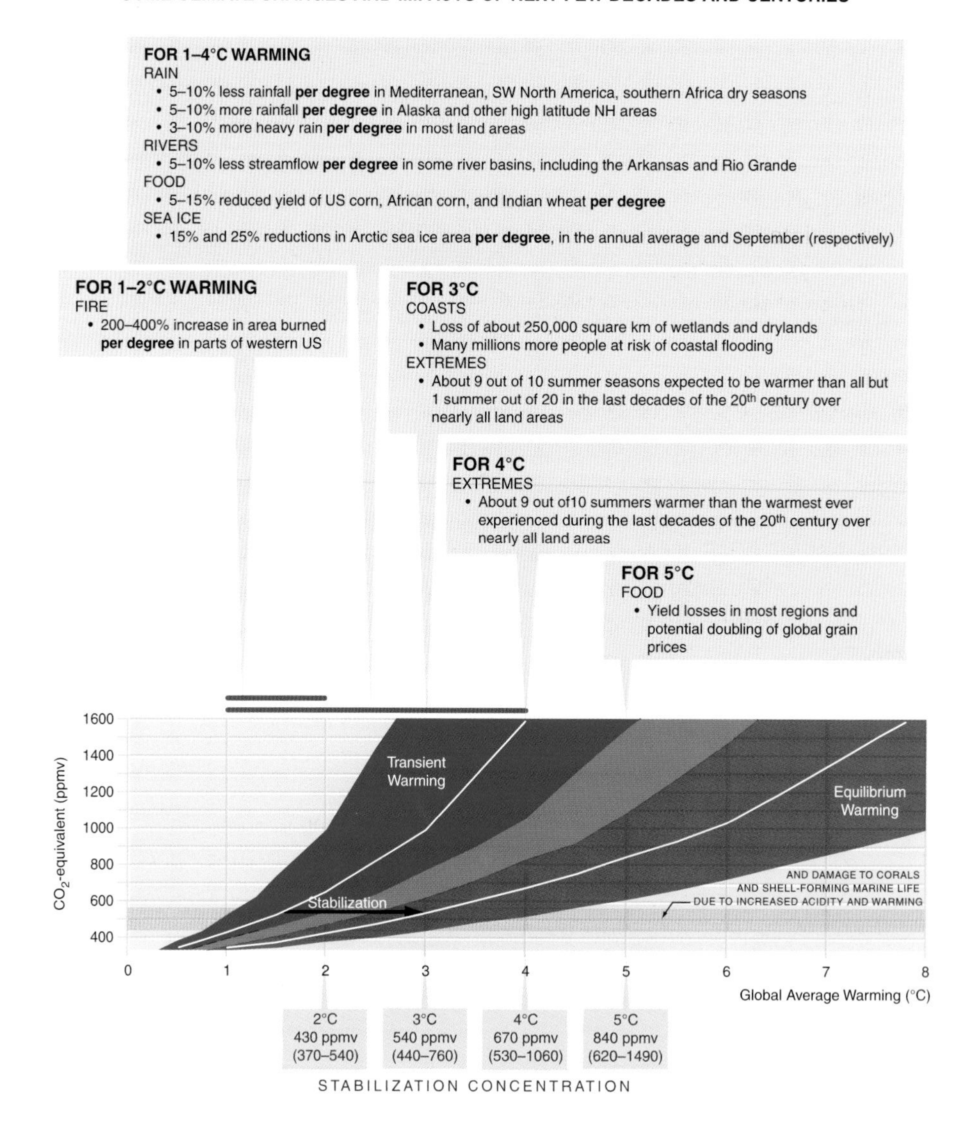

FIGURE S.5 Climate changes and climate impacts as a function of global warming (not in priority order or implied importance). These anticipated effects are projected to occur in the 21st century following the transient warming for a given CO_2 equivalent concentration, followed by further warming to the equilibrium value for stabilization at a given target concentration. As in previous figures, for discussion of transient and equilibrium warming see Chapter 3, where it is noted that the probability distribution of climate sensitivity is uncertain; larger or smaller warmings than the estimated likely value for a given carbon dioxide equivalent concentration cannot be ruled out. Ranges are shown for climate impacts over the globe or over large regions; specific regions, crops, river basins, etc. and their uncertainties are discussed in detail later in the report. {3.2, 3.3, 4.2, 4.5, 4.6, 4.7, 4.8, 4.9, 5.1, 5.2, 5.3, 5.4, 5.7, 5.8}

- 200-400% increases per degree in wildfire area burned in several western North American regions for 1-2°C
- Increased coral bleaching and net erosion of coral reefs, due to warming and changes in ocean acidity (pH) for carbon dioxide levels corresponding to about 1.5-3°C of global warming.
- Sea level rise in the range of 0.5 to 1.0 m in 2100 (reached in a scenario corresponding to about 3±1°C of global warming) and an associated increase in the number of people at risk from coastal flooding by 5-200 million[11] as well as global wetland and dryland losses of more than 250,000 square kilometers. {4.1, 4.2, 4.5, 4.6, 4.7, 4.8, 4.9, 5.1, 5.2, 5.3, 5.4}

Many important impacts of climate change are difficult to quantify for a given change in global mean temperature, but the risk of adverse impacts is likely to increase with global mean temperature change.

For some impacts, this difficulty arises because temperature is a primary, but not necessarily the only, driver of change. Quantification can also be difficult due to uncertainty in observing and modeling the response of a given system to temperature changes or other climate and non-climate factors, and additional complexity due to the influence of multiple environmental and other anthropogenic factors. It is clear from many scientific studies documenting projected impacts across numerous sectors and regions, however, that a number of impacts do scale approximately with global temperature. Hence, these are expected to intensify in response to a greater temperature change. An illustrative set of temperature-dependent impacts are summarized in Figure S.6. These include shifts in terrestrial and marine species ranges and abundances (including die-off in some cases), increased risk of heat-related human health impacts, loss of infrastructure in coastal regions (due to sea level rise) and the Arctic (due to sea level rise, retreat of sea ice and associated coastal erosion, and permafrost loss). This summary of temperature-related impacts is intended to be indicative rather than comprehensive. Figure S.6 does not include all possible temperature-sensitive impacts, such as projected extinctions due to climate change and increased risks to national security. {4.7, 4.9, 5.5, 5.6, 5.7, 5.8}

Uncertainty in the cumulative carbon or stabilized carbon dioxide concentration that corresponds to a given temperature target is large. It follows

[11]With the range depending mainly on uncertainty in adaptation measures undertaken.

IMPACT AREA	RISK OF IMPACTS INCREASES WITH TEMPERATURE	PRIMARY CLIMATE DRIVER	CONFOUNDING FACTORS
AGRICULTURE	Crop pests, weeds, and disease: shifts in geographic range and frequency	Extreme cold temperature, precipitation	Agricultural practices; herbicides & pesticides
TERRESTRIAL ECOSYSTEMS	Individual species: shifts in timing of flowering & breeding cycles, in geographic ranges, and in populations	Temperature: averages, extremes, degree-days	Landscape fragmentation
	Disturbances: changes in the frequency and timing of fire, pests, and disease	Temperature, precipitation, drought stress	Land management practices
	Forests: shifts in primary processes including nutrient cycling,transpiration, and respiration	Temperature, precipitation, carbon dioxide levels	?
COASTAL AND MARINE ECOSYSTEMS	Individual species: shifts in geographic ranges and die-off	SLR, saline and freshwater inputs, water temperature	Local coastal practices, such as subsidence
	Corals and mollusks: declining calcification rates, more frequent bleaching events	Maximum temperature, ocean acidification	?
	Coastal upwelling zones: shifts in nutrient availability	Shifts in surface winds & ocean circulation	?
	Oxygen minimum (dead) zones: expanding geographic area and duration	Ocean temperature, circulation, and nutrient availability	Water pollution (nitrates?)
ENERGY	Increasing demand for air conditioning and decreasing demand forwinter heating	Accumulated and extreme temperatures	Penetration & efficiency of air conditioning technology
INFRASTRUCTURE	In the Arctic: shortening of land, lengthening of marine transportation season	Permafrost melt, sea ice duration & extent	?
	Risk of impacts from extreme temperature, precipitation, and storms	Temperature and precipitation extremes	Shoreline development and protection
HEALTH	Increased risk of heat-related illness and death	Frequency, intensity & duration of heat waves	Extreme cold temperature, precipitation
	Shifts in timing and geographic range of allergens and vector-borne diseases	Average and extreme temperature & precipitation	Human spread, cultivation practices
WATER	Earlier peak streamflow, longer summer dry periods across much of the U.S.	Precipitation, temperature, snowpack and melt timing	Water management and demand

FIGURE S.6 Our understanding of the impacts of climate change is still evolving, and quantitative information is currently too limited to provide numerical estimates of the scale, scope, and timing of some impacts. This figure illustrates a number of such possible impacts along with their primary drivers, as well as available information on confounding factors. {2.4, 5.1-5.8}

that choices about stabilization targets depend upon judgments regarding the degree of acceptable risk.

The likely range of cumulative carbon emissions corresponding to a given warming level is estimated to lie between –30% to +40% of the best estimate. This range is due mainly to uncertainties in the carbon cycle response to emissions and the climate response to increased radiative forcing. For a cumulative anthropogenic emission of 1,000 GtC, our best estimate of the warming remains below 2°C, but there is an estimated 17% probability that the warming could exceed 2°C for more than 1,500 years. When cumulative emissions are increased to 1,500 GtC, the best estimate of the anthropogenic warming remains above 2°C for more than 3,500 years, and the very likely upper end warming is still above 2.5°C for more than 10,000 years. Higher values cannot be excluded, implying additional risk that cannot presently be quantified. On the other hand, at the lower end of carbon-climate likely uncertainty range, there may be about a 17% chance that warming could remain below 2°C even if as much as 1,700 GtC are emitted. Figures S.3 and S.5 provide some scientific reasons why global warming of a few degrees could be considered dangerous to some aspects of nature and society, but the corresponding uncertainty ranges should be emphasized here. For example, while the best estimate of a stabilization target corresponding to a long-term warming of 2°C is 430 ppm, the likely uncertainty range for this value spans from 380 ppm (below current observed levels) to 540 ppm (almost a doubling of carbon dioxide relative to pre-industrial times). {3.4, 6.1}

Many important aspects of climate change and its impacts are expected to be approximately linear and gradual, slowly becoming larger and more significant relative to climate variability as global warming increases.

This report highlights the importance of 21st century choices regarding stabilization targets and how they can be expected to affect many aspects of Earth's future. Progressively warmer temperatures are expected to slowly lead to larger and more significant changes for impacts including wildfire extent, decreases in yields of some (but not all) crops, streamflow changes, decreased Arctic sea ice extent, increases in heavy rainfall occurrence, and other factors presented. However, it should be noted that many climate changes and impacts remain poorly understood at present. For example, the record of past climates suggests that major changes such as dieback of the Amazon forests or substantial changes in El Niño behavior can occur. This report identifies some areas where recent science suggests reduced effects compared to earlier studies (including e.g., projected future changes in hurricane activity). This report does not identify any specific projections of abrupt climate changes that the committee considers to be robustly established, e.g., based on clear physical understanding of processes and multiple models. However, it is clear that the risk of surprises can be expected to increase with the duration and magnitude of the warming. Finally, this report shows throughout that present emissions represent commitments to growing current and future impacts, including the very long-term future over many thousands of years. {2.4, 3.4, 4.3, 4.4, 5.8, 6.1, 6.2}

Overview of Climate Changes and Illustrative Impacts

The following section provides an overview of a number of climate changes and impacts that can now be identified and estimated at different levels of warming. Highlighted in bold are the key impacts followed by supporting details.

CHANGES IN RAINFALL AND STREAMFLOW

Increases of precipitation in high latitudes and drying of the already semi-arid regions at lower latitudes are projected with increasing global warming, with seasonal changes in several regions expected to be about 5-10% per degree of warming. However, patterns of precipitation change show much larger variability across models than patterns of temperature. The basic large-scale pattern and magnitude of precipitation responses across the tropics, subtropics, and mid-latitude and high-latitude regions can be understood largely as the result of increasing water vapor in the atmosphere; these are broadly consistent with observed trends and physical understanding, and represent a very robust prediction across models. Precipitation in many of the world's monsoon regions is expected to increase during the rainy season. Precipitation associated with mid-latitude storms is also expected to increase. For some areas, particularly those near transitions between regions that become wetter and those that become drier, model disagreement is large. The continental U.S. region straddles changes that are both positive (over the northernmost areas) and negative (over the southwest areas) changes in both annual and Dec-Jan-Feb average precipitation. A large portion of the contiguous 48 U.S. states is in a transition zone where future rainfall changes cannot be projected with confidence at present. Models agree in projecting increases in precipitation on the order of 5-10% per degree C of warming in high latitudes in all seasons, including over Alaska, and they also project drying in the dry season in the south and southwest United States and Mexico (see Figure O.1). {4.2}

Streamflow Changes

Widespread changes in streamflow are expected in a warmer world, with many regions experiencing changes of the order of 5-15% per degree of warming. Streamflow is a key index of the availability of freshwater, a quantity that is essential for human and natural systems. Changes in streamflow depend upon both evaporation (and hence warming) as well as precipitation. In regions where decreases in precipitation are predicted, these decreases usually will be accompanied by larger decreases in streamflow.

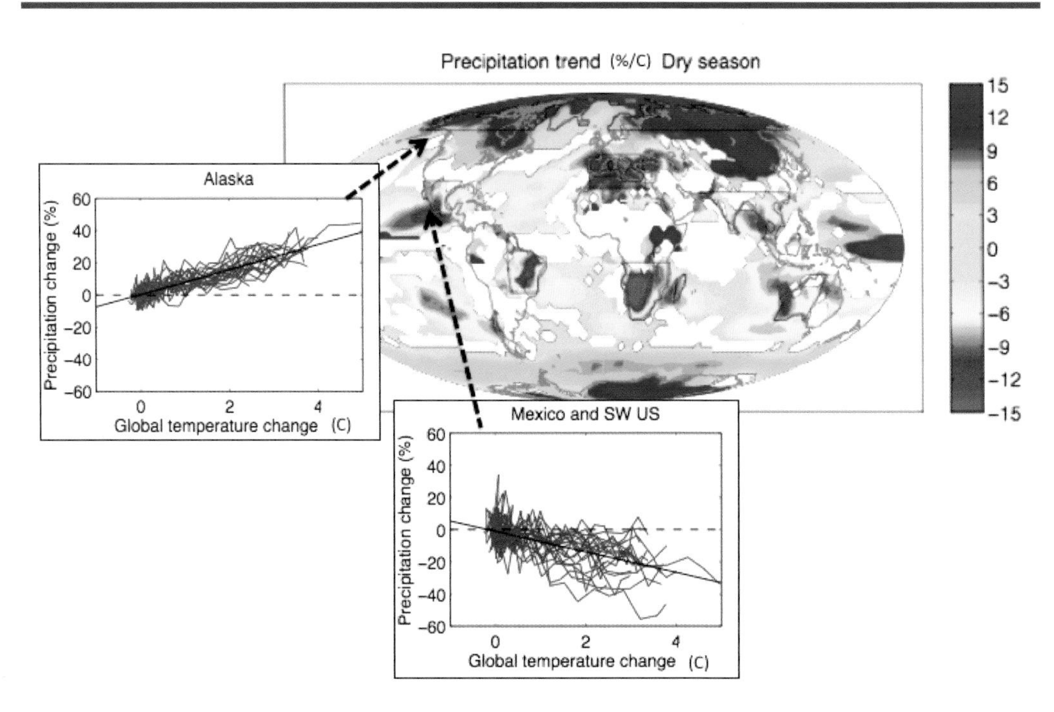

FIGURE O.1 Estimated changes in precipitation per degree of global warming in the three driest consecutive months at each grid point from a multi-model analysis using 22 models (relative to 1900-1950 as the baseline period). White is used where fewer than 16 of 22 models agree on the sign of the change. One ensemble member from each model is averaged over the dry season and decadally in several indicated regions including southwestern North America and Alaska, as shown in the inset plots. Adapted from Solomon et al. (2009), with additional inset panel for Alaska (courtesy R. Knutti) provided using the same datasets and methods as in that work. {4.2}

Streamflow is expected to decrease in many temperate river basins as global temperature increases. The greatest decreases are expected in areas that are currently arid or semi-arid. Most models project decreases in the southwest United States, while slight increases are projected in the northeast and northwest. There is strong agreement among models that runoff in the Arctic and other high-latitude areas, including Alaska, will increase. The greatest decreases per degree within the United States are projected for the Rio Grande Basin (about 12% per degree) and increases of about 9% per degree are expected in Alaska; see Figure O.2. Thus, warming of a few degrees can be expected to lead to large perturbations to water resources, especially in the southwestern and southern parts of the United States, many of which are already facing water resources challenges due to growing population and environmental issues. {5.3}

CHANGES IN EXTREME TEMPERATURE, PRECIPITATION, AND CLIMATE DYNAMICS

Temperature Extremes

Extreme temperatures are expected to increase in a warmer world. For example, for about 3°C of global warming, 9 out of 10 Northern Hemisphere summers are projected to be "exceptionally warm" in nearly all land areas, and every summer is projected to be "exceptionally warm" in nearly all land areas for about 4°C, where an "exceptionally warm" summer is defined as one that is warmer than all but about 1 of the 20 summers in the last decades of the 20th century. A complete review of the many studies evaluating changes in extremes in various regions is beyond the scope of the present study. Here we use as an illustrative example the effect of warming on seasonal extremes, based on a simple shift in the distribution of temperatures, using pattern scaling. Some studies have indicated the possibility that the variance of the distribution of temperature will increase, especially in regions that are projected to become drier (e.g., the Mediterranean basin), further enhancing the chances of extreme seasonal temperatures beyond that estimated here (see Figure O.3). {4.5}

Extreme Precipitation

Extreme precipitateon (heaviest 15% of daily rainfall) is likely to increase by about 3-10% per degree C as the atmospheric water vapor content increases in a warming climate, with changes likely to be greater

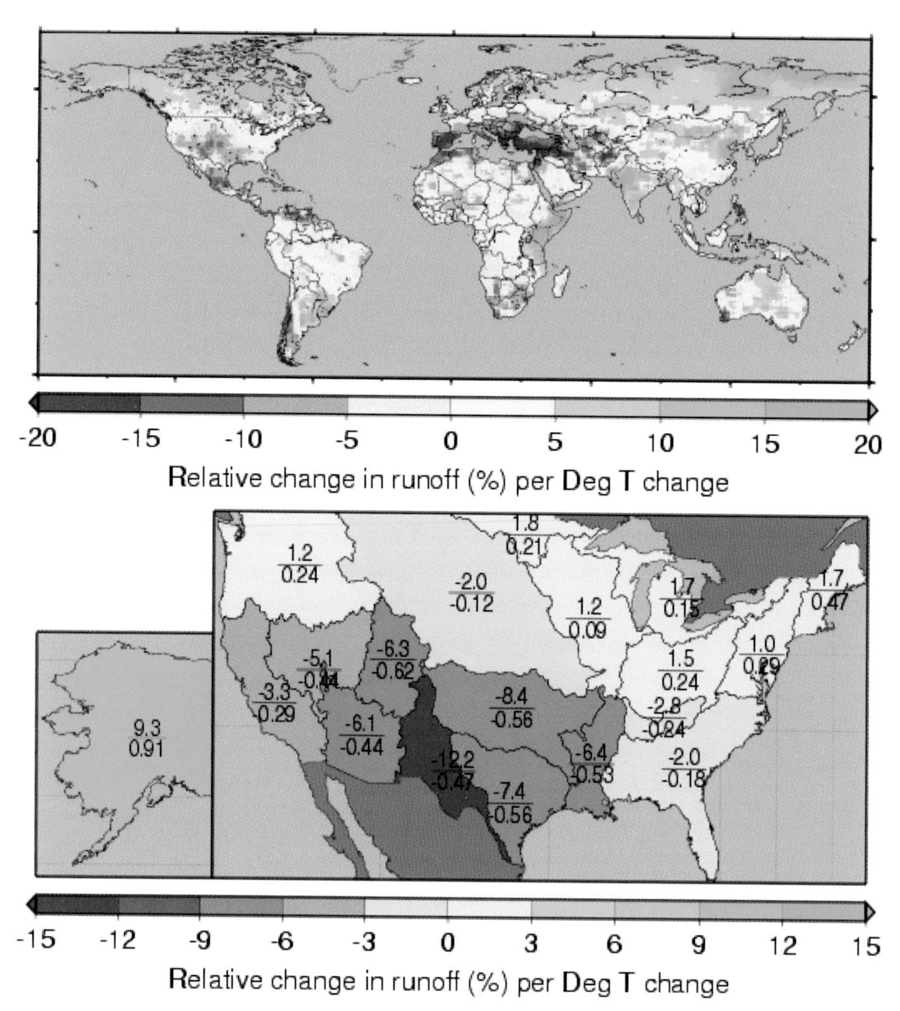

FIGURE O.2 Median runoff sensitivities (per degree of global warming) relative to 1971-2000 over 68 model pairs. Each pair consisted of an average over A2, A1B, and B1 global emissions scenarios of one IPCC Assessment Report 4 (AR4) GCM model's output, derived from 30-year runoff averages centered on the years for which the global average temperature increases were 1.0°C, 1.5°C, and 2.0°C, minus the 30-year model average runoff for 1971-2000, divided by the global temperature change. This analysis was performed for 23 models for 1.0°C and 1.5°C increases, and 22 models for a 2.0°C increase. Results are shown as averages over GCM grid cells (upper plot) or U.S. hydrological regions (lower plot). The notation a/b in each river basin denotes the mean change in percent (a), and the agreement among models (b); expressed as the fraction with positive changes minus the fraction with negative changes (FPN); see also Table 5.3. Table 5.3 contains standard errors and consistency across models, as indicated by FPN.

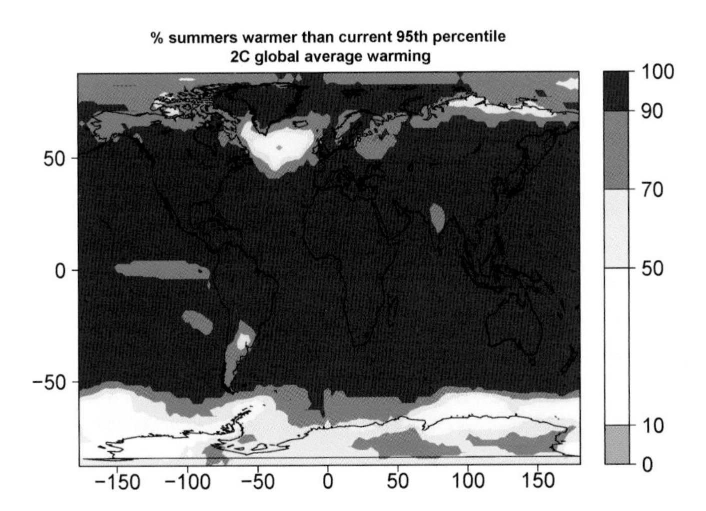

FIGURE O.3 Percentage of northern summers (June-July-August) warmer than the warmest 95th percentile (1 in 20) for 1971-2000, for 2°C global average warming above the level of 1971-2000, or about 3°C total warming since pre-industrial times, from an analysis of the multi-model CMIP3 (Coupled Model Intercomparison Project phase 3) ensemble. {4.5}

in the tropics than in the extratropics (this sensitivity may decrease somewhat as the warming progresses). While changes in precipitation extremes could lead to changes in flood frequency, the linkage between precipitation changes and flooding will be modulated by interactions between precipitation characteristics and river basin hydrology, the nature of which are not yet well understood. {4.6}

Hurricanes and Typhoons

Averaged over the tropics as a whole, the number of tropical cyclones is expected to decrease slightly or remain essentially unchanged. Models suggest that the average intensity of tropical cyclones (as measured by the wind speed) is likely to increase roughly by 1-4% per degree C global warming, or by 3-12% per degree C for the cube of this wind speed, often taken as a rough measure of the destructive potential of storm winds. For the North Atlantic, the changes in hurricane statistics are more uncertain than the global values, depending in large part on the spatial structure of the warming of the tropical oceans, and not just on the local warming in the Atlantic. Recent models project future changes in the number of Atlantic hurricanes ranging from –25% to +25% per degree C of global warming; thus, the sign of future changes in number of storms is uncertain. {4.3}

Ocean Circulation

The Meridional Overturning Circulation (MOC) in the Atlantic Ocean is expected to slow down in the 21st century due to warming associated with increased greenhouse gases and associated increased ocean stratification. As a result, warming in the northern North Atlantic Ocean and surrounding maritime regions is expected to be smaller than other oceanic regions. Changes in fisheries and marine ecosystems could also result from a MOC slow-down, but these impacts are poorly understood. {5.8}

CHANGES IN ICE, SNOW, AND FROZEN GROUND

Sea Ice

Arctic and Antarctic sea ice extent and volume are projected to decrease over the 21st century if greenhouse gas emissions continue to increase. Models project a clearly defined linear relationship between annual Arctic sea ice area decreases and global averaged surface air temperature. According to an analysis of an ensemble of models, annually averaged Arctic sea ice area reductions of about 15% are expected per degree C of global average warming (see Figure O.4). Greater reductions are expected for summer

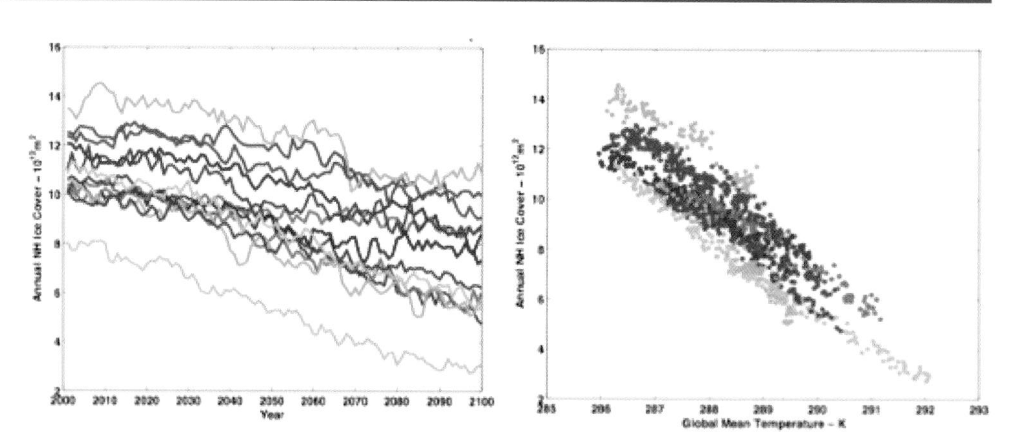

FIGURE O.4 Changes in annually averaged Arctic sea ice extent versus time from 13 CMIP3 models (left). The same information is plotted versus global mean temperature in the righthand panel. {4.7}

compared to winter. Figure O.4 illustrates that the retreat of Arctic sea ice is more compact when plotted versus global mean temperature rather than time, and this aids the understanding of the effect of stabilization at various target levels. Published studies of the range of the date when late summer Arctic sea ice is expected to disappear range from 2037 to beyond 2100. Models suggest that late summer Arctic sea ice decreases rapidly if warming exceeds about 2°C. By the end of the 21st century (global warming of about 3-5°C relative to pre-industrial conditions) an ice-free Arctic ocean in late summer is predicted by most models. In the decades after 2100, two models suggest that ice-free conditions may occur in winter if polar temperatures reach 13°C above present-day values. In the Antarctic, models predict a loss in ice cover ranging from 10-50% in winter and 33-100% in summer for a warming of about 1.7-4.4°C. The relationship between annual average sea ice area and the global average temperature suggests that ice recovery may occur if temperatures decrease. {4.7}

Snow Cover and Snowpack

Current trends in snow cover over the Northern Hemisphere suggest that the snow cover season has shortened and spring melt is occurring earlier compared to the past 50-100 years. Modelled changes in Northern Hemisphere snow cover are similar to the observations. Future decreases are consistent across the models and may reach −18% by 2090 (or a global warming of about 2-3°C). Snowpack has decreased over much of western North America since 1925, and this decrease has been linked to increasing temperatures over the West. While regional responses to the warmer surface temperatures may vary, the overall response suggests a significantly shorter snow season, smaller areal coverage, an earlier start to the melt season, a later start to the accumulation season, and decreased snowpack as the climate warms. The regions that show the most sensitivity to warming conditions are in maritime areas (both at low elevation and mountainous) while the continental interiors respond more slowly. The percentage change is largest in summer, but the greatest areal reductions are expected in spring. {4.7}

Permafrost

Northern Hemisphere permafrost is expected to degrade under global warming. Permafrost extent is expected to shrink, the region to retreat poleward, and the active layer to deepen as its temperature increases. These changes in the permafrost are linked to increasing global temperatures, and rates of degradation change with different emission scenarios such that higher emissions promote faster degradation. Damage to infrastructure over a region 1-4 million km^2 would be expected by the end of the 21st century if average Arctic air temperature increased by 5.5°C above the year 2000 average. {4.7}

Ice Sheets and Glaciers

Ice mass loss is occurring in some parts of Greenland and Antarctica, but contributions of the great ice sheets to sea level rise of coming decades and the next century remain uncertain. For 1993-2003, the estimated contributions to sea level rise (SLR) integrated across the Greenland and Antarctic ice sheets are 0.21±0.07 and 0.21±0.35 mm y^{-1}, respectively (0.021±0.007 and 0.021±0.035 m if continued for a century). Greenland lost roughly 180±50 Gt y^{-1} (0.5±0.14 mm y^{-1} SLR) for the time period 2003-2007. Recent observations have shown that changes in the rate of ice discharge into the sea can occur far more rapidly than previously suspected. The pattern of ice sheet change in Greenland is one of near-coastal thinning, primarily in the south and west along fast-moving outlet glaciers, and increased ice melt in the marginal region. However, the interior of the ice sheet is expected to be less vulnerable to future changes than these edge regions. Furthermore, current discharge rates may represent a transient instability, and whether they will increase or decrease in the future is unknown. A doubling in ice discharge along with a continued increase in surface melt using a "medium" emission scenario (AR4 A1B) would increase the global sea level by about 0.16 m by 2100, with 0.09 m contribution from ice dynamics, and 0.07 m from surface melt, respectively. The Antarctic ice sheet shows a pattern of near balance for East Antarctica, and greater mass loss from West Antarctica (including the Antarctic Peninsula) for the past few years; however, there is no strong evidence for increasing Antarctic loss over the past two decades. The Amundsen Coast basin (Pine Island ice and Thwaites Glacier) represents a potential for up to 1.5 m of equivalent sea level if it were to be entirely melted; doubling the current ice stream velocities in this region along with accelerated ice loss for the Antarctic Peninsula could raise sea level globally by 0.12 m in 2100. An increase in ice discharge has already been observed in several regions in Greenland. Assuming a doubling in ice discharge for outlet glaciers in Greenland and the Amundsen Coast basin in Antarctica, both ice sheets together could contribute up to about 0.28 m sea level by 2100 under the AR4 A1B warming scenario. {4.8}

Glaciers and small ice caps are losing mass and contributing to sea level rise. Glaciers and small ice caps are estimated to have contributed 0.8±0.17 mm y^{-1} SLR for the 1993-2003 time period, but they are not in balance with the present climate. The total sea level equivalent of glaciers and ice caps is 0.7 m. On average, glaciers and ice caps need to decrease in volume by about 26% to attain equilibrium with current warming, resulting in a minimum contribution to total change in sea level of about 0.18±0.03 m. For the AR4 A1B warming scenario, a contribution of 0.37±0.02 m SLR from glaciers by 2100 is projected. {4.8}

20th and 21st Century Changes in Sea Level

Global sea level has risen by about 0.2 m since 1870. The sea level rise by 2100 can be expected to total at least 0.60±0.11 m from thermal expansion (0.23±0.09 m) and loss of glaciers and ice caps (0.37±0.02 m). This lower limit is higher than previous studies due especially to improved information about glaciers. Additional contributions from Greenland and Antarctica are expected. Assuming ice loss from Greenland at the current rate, the total global sea level rise would be 0.65±0.12 m by 2100. Assuming a doubling in ice discharge from both Greenland and Antarctica, the total global average sea level rise would be 0.88±0.12 m by 2100. We therefore estimate a range of total global sea level rise in 2100 of about 0.5 to 1.0 m. Global sea level rise is a consequence of global warming and is caused by ocean water expansion and loss of ice stored on land (glaciers, small ice caps, and ice sheets). Satellite measurements show sea-level is rising at 3.1±0.4 mm y^{-1} since these records began in 1993 through 2003. This rate has decreased somewhat in the most recent years (2003-2008) to 2.5±0.4 mm y^{-1} due to a reduction of ocean thermal expansion from 1.6±0.3 mm y^{-1} to 0.37±0.1 mm y^{-1}, whereas contributions from glaciers, small ice caps, and ice sheets increased from 1.2±0.4 mm y^{-1} to 2.05±0.35 mm y^{-1}. Oceans respond slowly to global warming. The planet is already committed to a further 0.05 m sea level rise through thermal expansion alone over the next several centuries as a response to the past warming. Thermal expansion alone is expected to contribute about 0.23±0.09 m to sea level rise for the A1B scenario by 2100. Some semi-empirical models predict sea level rise up to 1.6 m by 2100 for a warming scenario of 3.1°C, a possible upper limit that cannot be excluded. {4.8}

IMPACTS ON NATURE AND SOCIETY

Food

Warming decreases yields of several crops in major growing regions, with ~5-10% yield loss per °C of local warming, or about 7-15% per °C of global warming (see Figure O.5). Crops tend to develop more quickly under warmer temperatures, leading to shorter growing periods and lower yields, and higher temperatures drive faster evaporation of water from soils. Increases in CO_2 levels can be beneficial for some crop and forage yields, for example by stimulating photosynthetic rates, but effects are much smaller for crops with a C4 photosynthetic pathway such as maize (corn). These direct effects of increased CO_2 compete with yield reductions linked to warming. {5.1}

Global climate change is expected to reduce yields of key food crops in some tropical regions by about 7-15% over about the next 20 years. This can be expected to make it more difficult to keep up with increasing food demand even if continuing advances in technologies and agricultural practices are as effective as in the past. As a point of comparison, the global

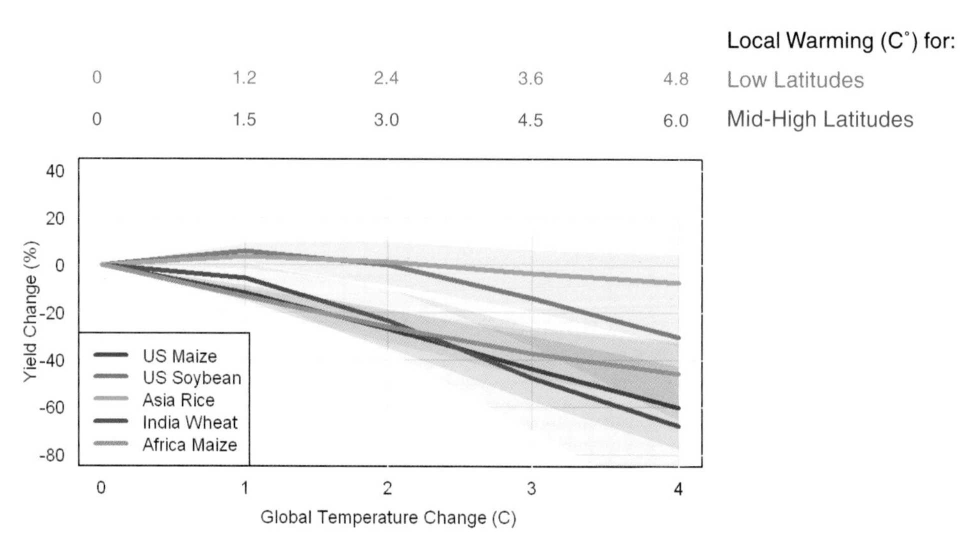

FIGURE O.5 Projected changes in yields of several crops worldwide as a function of global warming (relative to pre-industrial temperatures) in the absence of adaptation. Best estimates and likely uncertainty ranges are shown. {5.1}

demand for cereal crops can be expected to rise by about 25% over the same period. {5.1}

Up to roughly 2°C global warming, studies suggest that crop yield gains and adaptation measures (especially in higher latitude areas) could balance yield losses in tropical and other regions, but warming above 2°C is likely to increase global food prices. Major increases in trade of food from temperate to tropical areas are expected as a result of warming and represent one form of adaptation. Temperate growers are also likely to shift to earlier planting and longer maturing varieties as climate warms. However, adaptations are expected to be less effective in tropical regions where soil moisture, rather than cold temperatures, limits the length of the growing season. Very few studies have considered the evidence for ongoing adaptations to existing climate trends and have quantified the benefits of these adaptations. Future development of new varieties that perform well in hot and dry conditions may also promote adaptation, but the extent to which this will be effective remains unclear. At the higher warming scenarios considered in this report, it will be increasingly difficult to generate varieties with a physiology that can withstand extreme heat and drought while still being economically productive. Without adaptation, studies suggest that food prices could more than double if global warming were to be 5°C. These estimates do not include additional losses due to weeds, insects, and pathogens, changes in water resources available for irrigation, effects of increased flood or drought frequencies, or responses to temperature extremes. {5.1}

Global warming of 2°C would be expected to lead to average yield losses of U.S. corn of roughly 25% (±16% very likely range) unless effective adaptation measures are discovered and implemented. Nearly 40% of global corn production occurs in the United States, much of which is exported to other nations. The future yield of U.S. corn is therefore important for nearly all aspects of domestic and international agriculture. Higher temperatures speed development of corn and increase soil evaporation rates; further warming above 35°C can compromise pollen viability, all of which reduce final yields. A major challenge in developing drought- and heat-tolerant varieties is that traits that confer these attributes often reduce yields in good years. {5.1}

Fire

Wildfire frequency and extent is expected to change in many countries as the global average temperature increases. Site-specific studies suggest that large increases in the area burned are expected in parts of Australia, western Canada, Eurasia, and the United States. The primary driver of these changes is the warming in most of the regions evaluated, with lesser contributions from changes in precipitation. {5.4}

Areas of the United States that are particularly vulnerable to increases in wildfire extent include the Pacific Northwest and forested regions of the Rockies and the Sierra. Studies are limited in number but suggest that warming of 1°C (relative to 1950-2003) is expected to produce increases in median area burned by about 200-400% (see Figure O.6). Some dry grassland and shrub regions (for example, in the southwestern United States) may experience a decrease in wildfires, because warming without increases in precipitation would reduce biomass production and hence limit the availability of fuel. Uncertainties include understanding of local soil moisture changes with global warming. Over time, extensive warming and associated wildfires could exhaust the fuel for fire in some regions, as forests are completely burned. {5.4}

Ocean Acidification

Rising atmospheric CO_2 alters ocean chemistry, leading to more acidic conditions (lower pH) and lower chemical saturation states for calcium carbonate minerals used by many plants, animals, and microorganisms to make shells and skeletons. Ocean acidification is documented clearly from ocean time-series measurements for the past two decades. Surface ocean pH has dropped on average by about 0.1 pH units from pre-industrial levels (pH is measured on a logarithmic scale, and a 0.1 pH drop is equivalent to a 26% increase in hydrogen ion concentration). Additional declines of 0.15 and 0.30 pH units will occur if atmospheric carbon dioxide reaches about 560 ppm and 830 ppm, respectively (see Figure O.7). Polar surface waters will become under-saturated with respect to aragonite, a key calcium carbon mineral, for atmospheric CO_2 levels of 400-450 ppm for the Arctic and 550-600 ppm for the Antarctic. In tropical surface waters, large reductions in calcium carbonate saturation state will occur, but waters will remain super-saturated for projected atmospheric CO_2 during the 21st century. Subsurface waters will also be affected, but more slowly, governed by ocean circulation; the fastest rates will occur in the upper few hundred meters globally and in polar regions where cold surface waters sink into the interior ocean. {4.9}

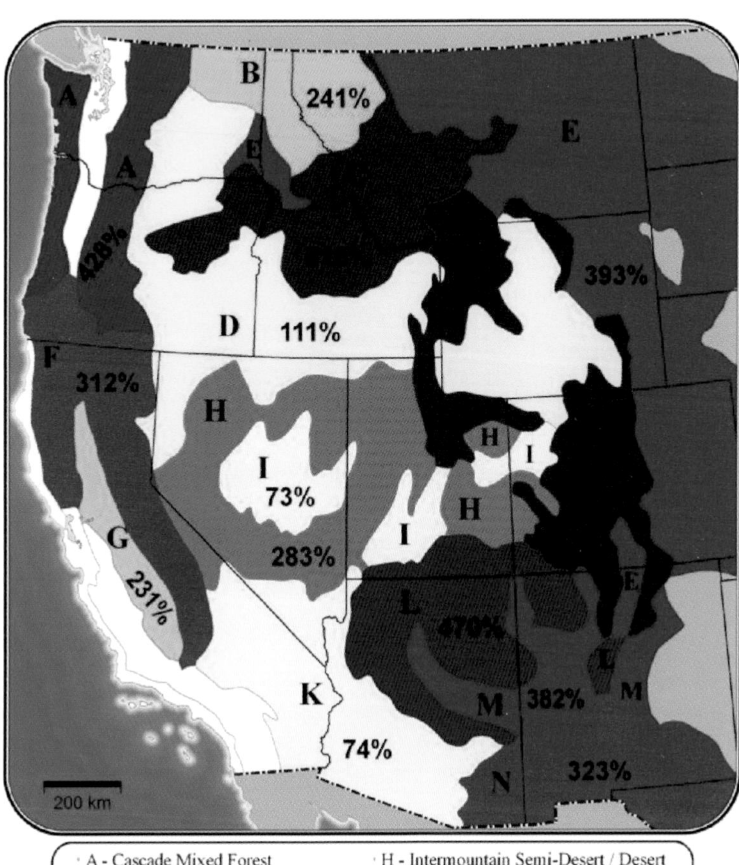

FIGURE O.6 Percent increase (relative to 1950-2003) in median annual area burned for ecoprovinces of the West with a 1°C increase in global average temperature. Changes in temperature and precipitation were aggregated to the ecoprovince level using the suite of models in the CMIP3 archive. Climate-fire models were derived from National Clmatic Data Center (NCDC) climate division records and observed area burned data following methods discussed in Littell et al. (2009). {5.4}

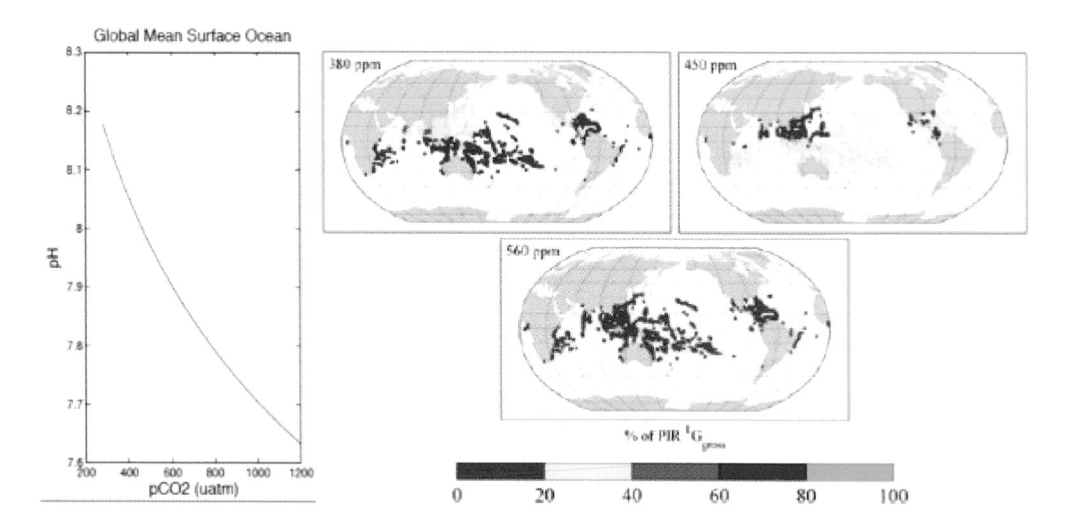

FIGURE O.7 (left panel) Variation in pH of global mean surface waters with CO_2. (right panel) Global coral reef distribution and their net community calcification; the biological production of calcium carbonate skeleton or shell material, relative to their pre-industrial rate (280 ppm), in percent, taking into account both ocean acidification and thermal bleaching; the loss of algal symbionts in response to warming and other stressors, for each reef location at CO_2 stabilization levels of 380, 450, and 560 ppm. Source: Silverman et al. (2009). {4.9; 5.8}

Impacts of CO_2, pH, and Climate Change on Ocean Biology

The patterns and rates of ocean primary production will change due to higher sea surface temperatures and increased vertical stratification, altering the base of the marine food-web. Observations indicate a strong negative relationship between marine primary productivity and warming in the tropics and subtropics, most likely due to reduced nutrient supply, and low-latitude primary production is projected to decline on basin-scales under future climate warming. Primary production in some temperate and polar regions is projected to increase due to warming, reduced vertical mixing, and reduced sea ice cover. Subsurface oxygen levels are projected to decline due to warmer waters and altered ocean circulation, leading to an enlargement of oxygen minimum zones. {5.8}

The geographic range of many marine species is shifting poleward and to deeper waters due to ocean warming. Individual marine species will change differentially, for example with the ranges of pelagic fish likely changed more than those of demersal fish. Few studies have looked comprehensively across many marine taxa and geographic regions, but a recent study suggests the potential for significant changes in community structure in the Arctic and Southern Ocean biodiversity due to invasion of warm water species and high local extinction rates in the tropics and subpolar domains. {5.8}

Coral bleaching events will likely increase in frequency and severity under a warmer climate. Over the past several decades, warmer sea surface temperatures have led to widespread tropical coral bleaching events and increased coral mortality, and warming and more local human impacts are associated with declines in the health of coral reef ecosystems worldwide. Bleaching can occur for sea surface temperature changes as small as +1-2°C above climatological maximal summer sea surface temperatures, which corresponds to global average warming of about 1.5-3°C (Figures S.5 and O.7). {5.8}

Rising CO_2 and ocean acidification will likely reduce shell and skeleton growth by marine calcifying species such as corals and mollusks. Some studies suggest a threshold of 500-550 ppm CO_2 whereby coral reefs would begin to erode rather than grow, negatively impacting the diverse reef-dependent taxa (see Figure O.7). Polar ecosystems also may be particularly susceptible when surface waters become undersaturated for aragonite, the mineral form used by many mollusks. Indirect impacts of ocean acidification on non-calcifying organisms and marine ecosystems as a whole are possible but more difficult to characterize from present understanding. {5.8}

Impacts of 21st Century Sea Level Rise

Depending on socioeconomic development, population growth, and intensity of adaptation, it has been projected that 0.5 m of sea level rise would increase the number of people at risk from coastal flooding each year by between 5 and 200 million; as many as 4 million of these people could be permanently displaced as a result. More than 300 million people currently live in coastal mega-deltas and mega-cities located in coastal zones. The corresponding projections for 1.0 m of sea level rise suggest that the number of people at risk of flooding each year would increase by 10 to 300 million. {5.2}

Coastal erosion is expected to occur as sea level rises with warming temperatures. Global aggregate estimates suggest that wetland and dry-land worldwide losses would sum to more than 250,000 km^2 with 0.5 m of sea level rise; more than 90% of these losses are projected to occur in developing countries. {5.2}

Rising sea levels will impact key coastal marine ecosystems, coral reefs, mangroves, and salt marshes, through inundation and enhanced coastal erosion rates. Regional impacts will be influenced by local vertical land movements and will be exacerbated where the inland migration of ecosystems is limited by coastal development and infrastructure. {5.2}

Infrastructure Impacts

Climate change impacts on infrastructure—including transportation, buildings, and energy—are primarily driven by changes in the frequency and intensity of temperature extremes and heat waves, heavy rainfall and snow events, and sea level rise. Many impacts are directly tied to changes in climate thresholds, such as number of days above or below a certain temperature, or amount of rainfall accumulated in a 24-hour period, rather than mean temperatures. Extreme events confront infrastructure with conditions outside the range for which they were built; to the extent that these extreme events increase in a given region, vulnerability of infrastructure will increase. Studies clearly document substantial economic damages from past extreme events, but it is currently difficult to generalize any relationships between temperature change and the magnitude and/or cost of impacts across regions and sectors. {5.5}

Local conditions can magnify the susceptibility of infrastructure to climate-related impacts. High-risk locations include the Arctic and low-lying coastlines. Climate change impacts have already been observed in high-latitude and high-elevation areas built on permanently frozen ground. Impacts include increasing coastal erosion and shoreline damage from storms as sea ice retreats; and land-based impacts including a shorter land travel season and formation of cracks and sinkholes in the ground from melting permafrost. A significant amount of infrastructure is located in low-elevation regions at risk of flooding due to sea level rise and storm surge. Infrastructure in coastal areas includes cities, power stations, water treatment plants, roads and highways, homes and buildings, and oil and gas lines. {5.5}

Climate change is expected to increase electricity demand and affect production and reliability of supply. Observed correlations between daily mean near-surface air temperature and electricity demand suggest warmer summer temperatures, and more frequent, severe, and prolonged extreme heat events could increase demand for cooling energy. Increases in peak demand could be most severe in already heavily air-conditioned regions. At the same time, high temperatures combined with drought can threaten the reliability of present-day electricity supply from hydropower and traditional generation sources, such as coal, gas, and nuclear, that require cooling water. {5.5}

Human Health

Heat-related illness and deaths occur as a direct result of sustained, elevated levels of extreme temperatures during heat waves, which are projected to increase with increasing temperatures. The frequency and severity of heat waves in Europe and North America are projected to increase under climate change. Under a 2°C increase in global mean temperature, for example, the average number of days per year with maximum temperatures exceeding 38°C or 100°F across much of the south and central United States could increase by a factor of 3 relative to the 1961-1990 average. Under a 3.5°C increase, the number of days is projected to increase by 5 to nearly 10 times. Some adaptation is inevitable as populations become accustomed to permanently different conditions. However, most research indicates that the public health impacts of climate change are likely to increase with temperature extremes; and new research highlights the potential that heat stress may impose hard limits on the inhabitability of some land areas under global temperature changes on the order of 10°C or more. {5.5}

Climate change is likely to affect the geographic spread and transmission efficiency of illnesses and disease carried by hosts and vectors, but complexity precludes any quantitative estimates of the relationship between incidence of a given disease and temperature change. Confounding factors—involving viral, bacterial, plant, and animal physiology, as well as sensitivity to changes in climate extremes, including precipitation intensity and temperature variability—challenge attempts to resolve the influence of temperature on observed trends in disease incidence. Most recent projections suggest that the ranges of malaria and other diseases may shift, but increases in some areas will likely be accompanied by decreases in others. {5.5}

Climate change may exacerbate existing stressors, such as air pollution, water contamination, and pollen production. Warmer temperatures increase production of ground-level ozone, which affects respiratory health. For a given level of ozone precursor emissions, background ozone levels and days with high ozone pollution levels above a defined safety standard (or "ozone exceedances") are projected to increase across much of the United States. Where heavy precipitation increases, risk of water contamination could also increase. Shifts in growing season, mean temperatures, and atmospheric

CO_2 levels affect the length of the pollen season and the characteristics of the plans themselves, with some species shown to increase their pollen-producing capacity and even their toxicity. {5.6}

Ecology and Ecosystems

For at least the past 40 years, many species have been and are currently shifting the phenology (timing) of spring events in concert with warming temperatures. Examining 542 species of plants and 19 of animals, a large phenology (e.g., timing of blooming, egg laying, migrating) study of 21 European countries for the last 30 years of the 20th century found a total of 78% of species were shifting their spring phenology earlier and only 3% were shifting later. When combining all species showing change along with those in the same areas not showing a measurable change, species on average were found to change ~2.5 days per decade. Throughout the Northern Hemisphere the similar change was reported to be 2.3 days per decade. The magnitude of the change occurring only in those species showing a change on average was ~4 to ~5 days a decade. {5.7}

As the climate has warmed many species have been and are continuing to track this warming by shifting their ranges into areas that before warming were less hospitable due to cooler temperatures. Terrestrial species are moving toward the poles and up in elevation, while marine species are generally moving down to deeper waters. The average shift over many types of terrestrial species around the globe was about 6 km per decade. {5.7}

Historically, extinctions of most species have been found to be due to various stresses, such as land-use change, invasive species, and hunting, but now the vulnerability of many species to extinction is enhanced with the added stress placed upon them by climate change. Those species more prone to becoming in danger of extinction include those that have a maximum dispersal distance shorter than the distance to the closest "cool refuge" and those that are not a good colonizer and hence fail to become established in these cooler locations. {5.7}

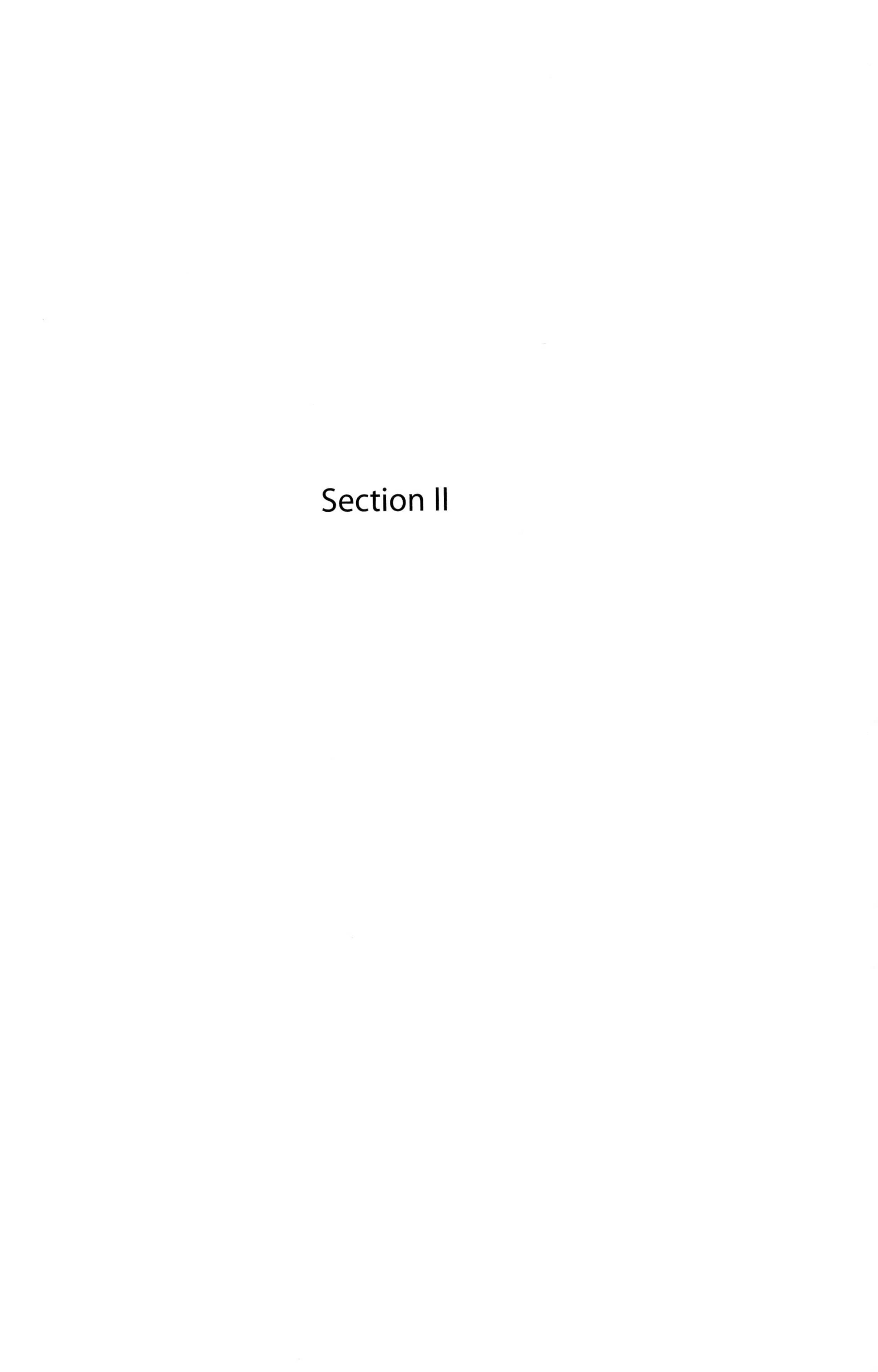

Section II

1

Introduction

1.1 THE FOCI OF THIS REPORT: CLIMATE CHANGES AND IMPLICATIONS FOR A RANGE OF ALTERNATIVE FUTURE WORLDS

The concentrations of several greenhouse gases (including carbon dioxide, methane, nitrous oxide, and other chemicals) have increased markedly since the start of the 20th century due to human activities. The changes in these gases very likely account for most of the globally averaged warming of the past 50 years (Hegerl et al., 2007). Carbon dioxide dominates the anthropogenic radiative "forcing" of Earth's climate due to manmade greenhouse gases. It has increased by more than 35% since 1750, now reaching the highest levels in at least 800,000 years (Forster et al., 2007; Luthi et al., 2008).

Human society faces important choices in the coming century regarding future emissions and the resulting effects that should be anticipated on Earth's climate, ecosystems, and people. One way of evaluating these choices is to consider the climate changes and impacts that are projected if human actions were to cause greenhouse gases to increase to particular concentration levels and then stabilize. The focus of this study is on the alternatives for the planet's future represented by stabilization of greenhouse gases at a broad range of "target" levels, hereafter referred to as stabilization targets. Transient climate changes and impacts experienced for increasing concentrations of greenhouse gases are also considered.

This report does not recommend or justify any particular target. Rather, our goal is to present the best scientific information available regarding the implications of different targets for human and natural systems. The charge to the committee was to evaluate the issue of serious or irreversible impacts of climate changes.

It should be emphasized that choosing among different targets is a policy issue rather than strictly a scientific one, because such choices involve questions of values, e.g., regarding how much risk to people or to nature might be considered too much. Some climate changes could be beneficial for some people or regions, while being damaging to others. For example, with global warming, fewer people may die in cold waves while more people die in heat waves; similarly, crops may be more productive in parts of Canada while less productive in the United States, raising issues of international food trade and transfer (see Section 5.1). Treatment of such effects as cancelling one another would generally be a value judgment and is not deemed to be appropriate here. Due to these considerations, we do not comprehensively cover possible benefits of climate change that could accrue to some people or regions. The study does not seek to evaluate the financial costs or feasibility of achieving any given stabilization target, nor to identify possible mitigation strategies to attain the targets.

The primary challenge for this study is to quantify insofar as possible the expected outcomes of different stabilization targets using analyses and information drawn from the scientific literature. Data available from publicly available archives were used in analyses carried out for this study, including, e.g., the CMIP3 climate model intercomparison project, the new representative concentration pathway (RCP) scenarios, etc. Some analyses and runs were also carried out, using published models and methods. The report covers emissions, concentrations, changes in the physical climate (such as temperature, rainfall, sea level, etc.) and their time scales, as well as associated impacts (such as food production, flooding, ecosystem damage, etc). In evaluating impacts, we seek to identify a baseline that includes expectations regarding adaptation to climate change where appropriate, but we also identify instances where adaptation is possible but where its feasibility, likelihood, or effectiveness is presently unknown. The report represents a brief summary of a vast scientific literature and seeks to be illustrative and representative rather than comprehensive.

Warming is the frame of reference for evaluation of impacts used in this report for both conceptual and practical reasons. Many key future climate impacts are dependent upon the amount of global warming. Indeed, available data and modeling suggest that the magnitudes of several key impacts can be evaluated with relatively good accuracy for given amounts of global warming. Much of the available literature and analysis of climate impacts can be tied to specific warming levels but not readily linked to CO_2-equivalent concentrations (due for example to lack of specification of aerosol forcing between studies). Indicated warming levels are related here to the corresponding best estimates and uncertainties in CO_2-equivalent concentrations as well as to cumulative carbon emissions.

The study presents best estimates for climate change and its impacts at varying warming levels and associated stabilization targets, and it also briefly describes the level of understanding of the processes involved (e.g., physical, biological, etc.) in such changes. Presentation of best estimates provides a clear view of the best current understanding that may otherwise be poorly communicated, and this is one aim of this study. However, we also balance descriptions of best estimates with their corresponding uncertainties, along with appropriate coverage of issues of uncertain risks. Climate changes and their impacts are discussed on a global basis, and specific regional examples within America and American territories are also presented for the purpose of illustration. We summarize as appropriate the following factors: (1) the extent to which multiple studies are available, and the robustness of findings across work by a range of authors, (2) the scientific confidence in understanding of the underlying processes, and (3) studies that already attribute a contribution of observed changes to anthropogenic effects where available (see Section 1.2). Where attribution is already possible for current levels of climate change, confidence in further future changes is generally strengthened.

Many climate changes or impacts currently are understood only in a qualitative manner, and thus are not quantifiable as a function of stabilization target. The report assesses and identifies these unquantified factors to the extent practical based upon the available literature. It should be emphasized that these should not be considered negligible; indeed some of these could be very important, or even dominant, in evaluating future risks due to anthropogenic climate change.

Many studies involve the use of "pattern scaling" whereby it is assumed that the spatial pattern of future climate change computed for one level of perturbation (i.e., radiative forcing) may be scaled to derive values for another test case such as one with stronger forcing and larger perturbations. A variety of studies have shown that such methods generally simulate the results of atmosphere-ocean general circulation models rather well (see Section 4.2), although results are generally less robust for precipitation than for warming (see Section 4.3), and near regions where strong feedbacks such as sea ice retreat take place. We employ pattern scaling for many of the estimates in this report.

Earth's history shows that climate changes can occur on time scales ranging from decades to centuries to millennia. All of these time scales are considered here. As there is abundant evidence that climate is already changing in part due to human activities, a focus of the report is on the next few decades to century, where the climate changes under increased human forcing of the climate system represent a transient climate change, linked to transient climate response (or TCR, see Section 3.2), and reflects a changing climate in which relatively fast-responding variables such as water vapor and cloudiness change but slower processes such as ocean warming and ice melt may lag behind. This implies that only a portion of the climate change is "realized" during a period of increasing greenhouse gas concentrations. If greenhouse gases are stabilized following a period of changing radiative forcing, the climate system response within a few centuries can be expected to reflect the equilibrium climate sensitivity (see Section 3.2). On very long time scales of multi-millennia, changes in factors such as ice sheets and vegetation can lead to different and generally stronger climate changes under stabilization (see Chapter 6).

The report considers a range of warming levels, which can be related to stabilization targets ranging from 350 to near 2,000 ppmv of total greenhouse gas forcing expressed as equivalent carbon dioxide concentration (see Box 1.1 for definition). We note that 1,000 ppmv is a level that a number of studies suggest may be reached by 2100 for a "business as usual" scenario with little or no mitigation of emissions. Even higher concentrations have precedent in the Earth's paleo history millions of years ago (see Chapter 6), and illustrate how the Earth may be changed if, for example, feedbacks from the Earth system trigger large releases of carbon from peat.

This report is organized as follows:

The next part of this introductory chapter summarizes the attribution of currently observed climate changes and impacts, as a starting point for discussion of the future.

Chapter 2 describes the relationship between greenhouse gas emissions and concentrations for a range of gases and cases. Aerosols and other forcings are also discussed.

Chapter 3 presents the current understanding of how the climate changes due to greenhouse gas concentrations on a range of time scales including

BOX 1.1 CARBON DIOXIDE-EQUIVALENT (CO_2-EQ) CONCENTRATIONS

Greenhouse gases differ in their warming influence (radiative forcing) on the global climate system due to their different radiative properties and lifetimes in the atmosphere. These warming influences may be expressed through a common metric based on the radiative forcing of CO_2.

- ***CO_2-equivalent concentration*** is the concentration of CO_2 that would cause the same amount of radiative forcing as a given mixture of CO_2 and other forcing components.

decades, centuries, and millennia. It discusses best estimates and uncertainties, including those relating to feedbacks (such as carbon cycle feedbacks whereby more carbon is emitted from the biosphere in a warming planet).

Chapters 4 through 6 comprise the core of this report. They present climate changes and impacts responses as a function of warming, that is, corresponding to stabilization targets of 350 to 2,000 ppmv CO_2-equivalent, considering time frames up to 2100, as well as very long time frames several thousand years in the future.

1.2 ATTRIBUTION

We offer here a brief summary of detection and attribution results as they provide part of the foundation of the discussion of future projected changes in the physical climate system and the impacts on natural and human systems that could ensue from them. This summary is not intended to be comprehensive of all detection and attribution studies to date, but will start from the IPCC Earth Assessment Report (AR4), updating it by those more recent results that are specifically relevant to this report.

Formal detection and attribution of an anthropogenic influence over the *physical climate system* is based on analysis of spatial and temporal patterns in observations of climate parameters and on comparison of their statistical characteristics with those of the same patterns as simulated by climate models. Because models can be integrated by applying the known external forcings in designed experiments (natural only, anthropogenic only, natural and anthropogenic jointly[1]) or in unforced mode (i.e., a control simulation), the behavior of the system subjected to different forcings as well as in control mode can be characterized, and the observed behavior of the real climate system can be compared to test consistency with a naturally varying process or with a process subjected to externally (especially manmade) forcings, to a given degree of statistical confidence. The progress of formal detection and attribution (D&A) is thus linked inextricably to the accumulation over time and space of quality observations that allows computation of robust statistics of the parameters under study and comparison to the ability of climate models to reliably simulate those parameters' natural and forced behavior. Importantly, formal D&A compares *spatial and/or temporal patterns* of change in observations and model simulations, not simply magnitudes of changes, seeking to test the consistency of the process-driven behaviors between models and observations. These behaviors are defined

[1]Each of these examples can be split into more complex designs with single natural/anthropogenic forcings administered in isolation or jointly.

by the direction and geographical shape of the change in the observed and simulated variables, not by the absolute values of trends and spatial anomalies. The best established and robust results in detection and attribution of anthropogenic influences are those that have been documented for temperature increases at global and continental scales, at the surface, in the troposphere, and in the oceans. When moving from temperature to other parameters and from global and continental scales to smaller regional scales, the confidence and robustness of the results diminish but are still high for many other measures of physical changes (e.g., sea level pressure, temperature extremes, zonally averaged precipitation, atmospheric moisture and surface specific humidity). The evaluation of the IPCC has also included detection and attribution of changes that have not been analyzed through the statistical machinery of formal D&A but are either closely related to changes that have been so analyzed, or have been proven consistent in a qualitative way with experiments including anthropogenic forcings and not consistent with model simulations forced by natural inputs alone. In all cases the scientific reasoning and understanding of the mechanisms linking anthropogenic forcings to these changes buttress the evidence from observational records and models.

In Table 1.1 we list the physical parameters of the climate system whose changes have been detected and attributed. We list first those D&A results that have been documented in IPCC AR4, along with a measure of confidence assigned, followed by a list compiled from more recent peer-reviewed literature.

Formal D&A, as just described, relies on model simulations, by which a treatment (human factors included) versus control experiment (only natural factors included) is set up for D&A of climate variables, but D&A of impacts requires an additional modeling step, by which the behavior of the climate system is translated into effects on the natural of human systems under study. The additional modeling adds limitations and uncertainties, but, where it is possible using current understanding, D&A of impacts is methodologically similar to D&A of physical parameters' change, and it is only the degree of significance or uncertainty in the results that will have to account for the compounding of modeling steps and the approximation errors thus added. For many types of impacts, however, a direct modeling, with the ability to include the confounding factors that are often required in addition to the climate changes, is not possible, and D&A has to follow a multi-step pathway:

TABLE 1.1 Physical Parameters of the Climate System Whose Changes Have Been Detected and Attributed to Greenhouse Gas Increases and Other Human Factors by IPCC AR4 and More Recent Peer-reviewed Literature

Phenomena Affected by Anthropogenic Contribution	Degree of Confidence	Reference
Global surface air temperature (SAT) increase in the past half century	Very Likely	IPCC AR4 WG1 Ch9 (IPCC, 2007a)
Larger than observed, in the absence of volcanoes and aerosols, Global Surface Air Temperature (SAT) increase in the past half century.	Likely	IPCC AR4 WG1 Ch9 (IPCC, 2007a)
Warming of continental-scale SAT since middle of 20th century	Likely	IPCC AR4 WG1 Ch9 (IPCC, 2007a)
Temperature extremes over land in the Northern Hemisphere and Australia	Likely	IPCC AR4 WG1 Ch9 (IPCC, 2007a)
Tropopause height increases in latter half of 20th century	Likely (with contribution of stratospheric ozone decrease)	IPCC AR4 WG1 Ch9 (IPCC, 2007a)
Tropospheric warming in latter half of 20th century	Likely	IPCC AR4 WG1 Ch9 (IPCC, 2007a)
Simultaneous tropospheric warming and stratospheric cooling in latter half of 20th century	Very likely (because of their happening in conjunction)	IPCC AR4 WG1 Ch9 (IPCC, 2007a)
Warming of the upper layer of the oceans	Likely	IPCC AR4 WG1 Ch9 (IPCC, 2007a)
Sea level rise in latter half of 20th century	Very Likely	IPCC AR4 WG1 Ch9 (IPCC, 2007a)
Sea level pressure trends in latter half of the 20th century (spatial patterns of increases and decreases in both hemispheres[a])	Likely	IPCC AR4 WG1 Ch9 (IPCC, 2007a)
Reduction of NH (Arctic) sea ice extent and global glacier retreat in latter half of 20th century	Likely	IPCC AR4 WG1 Ch9 (IPCC, 2007a)
Increases in intense tropical cyclone activity	More Likely Than Not[b]	IPCC AR4 WG1 Ch9 (IPCC, 2007a)
Increases in heavy rainfall on global land areas during latter half of 20th century	More Likely Than Not[b]	IPCC AR4 WG1 Ch9 (IPCC, 2007a)
Increased risk of drought in latter half of 20th century	More Likely Than Not[b]	IPCC AR4 WG1 Ch9 (IPCC, 2007a)
Zonal mean precipitation changes (increases in the Northern Hemisphere mid-latitudes, drying in the Northern Hemisphere subtropics and tropics, and moistening in the Southern Hemisphere subtropics and deep tropics	NA	Zhang et al., 2007c
Water vapor and surface specific humidity increases	NA	Santer et al., 2007; Willett et al., 2007.

continued

TABLE 1.1 Continued

Higher temperatures/early snowmelt/early run-off in U.S. Southwest	NA	Maurer et al., 2007; Barnett et al., 2008; Bonfils et al., 2008, Hidalgo et al., 2009; Pierce et al., 2008
Arctic and Antarctic temperature increases; Arctic seaice decrease; Arctic precipitation increase	NA	Gillett et al., 2008a; Min et al., 2008.
Increased atmospheric winter (JFM and JAS) storminess in high latitudes	NA	Wang et al., 2009
SSTs warming in cyclogenesis regions of Atlantic and Pacific oceans	NA	Gillett et al., 2008b
Increased ocean salinity in Atlantic Ocean	NA	Stott et al., 2008

NOTE: The degree of confidence is noted when available from the expert judgment by the IPCC AR4 authors. More recent results (indicated by the light blue background) have appeared in the peer-review literature, but their degree of confidence has not been assessed and thus does not appear in our table.

[a]See Figure 9.16 of IPCC AR4 WG1 (Chapter 9).

[b]Consistent with theoretical expectations of the effects of anthropogenic forcings but have not been detected and/or attributed, due to modeling uncertainties or lack of skill, and poor quality/limited extent of the data record.

1. a change in the system impacted is first associated with climatic factors, and, separately,
2. the climatic factors are shown to be attributable to anthropogenically caused changes.

In order to link the two steps formally, a measure of significant positive spatial correlation between changes in the impacted system and changes in climate was primarily used in the IPCC WG2 report to summarize many studies of impacts all over the globe. This correlation-based argument was also supported by the existence of several studies that performed formal D&A through modeling of specific systems or domains, and, importantly, by scientific understanding of the reasons why the impacts are consistent with warming of the climate system.

In summary, most of the attribution of impacts that has taken place thus far has relied on the documented attribution of the warming patterns of the physical system coupled to the scientific understanding of the effects of such warming on natural and human systems, together with significantly

TABLE 1.2 List of Observed Impacts Attributed to Global Warming by IPCC AR4 WG2 with Results from Two Studies Using Formal D&A Methodology

Driving Change in Physical Climate	Impacts	Reference
Changing snow/ice/frozen ground	Increase in number and extension of glacial lakes, increase in avalanches and instability of permafrost, and changes in polar ecosystems	IPCC AR4 WG2 (IPCC, 2007b)
Warmer temperatures	Increase in runoff and anticipated snow melt in spring; increase in temperatures in lakes and rivers	IPCC AR4 WG2 (IPCC, 2007b)
Warmer temperatures	Earlier timing of spring event (foliage, migrations, and egg-laying) and range shifts (poleward and upward) of terrestrial biological systems	IPCC AR4 WG2 (IPCC, 2007b)
Rising ocean water temperatures/ changes in ice cover, salinity, oxygen levels, and circulation	Shifts in the ranges of algae/plankton/fish abundance in high-latitude oceans and high-latitude/high-elevation lakes	IPCC AR4 WG2 (IPCC, 2007b)
Increase in atmospheric concentration of CO_2	Ocean acidification (no effect on ocean life documented yet	IPCC AR4 WG2 (IPCC, 2007b)
Summer temperature warming in Canada	Increase in area burned by forest fires in Canada	Gillett et al., 2004
Increase in temperature	Increase in growing season length	Christidis et al., 2007

NOTE: Confidence levels are not available because the IPCC AR4 WG2 did not supply them.

positive spatial correlation between said impacts and regionally differentiated warming.

This reasoning is at the basis of the list of impacts attributed to global warming by IPCC AR4 WG2 in its Summary for Policy Makers, which we list, together with results from two studies that applied formal D&A methodology,[2] in Table 1.2.

[2]Of these two studies, the first is reviewed by WG,2 but its results are not explicitly listed in its Summary for Policy Makers, SPM; the second appeared later than the release of the report.

2

Emissions, Concentrations, and Related Factors

2.1 CONTRIBUTION OF DIFFERENT CHEMICALS TO CO_2 EQUIVALENT LEVELS AND CLIMATE CHANGES

A range of anthropogenic chemical compounds contribute to changing Earth's energy budget, thereby causing the planet's global climate to change. For example, increases in greenhouse gases absorb infrared energy that would otherwise escape to space, acting to warm the planet, while some types of aerosol particles can contribute to cooling the planet by reflecting incoming visible light from the Sun. These components of our atmosphere are emitted from a variety of human activities, including for example fossil fuel burning, land-use change, industrial processes such as cement production, and agriculture. The gases and particles involved are frequently referred to as drivers of climate change, or radiative forcing agents. Detailed reviews of radiative forcing is presented in Forster et al. (2007) and Denman et al. (2007). Radiative forcing due to various climate change agents can be converted to equivalency with the concentration of CO_2 (CO_2 equivalent), one frame of reference for this report (see Figure 2.1). Here we briefly summarize how major forcing agents contribute to current and future CO_2-equivalent target levels and explore implications for global mean temperature increases.

Some greenhouse gases and aerosols are retained for days to years in the atmosphere after emission. The concentrations of such compounds in the atmosphere are tightly coupled to the rate of emission. Their concentrations would drop rapidly if emissions were to cease. Increasing emissions lead to increases in concentrations of such gases, while constant emissions are required for their concentrations to be stabilized. Methane is a key greenhouse gas with an atmospheric lifetime of about 10 years whose concentration has approximately doubled since the pre-industrial era (1750), and it is the second most important greenhouse gas, currently contributing about

25 ppmv of CO_2 equivalent (see Figure 2.1). From about 1998 to 2007, methane concentrations remained nearly constant (Forster et al., 2007). However, methane began to increase after 2007. In the absence of mitigation, methane is expected to continue to make significant contributions to climate change during the 21st century (see Section 2.2).

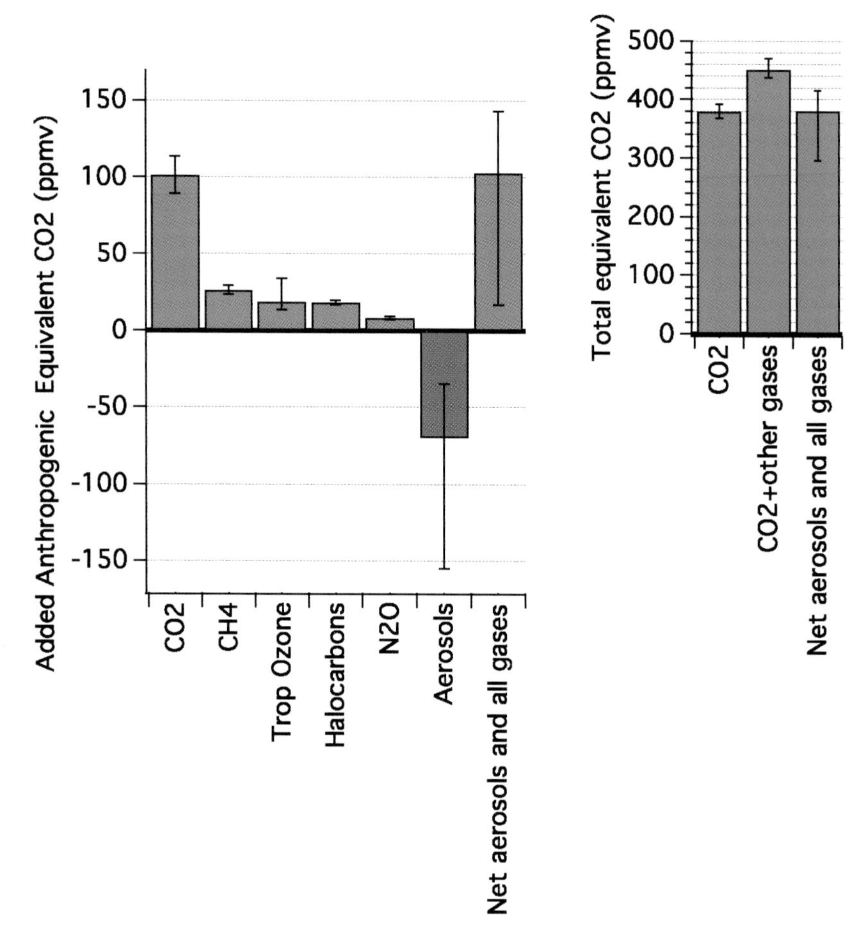

FIGURE 2.1 (left) Best estimates and very likely uncertainty ranges for aerosols and gas contributions to CO_2-equivalent concentrations for 2005, based on the radiative forcing given in Forster et al. (2007). All major gases contributing more than 0.15 W m^{-2} are shown. Halocarbons including chlorofluorocarbons, hydrochlorofluorocarbons, hydrofluorocarbons, and perfluorocarbons have been grouped. Direct effects of all aerosols have been grouped together with their indirect effects on clouds. (right) Total CO_2-equivalent concentrations in 2005 for CO_2 only, for CO_2 plus all gases, and for CO_2 plus gases plus aerosols.

In sharp contrast, some greenhouse gases have biogeochemical properties that lead to atmospheric retention times (lifetimes) of centuries or even millennia. These gases can accumulate in the atmosphere whenever emissions exceed the slow rate of their loss, and concentrations would remain elevated (and influence climate) for time scales of many years even in the complete absence of further emission. Like the water in a bathtub, concentrations of carbon dioxide are building up because the anthropogenic source substantially exceeds the natural net sink. Even if human emissions were to be kept constant at current levels, concentrations would still increase, just as the water in a bathtub does when the water comes in faster than it can flow out the drain. The removal of anthropogenic carbon dioxide from the atmosphere involves multiple loss mechanisms, spanning the biosphere and ocean (see Section 2.4), and carbon dioxide removal cannot be characterized by any single lifetime. Although some carbon dioxide would be lost rapidly to the terrestrial biosphere and to the shallow ocean if human emissions cease, some of the enhanced anthropogenic carbon will remain in the atmosphere for more than 1,000 years, influencing global climate (Archer and Brovkin, 2008). The warming induced by added carbon dioxide is expected to be nearly irreversible for at least 1,000 years (Matthews and Caldeira, 2008; Solomon et al., 2009), see Section 3.4.

Figure 2.1 shows that carbon dioxide is the largest driver of current anthropogenic climate change. Other gases such as methane, nitrous oxide, and halocarbons also make significant contributions to the current total CO_2-equivalent concentration, while aerosols (see Section 2.3) exert an important cooling effect that offsets some of the warming. The best estimate of net total CO_2 equivalent concentration of the sum across these forcing agents in the year 2005 is about 390 ppmv (with a very likely range from 305 to 430 ppmv). Global carbon dioxide emissions have been increasing at a rate of several percent per year (Raupach et al., 2007). If there were to be no efforts to mitigate its emission growth rate, scenario studies suggest that carbon dioxide could top 1,000 ppmv by the end of the 21st century. Carbon dioxide alone accounts for about 55% of the current total CO_2-equivalent concentration of the sum of all greenhouse gases, and it will increase to between 75 and 85% by the end of this century based on a range of future emission scenarios (see Section 2.2). Thus carbon dioxide is the main forcing agent in all of the stabilization targets discussed here, but the contributions of other gases and aerosols to the total CO_2-equivalent remain significant, motivating their consideration in analysis of stabilization issues.

How large a reduction of emissions is required to stabilize carbon dioxide concentrations, and does it depend upon when it is done or on the chosen target stabilization concentration? Studies over the past five years of so using many different carbon cycle models have improved our understanding of requirements for carbon dioxide stabilization. This is because of more detailed treatments of carbon-climate feedbacks, including the ways in which warming decreases the efficiency of carbon sinks as compared to earlier work (e.g., Jones et al., 2006; Matthews, 2006). Figure 2.2 shows an example of stabilization for two different Earth Models of Intermediate Complexity (EMICs), the University of Victoria (UVIC) model and the Bern model (see Methods section for descriptions of these two models; see also Plattner et al., 2008, and references therein for a model intercomparison study). In this example test case, carbon dioxide emissions increase at current growth rates of about 2% per year to a maximum of about 12 GtC per

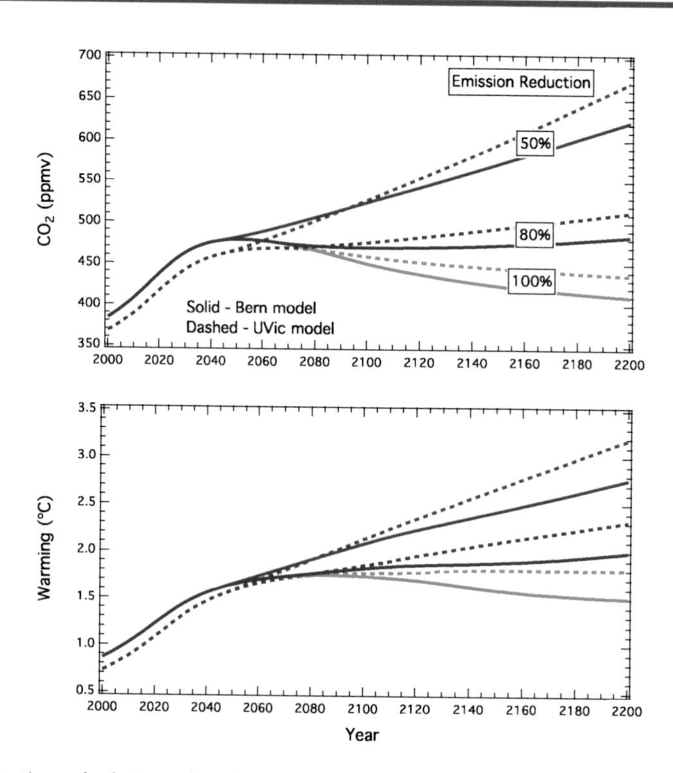

FIGURE 2.2 Illustrative calculations showing CO_2 concentrations and related warming in two EMICS (the Bern model and the University of Victoria model, see Methods) for a test case in which emissions first increase, followed by a decrease in emission rate of 3% per year to a value 50%, 80%, or 100% below the peak. The test case with 100% emission reduction has 1 trillion tonnes of total emission and is also discussed in Section 3.4.

year, followed by a decrease of 3% per year down to a selected total reduction of 50, 80, or 100%. The rate of decrease of 3% per year used here is derived from scenario analysis described in the next section. This section together with the next section aim to probe what plausible rates of emissions reduction based upon scenario studies imply for the future evolution of carbon dioxide concentrations. The rate of possible emissions reductions of carbon dioxide depends upon factors including e.g., commitments to existing infrastructure and development of alternatives, see Section 2.2. It is interesting to note that even in the case of the phaseout of ozone-depleting substances under the Montreal Protocol, emissions reductions were about 10% per year initially but stalled at a total reduction of about 80% of the peak, with some continuing emissions of certain gases occurring due for example, the challenge of finding alternatives for fire-fighting applications (see IPCC, 2005).

Figure 2.2 shows that carbon emission reductions of 50% do not lead to long-term stabilization of carbon dioxide, nor of climate, in either of these models, as has also been shown in previous studies (e.g., Weaver et al., 2007). It is noteworthy that the Bern model has weaker carbon-climate feedbacks than the UVIC model; nevertheless both models show the need for emissions reductions of at least 80% for carbon dioxide stabilization even for a few decades, while longer-term stabilization requires nearly 100% reduction. Very similar results were obtained in other test cases run for this study considering peaking at higher values, or decreasing at rates from 1 to 4% per year (see also Meehl et al., 2007; Weaver et al., 2007). Figure 2.3 shows sample calculations evaluated in Meehl et al. (2007) using three different models for various stabilization levels. Figure 2.3 shows that stabilization levels of 450, 550, 750, or 1,000 ppmv require eventual emission reductions of 80% or more (relative to whatever peak emission occurs) in all of the models evaluated. Thus current representations of the carbon cycle and carbon-climate feedbacks show that anthropogenic emissions must approach zero eventually if carbon dioxide concentrations are to be stabilized in the long term (Matthews and Caldeira, 2008). This is a fundamental physical property of the carbon cycle and is independent of the emission pathway or selected carbon dioxide stabilization target. Box 2.1 discusses how emissions of non-CO_2 greenhouse gases could affect attainment of stabilization targets.

Figures 2.2 and 2.3 illustrate a fundamental change in understanding stabilization of climate change that has been prompted by the scientific literature of the past two years or so (see Jones et al., 2006; Matthews and Caldeira, 2008). Early work on stabilization using relatively simple models

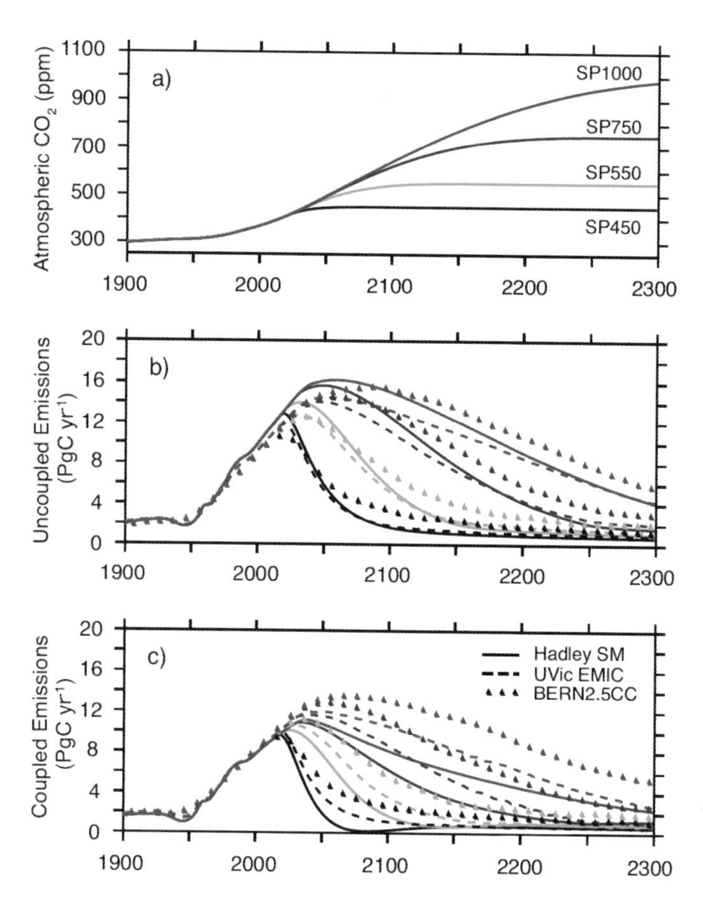

FIGURE 2.3 (a) Illustrative atmospheric CO_2 stabilization scenarios for 1,000, 750, 550, and 450 ppmv; SP1000 (red), SP750 (blue), SP550 (green) and SP450 (black), from Meehl et al. (2007). (b) Compatible annual emissions calculated by three models, the Hadley simple model (solid), the UVIC EMIC (dashed), and the BERN2.5CC EMIC (triangles) for the three stabilization scenarios. Panel (b) shows emissions required for stabilization without accounting for the impact of climate on the carbon cycle, while panel (c) included the climate impact on the carbon cycle, showing that emission reductions in excess of 80% (relative to peak values) are required for stabilization of carbon dioxide concentrations at any of these target concentrations.

suggested that slow reductions in emissions could lead to eventual stabilization of climate (e.g., Wigley et al., 1996). But recent studies using more detailed models of key feedbacks in the ocean, biosphere, and cryosphere, have underscored that although a quasi-equilibrium may be reached for a limited time in some models for some scenarios, stabilizing radiative forcing at a given concentration does not lead to a stable climate in the long run. Cumulative emitted carbon can more readily be linked to climate stabiliza-

BOX 2.1 STABILIZATION AND NON-CO_2 GREENHOUSE GASES

Because carbon emissions reductions of more than 80% are required to stabilize carbon dioxide concentrations, small continuing emissions of carbon dioxide, or emissions of CO_2-equivalent through other gases, could have surprisingly important implications for stabilizing climate change. For example, emissions of the hydrofluorocarbons (HFCs) currently used as substitutes for chlorofluorocarbons make a small contribution to today's climate change. However, because emissions of these gases are expected to grow in the future if they are not mitigated, and because of the stringency of the requirement of near zero emissions of CO_2-equivalent, these gases could represent a significant future impediment to stabilization efforts. For example, Figure 2.4 below shows that in the absence of mitigation, the HFCs could represent as much as one-third of the allowable CO_2-equivalent emissions in 2050 required for a stabilization target of 450 CO_2-equivalent. Thus, the analysis presented here underscores that stabilization of climate change requires consideration of the full range of greenhouse gases and aerosols, and of the full suite of emitting sectors, applications, and nations.

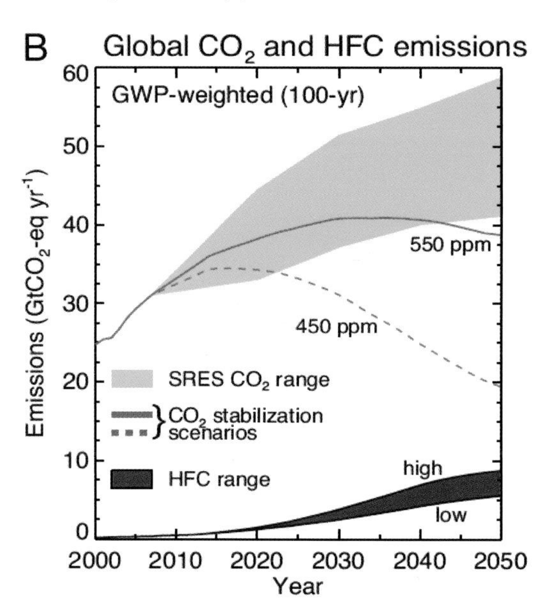

FIGURE 2.4 Global CO_2 and HFC emissions expressed as CO_2-equivalent emissions per year for the period 2000-2050. The emissions of individual HFCs are multiplied by their respective GWPs (direct, 100-year time horizon) to obtain aggregate emissions across all HFCs expressed as equivalent $GtCO_2$ per year. High and low estimated ranges based on analysis of likely demand for these gases and assuming no mitigation of HFCs are shown. HFC emissions are compared to emissions for the range of SRES CO_2 scenarios, and two 450- and 550-ppm CO_2 stabilization scenarios. The estimated CO_2-equivalent emissions due to HFCs in the absence of mitigation reach about 6 $GtCO_2$-equivalent in 2050, or about a third of the emissions due to CO_2 itself at that time in the 450 ppm stabilization scenario. Source: Velders et al. (2009).

tion, due to the irreversible character of the induced warming driven by carbon dioxide (see Section 3.4).

2.2 INFORMATION FROM SCENARIOS

Figure 2.5 shows the emissions of manmade greenhouse gases from various sectors of the U.S. economy (U.S. EPA, 2008). For highly industri-

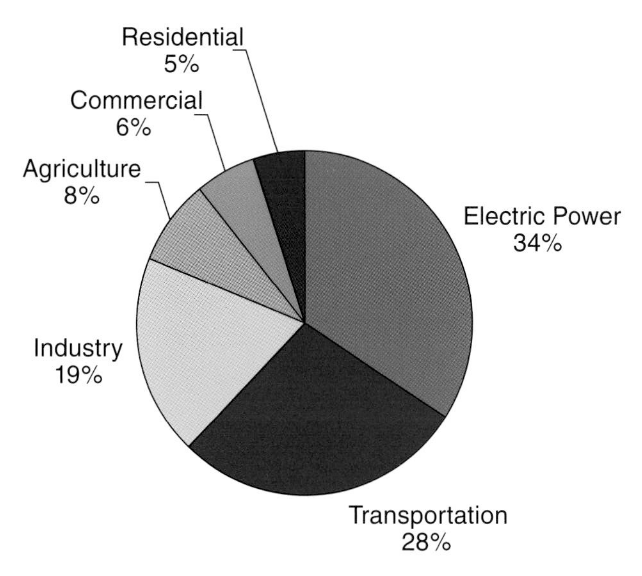

FIGURE 2.5 U.S. greenhouse gas emissions by sector in 2006. Source: U.S. EPA (2008).

alized countries such as the United States, the difficulty in reducing emissions will depend in large part on the lifetimes of the *existing* capital stock associated with the major emitting sectors. The electric sector is the largest source of manmade emissions in the United States, primarily due to the carbon dioxide emitted during the combustion of fossil fuels. The lifetime of coal-fired power plants is measured in decades. The next largest source of U. S. greenhouse gases is the transportation sector, again due to the combustion of fossil fuels. Here the lifetime of the capital stock is typically a decade or two.

Although developed countries historically have been the major emitter of greenhouse gases, developing countries are on track to overtake them in the next few years. In their case, the issue becomes one of the capital stock put in place in the *future* to support their industrialization process. With the huge economic growth projected for developing countries and in the absence of incentives to act otherwise, these countries will likely turn to the cheapest energy sources to fuel their growth. These fuels currently are fossil based: coal, oil, and gas. A recent study by the Energy Modeling Forum, based on eight Energy-Economy models, projected an annual growth rate of CO_2 emissions globally from the burning of fossil fuels and industrial uses to be of the order of 1 to 2 percent per year over the remainder of the century, in the absence of intervention (EMF 22, 2009). The study attributes much of the growth to developing countries.

Even if wealthier countries like the United States were to reduce their emissions to zero immediately, it is unlikely that global CO_2 emissions would be stabilized, much less global atmospheric concentrations (Blanford et al., 2009). Being in their post-industrial phase of development, the economic growth rates in developed countries are expected to be lower than those of developing countries and their mix of goods and services less carbon intensive. The cumulative reductions of developed countries, even with aggressive emissions reduction programs, are expected to be low when compared to those of developing countries.

One important contribution that developed countries can make to global emissions reductions is to develop the technological wherewithal that would not only be necessary for their own emission reductions, but also would be essential for developing countries to meet their economic development goals with affordable climate-friendly technologies.

As noted above, both the existing capital stock and that put in place in the future are critical to understanding the difficulty of transitioning away from the current path of growth in greenhouse gas emissions. Figure 2.6 shows representative carbon pathways (RCPs) for limiting radiative forcing (watts per m^2) at two alternative levels. These are referred to as the RCP 2.6[1] and RCP 4.5 scenarios. These are among a suite of pathways being developed for use in the IPCC 5th Assessment. The pathways shown in the figure were developed by the IMAGE and MiniCAM models, respectively (Moss et al., 2010).

Figure 2.6 highlights the importance of the carbon budget, that is, the area under the allowable emissions curve associated with a particular radiative forcing target. Being much lower in the RCP 2.6 scenario than in the RCP 4.5 scenario, we see the rate of growth first slows and then rapidly decline beginning in 2020. In the case of the higher CO_2 budget, emissions rise for another two decades before peaking. Notice that the maximum rate of decline is comparable in the two scenarios (about 3.5% per year); however, in the latter it is shifted out in time. The reasons for this shift are both the higher

[1]Although Moss et al. (2010) refers to this as the RCP 2.6 scenario, this is the one RCP scenario that peaks and then declines. For this reason it is also referred to as the RC P3-PD scenario. The RCP 3-PD has a unique shape. The radiative forcing of RCP 3-PD peaks and declines (PD), while the radiative forcing of the other RCPs stabilize or rise towards their higher 2100 levels. Specifically, the final RCP 3-PD prepared for climate modeling peaks at 2.99 W/m^2 in 2050 and then declines to 2.71 W/m^2 in 2100 with the decline continuing beyond 2100. The decline is due to the availability later in the century of a negative-emitting technology, biomass with carbon capture and storage (BECs).

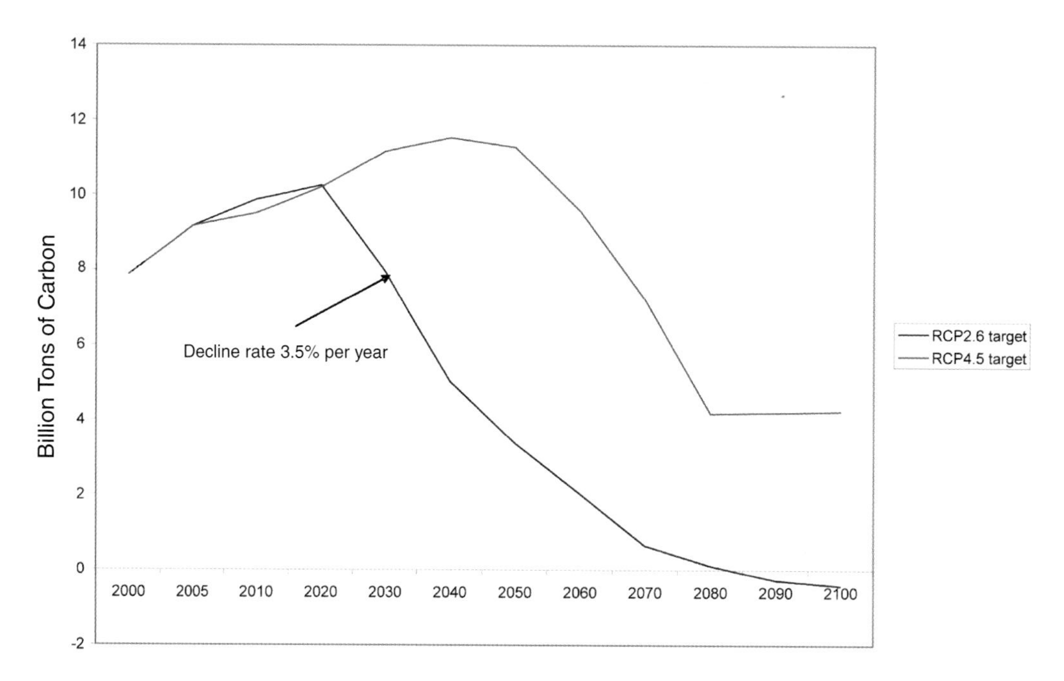

FIGURE 2.6 shows representative carbon pathways (RCPs) for limiting radiative forcing (W per m^2) at two alternative levels. The tighter the limit, the earlier the reductions must take effect. With the RCP 2.6 scenario, the rate of growth first slows and then rapidly declines beginning in 2020. In the case of the less stringent constraint, emissions rise for another two decades before peaking. Here the decline is shifted out in time.

carbon budget and a greater array of low-carbon, economically competitive alternatives, which are assumed to become available in the future.

We stress that there is a great deal of flexibility regarding the rate at which new technologies are substituted for existing ones, both on the supply and demand sides of the energy sector. The rate of retirement of existing carbon-intensive plant and equipment and their replacement with more climate-friendly alternatives will depend upon a number of factors. These include the stabilization target, reference case emissions in the absence of a price on CO_2 (either explicit or implicit), the availability and costs of alternatives, and the *willingness* to pay the costs of the transition to a low-carbon economy. The latter will depend on society's perception of the benefits (reduction in damages due to climate change). From a purely physical perspective, decline rates much higher than those shown here are feasible. It is a matter of the perceived urgency and the motivation to decarbonize.

Figure 2.7 shows the CO_2-equivalent concentrations for these two scenarios at three points in time. Notice that in the case of the tighter radiative forcing goal, there is some "overshoot". That is, the target is exceeded in the middle part of the century and then gradually approached. This is due to the assumption that there will be a "negative" emitting technology, Bioenergy with Carbon Capture and Sequestration (BECS). Otherwise a faster decline rate of the capital stock would be required.

2.3 SHORT-LIVED RADIATIVE FORCING AGENTS: PEAK TRIMMING VERSUS BUYING TIME

The role of CO_2 emissions in Earth's climate future is unique among the major radiative forcing agents, because the impact of the CO_2 emitted into the atmosphere will continue to alter Earth's energy budget for millennia to come. As noted above, Earth's energy budget is also subject to the influence of a number of short-lived radiative forcing agents whose radiative effect would decay to zero on a time scale of weeks to decades if their sources where shut off. These agents include aerosols, black carbon on snow or ice, and methane.

Aerosols are produced by burning biomass and fossil fuels, but unlike CO_2 they are not an inevitable by-product of combustion. Some aerosols reflect energy to space, but other aerosols such as black carbon absorb sunlight. Reflecting aerosols unambiguously lead to cooling of the surface. Their effect has offset a portion of the radiative forcing from the anthropogenic increase of greenhouse gases so far, and any action that reduces reflecting aerosol emissions will lead to a nearly immediate warming. The effect of airborne absorbing aerosols is more subtle, because they primarily act to shift the absorption of solar radiation from the surface to the interior of the atmosphere, leaving the top-of-atmosphere energy budget largely unchanged (Randles and Ramaswamy, 2008). The global mean effect of surface black carbon is difficult to quantify but is unambiguously a warming, amplified further by the albedo feedback of melting snow and ice (Flanner et al., 2007; Hansen and Nazarenko, 2004; McConnell et al., 2007). Aerosol effects can include direct damage to human health and agriculture, implying that they should be cleaned up for reasons independent of climate (Agrawal et al., 2008; Ramanathan et al., 2008). A key question is whether the effort to do so will help or hurt other efforts to keep warming in check. Although the discussion of aerosol effects will be based primarily on temperature changes, it should be kept in mind that the spatially inhomogeneous radiative forcing from aerosols can lead to regional effects such as changes in clouds and the

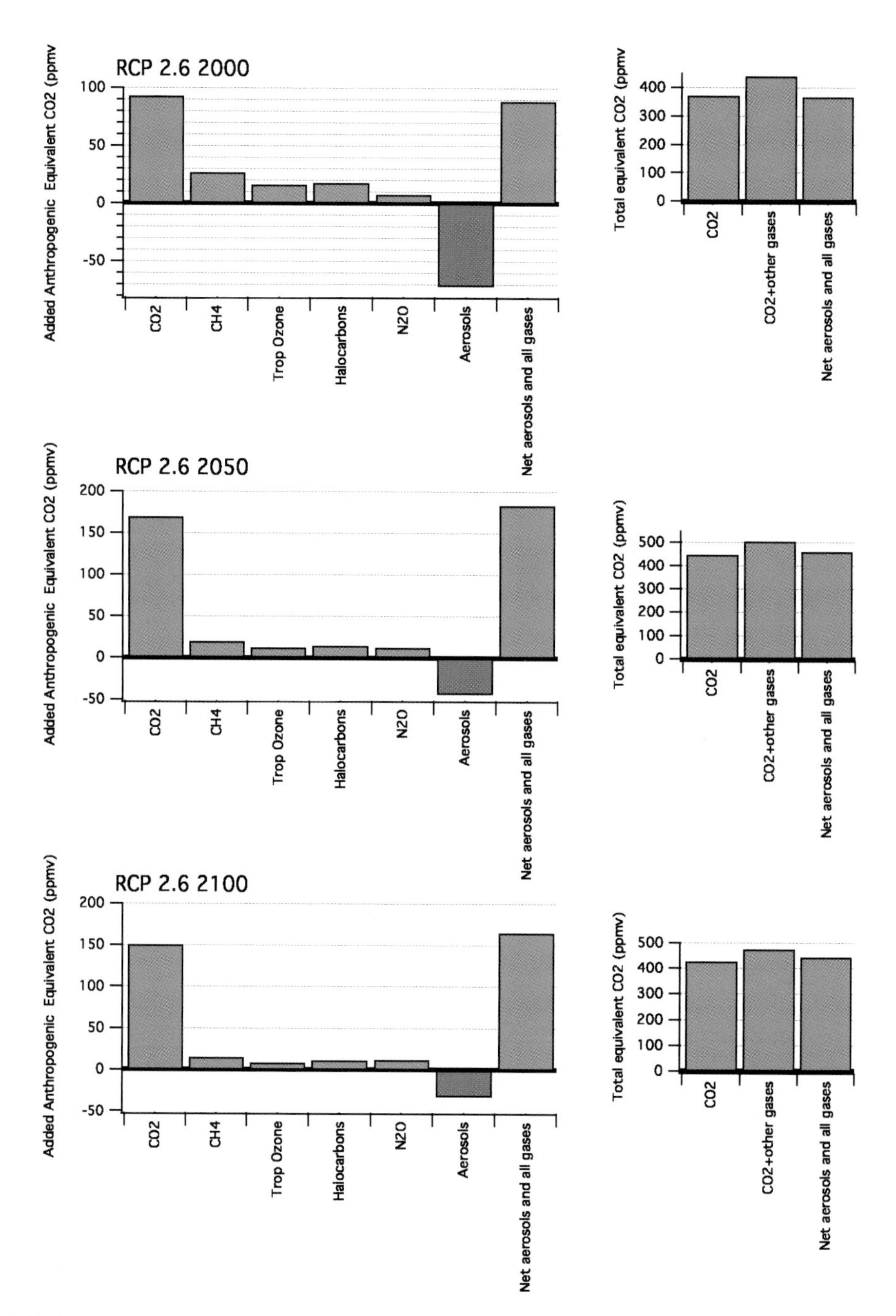

FIGURE 2.7 This figure illustrates components of radiative forcing (in CO_2-equivalent concentration units) for the RCP 2.6 (a) and RCP 4.5 (b) scenarios (see Moss et al., 2010). RCP 2.6 peaks at 3 W/m^{-2} before 2100 and then declines. There is some "overshoot" where the target is exceeded and is then gradually approached (see footnote 1). RCP 4.5 stabilizes at 4.5 W/m^{-2} after 2100.

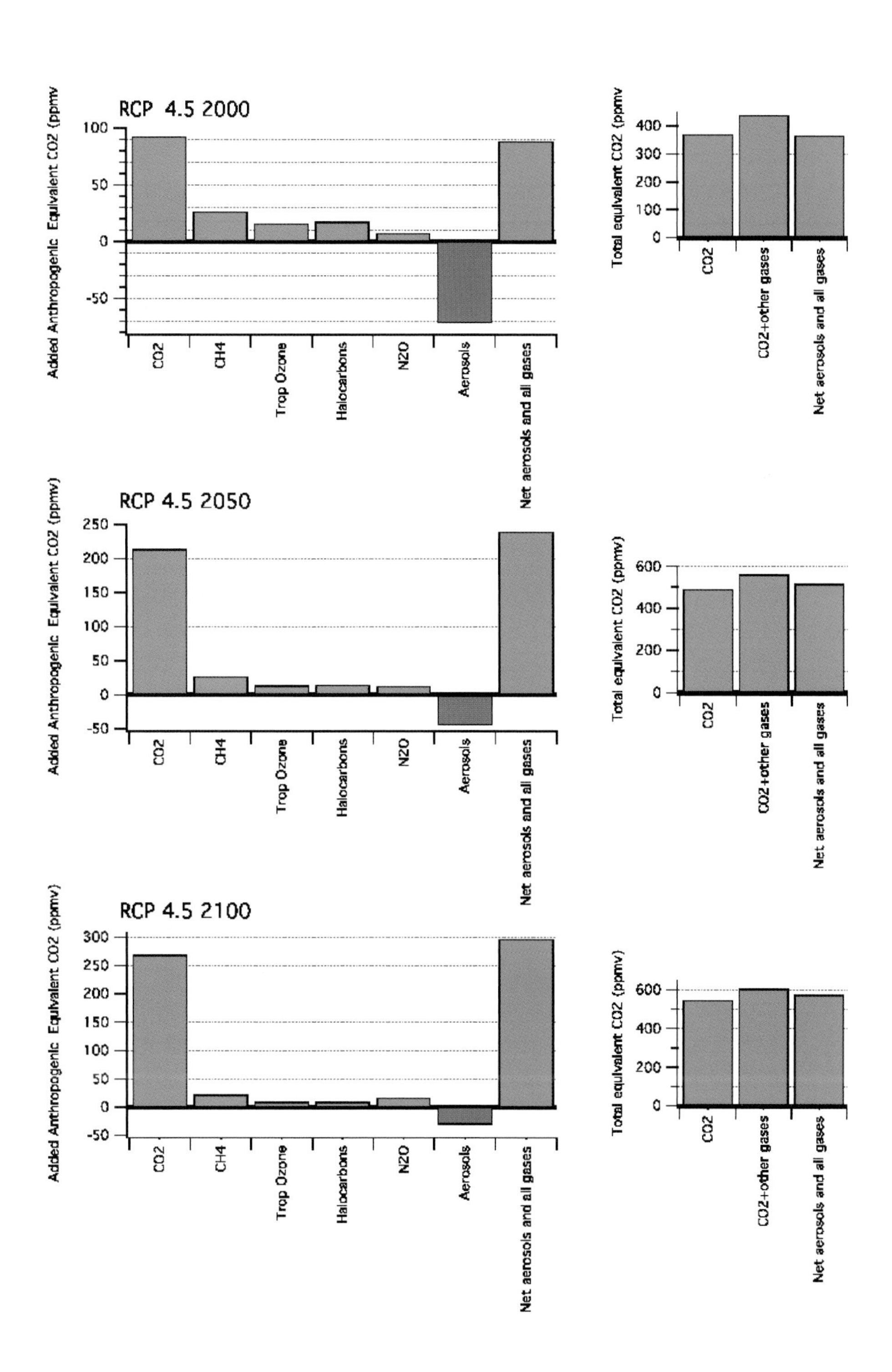
RCP 4.5 2000
Added Anthropogenic Equivalent CO2 (ppmv
100
50
0
-50
CO2
CH4
Trop Ozone
Halocarbons
N2O
Aerosols
Net aerosols and all gases
Total equivalent CO2 (ppmv
400
300
200
100
0
CO2
CO2+other gases
Net aerosols and all gases
RCP 4.5 2050
Added Anthropogenic Equivalent CO2 (ppmv)
250
200
150
100
50
0
-50
CO2
CH4
Trop Ozone
Halocarbons
N2O
Aerosols
Net aerosols and all gases
Total equivalent CO2 (ppmv)
600
400
200
0
CO2
CO2+other gases
Net aerosols and all gases
RCP 4.5 2100
Added Anthropogenic Equivalent CO2 (ppmv)
300
250
200
150
100
50
0
-50
CO2
CH4
Trop Ozone
Halocarbons
N2O
Aerosols
Net aerosols and all gases
Total equivalent CO2 (ppmv)
600
400
200
0
CO2
CO2+other gases
Net aerosols and all gases

hydrological cycle. These changes constitute an anthropogenic imprint on climate that is distinct from that associated with the overall warming due to greenhouse gases.

Methane is currently emitted by a diverse set of anthropogenic sources, many of which are related to agriculture. Once in the atmosphere, methane oxidizes to CO_2 with a time scale of about a decade. Since one molecule of CO_2 has much less radiative effect than a molecule of methane, for methane concentrations in the present low range, the resulting CO_2 is negligible in comparison to the CO_2 emitted by deforestation and fossil fuel burning. In contrast to CO_2, the climate effect of current anthropogenic methane emissions would decay relatively rapidly if emissions were to cease, as depicted in Figure 2.8. But it seems difficult to bring anthropogenic methane emissions to zero in the long term, given the continuing need for agriculture to feed the world's population. A continued long-term warming contribution from methane should therefore be anticipated, not because of persistence of methane in the atmosphere, but because of likely persistence of the source. For the much more massive methane release that could come from clathrate destabilization (see Section 6.1), the CO_2 produced by oxidation could have an important effect on climate.

The climate effects of short-lived radiative forcing agents are thus more reversible than those of CO_2, and therefore actions reducing emissions of short-lived agents have different implications for Earth's climate future than actions that affect CO_2 emissions. Insofar as it is perceived that control of methane or black carbon may be technically easier or less economically disruptive than controlling CO_2 emissions, mitigation of the short-lived warming influences has sometimes been thought of as a way of "buying time" to put CO_2 emission controls into place. This is a fallacy. While one does buy a rapid reduction by reducing methane or black carbon emissions, this has little or no effect on the long-term climate, which is essentially controlled by CO_2 emissions because of the persistence of CO_2 in the atmosphere. The situation is illustrated schematically in Figure 2.8. The time course of warming produced by CO_2 emissions alone is given schematically by the black line. If one adds short-lived radiative forcing agents with an aggregate warming effect into the mix, the effect will be to add to the temperature increase until such time as the emissions are brought under control, where after the temperature will quickly drop back to the CO_2-only curve (the blue and red solid lines on the curve, representing early or delayed mitigation of short-lived forcing agents). The effect of mitigation of methane and black carbon is thus to trim the peak warming rather than limit the long-term warming to which Earth is subjected. If the early action to mitigate methane emissions

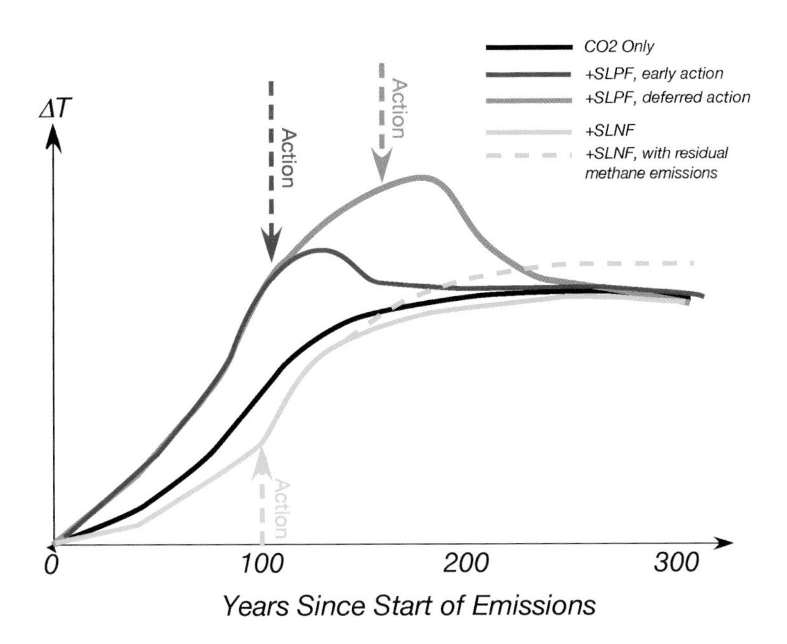

FIGURE 2.8 Qualitative sketch of the time-course of future temperature under various scenarios for control of emissions of short-lived radiative forcing agents. The time axis is given as time since the beginning of significant anthropogenic emissions of greenhouse gases. It is assumed that CO_2 emissions are brought to zero after 200 years. SLPF refers to short-lived positive forcing agents, like methane or black carbon on snow or ice. SLNF refers to short-lived negative forcing agents, primarily aerosols. "Early Action" refers to a scenario in which early, aggressive action is taken to mitigate emission of short-lived radiative forcing agents, while "Deferred Action" refers to a scenario in which such actions are delayed. The green line shows what happens if the aggregate of all short-lived forcings brought under control originally added up to a cooling effect (so that reducing them warms the climate). The dashed green line is similar, except that it assumes there is a residual methane emission that cannot be reduced to zero. The cumulative CO_2 emissions are assumed to be the same in all of these scenarios.

was done *instead of* actions that could have reduced net cumulative carbon emissions, the long-term CO_2 concentration would be increased as a consequence. Peak trimming in that case would come at the expense of an increased warming that will persist for millennia. Carbon emission control and short-term forcing agent control are two separate control knobs that affect entirely distinct aspects of Earth's climate and should not be viewed as substituting for one another.

It would be unrealistic to contemplate policies that would reduce black carbon emissions while leaving reflecting aerosol emissions intact, given that the diverse sources of emission yield an interlinked stew of absorbing and reflecting aerosols (Ramanathan et al., 2008). The green curve in Figure 2.8 shows what happens if the aggregate of all aerosols brought under control sums to a cooling effect before mitigation; the mitigation in this case accelerates the approach to the CO_2-only curve, as the masking effect of the aerosols is eliminated. If the long-term situation instead includes a recalcitrant methane emission rate that is stabilized but not brought to zero, then the long-term warming is brought above the CO_2-only case for a period as long as the methane emissions continue.

2.4 CARBON CYCLE

The evolution of atmospheric carbon dioxide concentrations depends on the balance of human emissions, natural processes that remove excess carbon dioxide from the atmosphere, and the sensitivity of land and ocean carbon reservoirs to climate change and land use (see Box 2.2). The historical atmospheric growth rate of carbon dioxide is well-constrained for the past 50 years from direct instrumental measurements and for periods prior to that from measurements of gases trapped in ice cores. Global average atmospheric CO_2 has risen from a pre-industrial level of about 280 ppm to about 390 ppm by the beginning of 2010. A definitive anthropogenic origin for the excess carbon dioxide can be assigned based on contemporaneous changes in carbon isotopes, a parallel decrease in atmospheric oxygen, and by the fact that the atmospheric carbon dioxide levels for the preceding several millennia of the Holocene had hovered within plus or minus 5 ppm of the pre-industrial value. Past fossil fuel combustion rates and carbon emissions from cement production and land-use change (e.g., deforestation, shifting land into pasture and agriculture) can be reconstructed, and net land and ocean carbon sources and sinks can be quantified from a combination of observations and numerical models. Figure 2.9 presents a recent synthesis for the global carbon system showing the fluxes between the atmosphere and various reservoirs versus time (Le Quéré et al., 2009). The terrestrial carbon fluxes are partitioned with carbon emissions from direct human land-use change including recovery from earlier human land use, separated from terrestrial carbon sinks in response to elevated CO_2 and climate.

The response of the global carbon cycle to human perturbations can be characterized by the airborne CO_2 fraction, the fraction of the cumulative carbon dioxide emitted by fossil fuel combustion and land-use change that remains in the atmosphere. The contemporary airborne fraction is currently slightly less than half (~0.45), and for any specified carbon emission trajectory, future atmospheric carbon dioxide concentrations depend on the

BOX 2.2 TIMES CALES FOR REMOVAL OF CO_2 FROM THE ATMOSPHERE

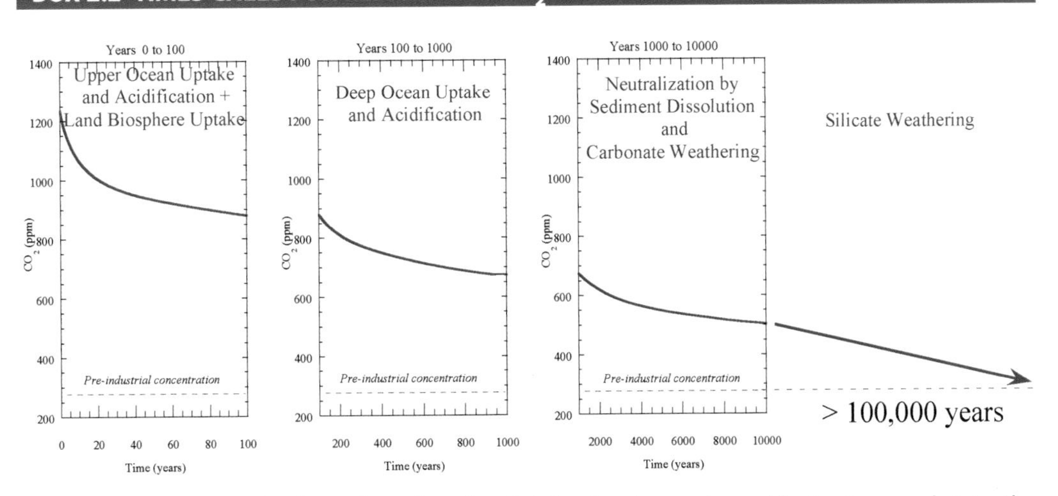

The figures show the fate of a pulse of 2,600 Gt of carbon released instantaneously into the atmosphere as CO_2. In the first hundred years CO_2 is absorbed into the upper ocean. The resulting acidification limits further uptake by the upper ocean waters. During this time period, there is also typically some uptake by the land biosphere. In the next 900 years, the saturated upper ocean waters mix with the deep ocean, allowing further uptake. Eventually, the deep ocean acidifies as well, limiting further uptake. Over the next 10,000 years the ocean becomes buffered by dissolution of carbonate sediments and by carbonates washed in from land, reducing the acidity and allowing the ocean to take up additional carbon. Over longer time scales spanning *more than* 100,000 years, most of the remaining CO_2 is removed by reacting with silicate minerals to form carbonates (e.g., limestone). We have not attempted to state the precise time required for silicate weathering to cause recovery to pre-industrial values, because of uncertainties in silicate weathering parameterization and uncertainties in the long term response of the glacial-interglacial cycle. The only long-term sink of CO_2 is silicate weathering, which is a very slowly increasing function of temperature. It would require over 20°C of warming to balance a steady state fossil fuel emission of only a half Gt of carbon per year, so that even an emission as low as this would lead to a steady accumulation of CO_2 in the atmosphere. This calculation does not allow for any long-term net release of carbon from land ecosystems or marine sediments, though it is known that the Earth system is capable of such releases. Any such release would increase the long term CO_2 concentrations and delay the recovery to pre-industrial values. (Data up to 10,000 years based on carbon cycle simulations of Eby et al. (2009). Silicate weathering time scale estimated from data given in Berner (2004). See Archer et al. (1997), and Archer (2005) for more details on the mechanisms of CO_2 removal.

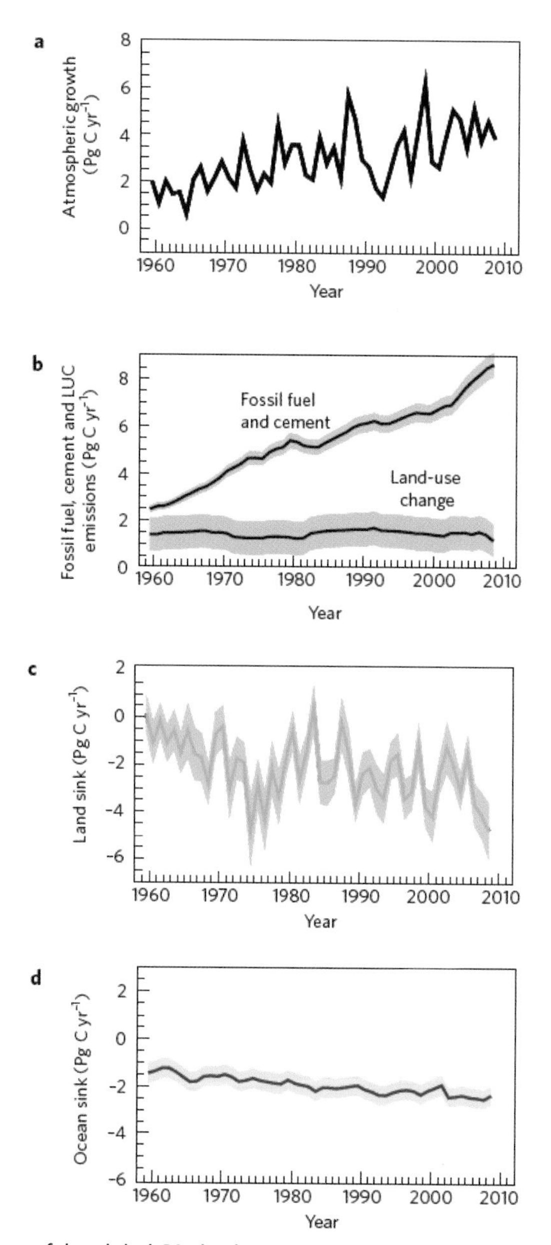

FIGURE 2.9 Components of the global CO_2 budget. (a) The atmospheric CO_2 growth rate. (b) CO_2 emissions from fossil fuel combustion and cement production and from land-use change. (c) Land CO_2 sink (negative values correspond to land uptake). (d) Ocean CO_2 sink (negative values correspond to ocean uptake). The land and ocean sinks (c,d) are shown as an average of several models normalized to the observed mean land and ocean sinks for 1990-2000. The shaded area is the uncertainty associated with each component. Adapted from Le Quéré et al. (2009).

airborne fraction and how it evolves with time. At present, the ocean and the land biosphere contribute about equally to the drawdown of excess carbon dioxide from the atmosphere. Theoretical arguments and numerical models suggest that the efficiency of both the land and ocean carbon sinks may decline in the future under warmer climate conditions, which would act to amplify climate warming (Fung et al., 2005; Friedlingstein et al., 2006). The magnitude of the climate-carbon cycle feedback, however, varies substantially across model simulations and is a substantial uncertainty in future climate projections.

Excess atmospheric carbon dioxide dissolves in surface seawater as inorganic carbon through well-known physical-chemical reactions. The distribution and global inventory of anthropogenic carbon dioxide in the ocean are well characterized based on global ship-based observations collected during the late 1980s and 1990s (Sabine et al., 2004). Ongoing measurements at time-series sites and along ocean sections constrain uptake over decadal time periods (IOC, 2009). Oceanic anthropogenic CO_2 uptake up to present has been governed primarily by atmospheric CO_2 concentrations and the rate of ocean circulation that exchanges surface waters equilibrated with elevated CO_2 levels with subsurface waters. In particular, key pathways include the ventilation of the wind-driven thermocline and deep and intermediate water formation. Ocean models constrained by field data provide estimates of the oceanic transport and air-sea flux of anthropogenic CO_2 as well as reconstructions of past ocean uptake and projections for the future (Matsumoto et al., 2004; Gruber et al., 2009; Khatiwala et al., 2009).

Future ocean uptake of anthropogenic carbon is expected to decrease in efficiency (i.e., absorb a smaller fraction of the emissions); the ocean CO_2 sink is expected to continue to increase, but more slowly than the emissions. Under elevated CO_2, the chemical buffer capacity of seawater decreases, lowering the amount of inorganic carbon absorbed when surface waters are equilibrated with the atmosphere. Upper-ocean warming reduces the solubility of carbon dioxide in seawater. Anthropogenic CO_2 uptake will be further reduced because of increased vertical stratification, reduced ocean ventilation rates, and reduced deep and intermediate water formation rates, which are expected due to warming in the tropics and subtropics and increased freshwater input in temperate and polar regions due to elevated precipitation and sea-ice melt (Sarmiento et al., 1998). In contrast, an increase in the strength of Southern Ocean winds, associated with a more positive phase of the Southern Annular Mode, may increase future uptake of anthropogenic CO_2 (Russell et al., 2006).

Ocean biogeochemistry plays an important role in ocean carbon storage, and biogeochemical responses to changing ocean circulation also need to be considered when assessing future net carbon uptake. The inorganic carbon concentration in subsurface ocean waters is generally elevated over surface concentrations because of the downward transport and subsequent respiration of organic water originally produced in the surface layer. In coupled carbon-climate models, biogeochemical feedbacks to a warmer climate tend to partially offset physical-chemical effects and act to reduce the overall strength of ocean climate-carbon cycle feedbacks. In the Southern Ocean, enhanced outgassing of natural CO_2 due to stronger winds and upwelling may more than compensate for increased anthropogenic CO_2 uptake, leading to a net reduction in ocean uptake (Le Quéré et al., 2008; Lovenduski et al., 2007, 2008). Recent observations of the air-sea difference in the partial pressure of carbon dioxide, the driving force for air-sea CO_2 exchange, indicate a weakening of oceanic uptake in a number of regions, although there remains some debate whether this signal should be attributed to climate change, ozone depletion, or primarily decadal climate variability (Le Quéré et al., 2009; Watson et al., 2009).

It is more difficult to directly constrain, on a global scale, the net fluxes of carbon into and out of the more heterogeneous terrestrial carbon reservoirs, and terrestrial uptake is often estimated from a combination of terrestrial biogeochemical models and satellite remote sensing approaches that have been assessed using process experiments, local CO_2 flux towers, etc. (Canadell et al., 2007; Raupach et al., 2007; Le Quéré et al., 2009). Land carbon uptake can be computed in a top-down fashion by difference from the estimated fluxes to the atmosphere, the ocean sink, and the growth rate in the atmosphere. Slightly more sophisticated approaches utilize the spatial and temporal variations in atmospheric CO_2 with transport models to infer land and ocean surface fluxes (Rödenbeck et al., 2003; Peylin et al., 2005). Atmospheric carbon isotope and oxygen/nitrogen ratios also provide critical constraints on the partitioning of carbon uptake between the ocean and land biosphere (Rayner et al., 1999).

The contemporary land carbon budget is governed by a combination of interacting natural and anthropogenic processes rather than any single mechanism (Pacala et al., 2001; Schimel et al., 2001). Deforestation and biomass burning result in net CO_2 fluxes to the atmosphere as high-carbon forests are turned into comparatively low-carbon pastures and croplands (Houghton, 2003). This process is now occurring mainly in the tropics and is partially countered by temperate regrowth on abandoned farm and pasture-land (Shevliakova et al., 2009). The impacts of land-use change can extend for decades after the initial disturbance, and contemporary land carbon fluxes may reflect as much aggregate events in the past as current conditions, complicating the task of deconvolving underlying mechanisms. Experimentally, fertilization of photosynthesis by higher atmospheric carbon dioxide levels increases plant growth, in many cases substantially, and contributes to carbon uptake. When included in global models, CO_2 fertilization leads to a modest current sink that may continue for many decades into the future. Some studies suggest, however, that this effect may be smaller than earlier thought perhaps due to other limitations such as nitrogen (Long et al., 2006; Thornton et al., 2009a). Many Northern Hemisphere ecosystems are also nitrogen limited, and deposition of reactive nitrogen mobilized by fossil fuel combustion may also cause increased carbon uptake (Lamarque et al., 2005; Thornton et al., 2009a). Land carbon storage varies on interannual time scales due to climate variability and in particular variations in temperature, precipitation, and water availability (Sitch et al., 2008). Wildfire also plays a major role in the global carbon cycle (Randerson et al., 2006), and increased tropical fire associated with El Niño droughts may contribute to increases in the growth rate of atmospheric carbon dioxide concentrations during recent El Niño years (Van der Werf et al., 2006).

Climate warming may cause land ecosystems to lose carbon because respiration is more temperature sensitive than photosynthesis, but there is a wide range of estimates for the climate sensitivity of land carbon stocks. Larger effects could come if future climates lead to additional disturbances, especially fire, pests, or widespread replacement of forests to grasslands, which could rapidly release large amounts of carbon. Current attention is focusing on both the role of human and climate-caused disturbance in controlling future ecosystem carbon storage and on physiological processes, such as carbon dioxide fertilization. Water availability to support photosynthesis and primary productivity is thought to be significant, with models (Fung et al., 2005) and observations (Angert et al., 2005) indicating that decreasing water balance (drier soil conditions) decreases carbon uptake. Global effects are a balance between warmer conditions favoring a longer growing season in the Northern Hemisphere temperate zone, increasing carbon uptake, and drier soils in the tropics, decreasing carbon uptake (Fung et al., 2005; Friedlingstein et al., 2006). Past and future land-use change will also affect land carbon storage and needs to be considered when projecting future atmospheric CO_2 levels.

The combined effects of ocean and land feedbacks have been explored in a series of coupled climate-carbon models reported by the international C^4MIP (Coupled Climate-Carbon Cycle Model Intercomparison Project; Friedlingstein et al., 2006; Figure 2.10). Simulations are conducted for the

20th and 21st centuries forced with specified trajectories for fossil fuel and land-use CO_2 emissions; carbon-climate sensitivity is assessed by comparing simulations with and without coupling between model atmospheric CO_2 and atmospheric radiation (and thus climate). Some simulations exhibit a large amplifying effect whereby climate effects lead to increased release of carbon dioxide, mostly from terrestrial processes. By the year 2100, simulations with strong feedbacks have higher atmospheric CO_2 levels as much as 200 ppm above the companion simulation without climate feedbacks. The airborne CO_2 fraction tends to increases in models where uptake processes are more sensitive to climate, where uptake processes are less sensitive to atmospheric CO_2, and where physical climate sensitivity to CO_2 is large.

The land modules in the C^4MIP simulations did not include interactions between carbon and other limiting nutrients such as nitrogen. When soil organic matter is respired due to warming, nitrogen is released, stimulating enhanced plant growth that can offset the CO_2 released from the soil matter. More recent coupled simulations that include this effect indicate a small negative climate-carbon feedback on atmospheric CO_2 (Thornton et al., 2009a). Interestingly, the carbon-nitrogen model had only a very weak CO_2 fertilization effect on land photosynthesis because of nitrogen limitation. As a result, the model land carbon uptake responded only weakly to rising atmospheric CO_2 and had as a result one of the largest atmospheric CO_2 levels at the end of the 21st century. The degree to which CO_2 fertilization is modulated by nitrogen is an important unresolved science question, and for the carbon cycle, sensitivity of the sinks to CO_2 is as important as sensitivity to climate. The C^4MIP simulations did not address the full suite of interactions between the carbon cycle and other changes in the Earth System; for example, ocean carbon storage can be influenced by ozone-driven changes in Southern Ocean winds (Le Quéré et al., 2008; Lovenduski et al., 2008; Lenton et al., 2009). Finally, the current generation of coupled climate-carbon cycle models neglect a number of carbon reservoirs, such as high-latitude peats, permafrost and methane clathrates that could be important on longer time scales (see Section 6.1).

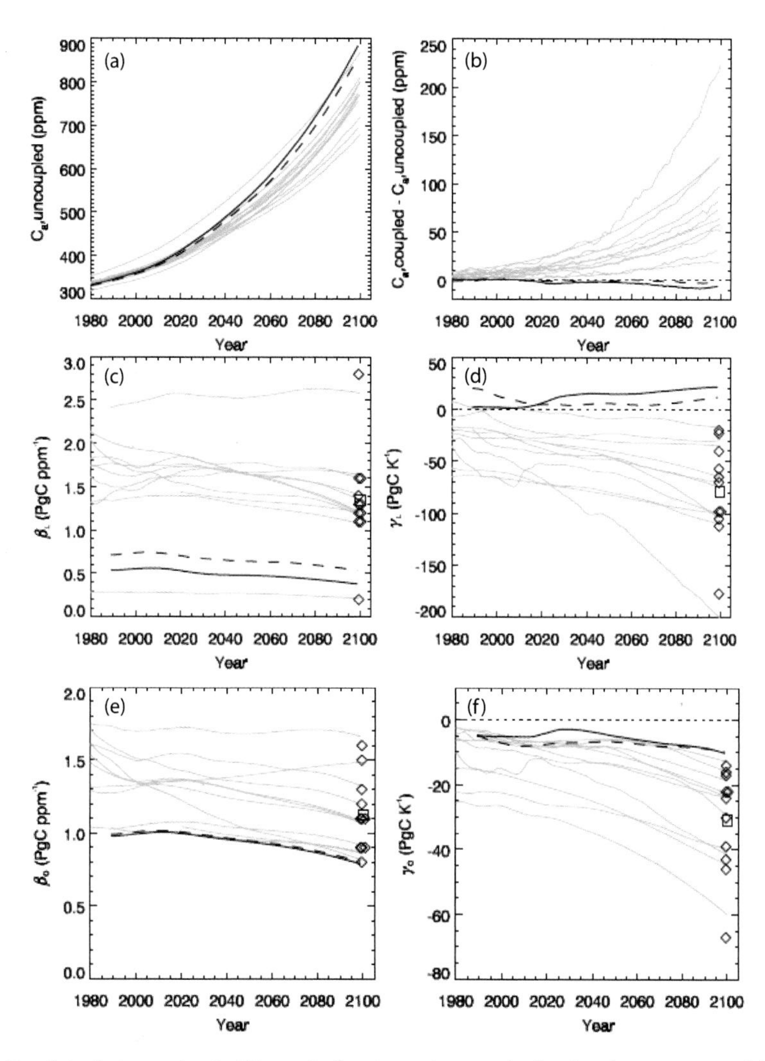

FIGURE 2.10 Predicted atmospheric CO_2 and climate-carbon cycle feedback parameters. (a) Atmospheric CO_2 trajectories for simulations with the same prescribed fossil fuel emissions and no CO_2-radiative feedbacks. (b) Difference in atmospheric CO_2 due to radiative coupling. (c) Land biosphere response to increasing atmospheric CO_2. (d) Land biosphere response to increasing temperature. (e) Ocean response to increasing atmospheric CO_2. (f) Ocean response to increasing temperature. Gray lines show results from carbon-only model studies from Friedlingstein et al. (2006) (also reported in IPCC, 4th assessment). Thick black lines are from a nitrogen-carbon model (Thornton et al., 2009a); thick solid line for pre-industrial nitrogen deposition and thick dashed line for anthropogenic nitrogen deposition. Diamonds show the feedback parameters estimated at year 2100 for previous studies (Friedlingstein et al., 2006), and squares show their mean. Thin dotted lines indicate zero response. (adapted from Thornton et al., 2009a).

3

Global Mean Temperature Responses

3.1 OVERVIEW OF TIME SCALES AND CLIMATE SENSITIVITY

The rapid addition of carbon dioxide and other greenhouse gases to the atmosphere by human activities throws Earth's energy budget substantially out of balance. For a time, Earth receives more energy from the Sun than it loses by emission of infrared radiation to space. The climate system rectifies this imbalance through processes acting over a range of different time scales. Restoring balance invariably results in a warmer climate, and the amount of warming associated with a prescribed addition of carbon dioxide is called the *climate sensitivity*. One obtains different climate sensitivities over different time scales, because additional processes come into play at long time scales, which are not important over shorter time scales.

In this report, we will deal with three kinds of climate sensitivity. The first kind characterizes the equilibrium response to changes in CO_2, including relatively fast feedbacks, primarily water vapor, clouds, sea ice, and snow cover. This equilibrium response assumes that the oceans have had time to equilibrate with the new value of carbon dioxide and with these fast feedbacks, an equilibration that is estimated to require multiple centuries to a millennium. When the term *climate sensitivity* is used without further qualifiers, it should be understood as meaning this form of relatively fast-feedback equilibrium climate sensitivity; it will be discussed in Section 3.2. The term "equilibrium" in this chapter will always refer to the equilibrium response incorporating only these feedbacks.

The second form of sensitivity we shall deal with—*Transient Climate Response*—characterizes the early stages of warming, when the deep ocean is still far out of equilibrium with the warming surface waters. Transient climate response is of great importance because it is appropriate to understanding climate variations and impacts of the 20th and 21st centuries. This form of sensitivity will be discussed in Section 3.3.

Our judgment as to the probable values of transient climate response and equilibrium climate sensitivity is summarized in Table 3.1. The values given in this table, used together with the pattern-scaling for regional climate discussed in Chapter 4, will form the basis for our evaluation of the impacts of climate change. The reasoning leading to these estimates, and an exposition of the physical processes involved, are given in Sections 3.2 and 3.3.

At the opposite end of the spectrum of time scales, the third kind of climate sensitivity, called *Earth System Sensitivity*, incorporates a range of slower feedback processes that can set in during the millennia over which anthropogenic CO_2 emissions are expected to continue to affect the climate. These include long-term carbon cycle feedbacks and partial or total deglaciation of Greenland and Antarctica. The human imprint on climate will outlast the fossil fuel era by millennia, because of the long atmospheric lifetime of CO_2, and perhaps because of the additional feedbacks the resulting warming may entrain. To future geologists, the fossil fuel era will appear as a boundary between the Holocene epoch and a substantially hotter epoch dubbed the *Anthropocene* (Crutzen and Stoermer, 2000)—which will be driven by different factors (human factors) than any climate states observed at any time in more than a million years. Will the great ice sheets of Greenland and Antarctica survive the Anthropocene? How much of the world's present biodiversity will survive the Anthropocene? How will agriculture in an Anthropocene climate feed whatever population may prevail

TABLE 3.1 Transient and Equilibrium Global Mean Warming as a Function of the Atmospheric CO_2 Concentration

CO_2 (ppm)	Trans. Low	Trans. Prob L	Trans. Med	Trans. Prob H	Trans. High	Eq. Low	Eq. Med	Eq. High
350	0.4	0.4	0.5	0.7	0.8	0.7	1.0	1.4
450	0.8	0.9	1.1	1.5	1.8	1.4	2.2	3.0
550	1.1	1.3	1.6	2.1	2.5	2.1	3.1	4.3
650	1.3	1.6	2.0	2.7	3.2	2.6	3.9	5.4
1000	2.0	2.4	3.0	4.0	4.8	3.9	5.9	8.1
2000	3.1	3.7	4.7	6.2	7.4	6.0	9.1	12.5

NOTE: Warming is given in degrees C relative to a pre-industrial climate with a CO_2 concentration of 280 ppm. The equilibrium values were extrapolated logarithmically from data given in Table 8.2 of IPCC, Working Group I, The Physical Science Basis (IPCC, 2007a). The "Eq. Low" column is based on the minimum sensitivity in the ensemble of models, the "Eq. High" is based on the maximum, and the "Eq. Med" is based on the median. The transient climate response estimates are discussed in Section 3.3; the "Prob. Low" column represents the low end of the probable range, the "Prob. High" column the high end of the probable range, and the "Med" column the median value of transient climate response. The stated warming reflects only the influence of CO_2 on the climate, but the table can be used to estimate the effect of other greenhouse gases by using the radiative-

in the coming centuries? The answer to these questions in large measure rests on the net amount of carbon dioxide released during the fossil fuel era, however long that may last. The discussion of Earth System Sensitivity and other considerations pertaining to very long-term climate change will be deferred to Chapter 6.

3.2 EQUILIBRIUM CLIMATE SENSITIVITY

Climate sensitivity is calculated by determining how much Earth's surface and atmosphere need to warm in order to radiate away enough energy to space to make up for the reduction in energy loss out of the top of the atmosphere caused by the increase of CO_2 or other anthropogenic greenhouse gases. In equilibrium, there is no net transfer of energy into or out of the oceans, so the equilibrium sensitivity can be treated in terms of the top-of-atmosphere energy balance.

The top-of-atmosphere balance is the rate at which energy escapes to space in the form of infrared minus the rate at which energy is absorbed in the form of sunlight, both expressed per square meter of Earth's surface. Figure 3.1 shows schematically how the balance depends on surface temperature, for a case which is initially in equilibrium at temperature T_0 (where the blue solid or dashed line crosses the horizontal axis). If CO_2 is increased, leading to a reduction in outgoing radiation by an amount Δ F, Earth must warm up so as to restore balance. The amount of warming required is determined by the slope of the line describing the increase in energy imbalance with temperature. A lower slope results in a higher climate sensitivity, as illustrated by the dashed slanted lines in Figure 3.1. Climate sensitivity is often described in terms of the warming $\Delta\ T_{2X}$ that would result from a standardized radiative forcing $\Delta\ F_{2X}$ corresponding to a doubling of CO_2 from its pre-industrial value. In the following we use the value $\Delta\ F_{2X}$ = 3.7 W/m^2 diagnosed from general circulation models to express the slope in terms of $\Delta\ T_{2X}$ (IPCC, 2007a).

The most basic feedback affecting planetary temperature is the black-body radiation feedback, which is the tendency of a planet to lose heat to space by infrared radiation at a greater rate as the surface and atmosphere are made warmer while holding the composition and structure of the atmosphere fixed (see Table 3.2). This feedback was first identified by Fourier (1827; see Pierrehumbert, 2004). For a planet with a mean surface temperature of 14°C (about the same as Earth's averaged over 1951-1980) the black-body feedback alone would yield $\Delta\ T_{2X}$ = 0.7°C if the planet's atmosphere had no greenhouse effect of any kind. Such a planet would have

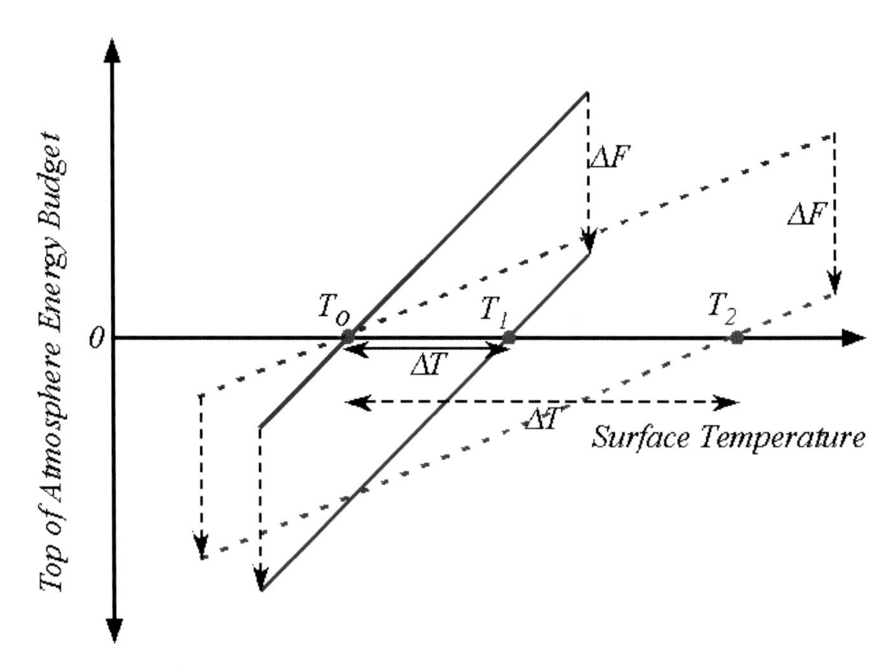

FIGURE 3.1 Determination of surface temperature response to a radiative forcing Δ F, in terms of the top-of-atmosphere energy budget. The top-of-atmosphere energy budget is the net of outgoing infrared radiation minus incoming solar radiation. The budget is zero when the system is in equilibrium. In these graphs, the budget is expressed schematically as a function of surface temperature. Equilibrium surface temperature is determined by the point where the line crosses the horizontal axis. If the system starts in equilibrium, but CO_2 is increased so that the line is shifted downward by an amount Δ F (via reduction in outgoing infrared), then the intersection point shifts to warmer values by an amount Δ T. When the slope of the energy budget line is smaller, a given Δ F causes greater warming, as indicated by the pair of lines with reduced slope. This connects the slope of the energy budget line with climate sensitivity. The slope of the line is like the stiffness of a spring, and the radiative forcing Δ F is like the force with which one tugs on the spring. When the spring is not very stiff (e.g., a spring made of thin rubber bands) a given force will make the spring stretch to a great length—analogous to a large warming. If the spring is very stiff (e.g., a heavy steel garage door spring) the same force will cause hardly any stretching at all—analogous to low climate sensitivity. A spring with no stiffness at all would represent a very special case, demanding a specific physical explanation, just as would a case of zero slope of the energy budget line, which corresponds to infinite climate sensitivity.

to be closer to the Sun than Earth is, in order to make up for the lack of a greenhouse effect. The same greenhouse effect that keeps Earth from freezing over in its actual orbit reduces the temperature at which Earth radiates to space, reduces the slope characterizing the black-body feedback, and hence increases the sensitivity. Taking into account the greenhouse effect

TABLE 3.2 Key Physics and Processes Contributing to Climate Sensitivity (warming expected if carbon dioxide doubles from an unperturbed value of 278 ppmv to 556 ppmv)

Black-body radiation alone, ignoring the greenhouse effect of the unperturbed atmosphere	0.7°C
Black-body radiation but also including the greenhouse effect in the unperturbed atmosphere	0.9°C
As above, but also including well-documented feedbacks due to tropospheric water vapor changing at fixed relative humidity and changes in the lapse rate (vertical structure of the atmosphere), no clouds	1.5°C
As above, also including clouds but keeping them fixed	1.8°C
As above, including clouds and allowing the clouds to vary, along with other feedbacks such as snow and sea ice retreat, from IPCC AR4 suite of models	Best estimate 3.2°C Likely range 2.1-4.4°C

of the unperturbed (background) atmosphere gives ΔT_{2X} of about 0.9°C in the absence of clouds.[1] But abundant evidence and basic physics shows that atmospheric water vapor must increase in a globally warmer world, and multiple lines of evidence confirm that both the atmosphere and general circulation models conform to a feedback that acts approximately as if the relative humidity is kept fixed (Held and Soden, 2000; Pierrehumbert et al., 2007; Dessler and Sherwood, 2009). When this result is used to incorporate the water vapor feedback into calculations of Earth's infrared emission to space, and the lapse rate feedback is also taken into account, we find that ΔT_{2X} increases to 1.5°C. This figure is helpful for understanding the physics contributing to climate sensitivity, but it is incomplete because it is only for clear sky conditions. In the real atmosphere, clouds contribute to Earth's background greenhouse effect, and their possible changes represent a key feedback.

The feedbacks that modify the basic black-body feedback are at the heart of predicting future climate. The combined water vapor and lapse rate feedback increases climate sensitivity by affecting the infrared emission side of the balance. Snow and sea-ice retreat work instead on the solar absorption side, but they also increase the sensitivity. Clouds work on both the infrared and solar side, and their net influence on sensitivity can go either way.

A quantitative treatment of cloud, snow, relative humidity, and sea-ice feedbacks requires the use of general circulation models. The estimates of equilibrium climate sensitivity in this report will be based on simulations employing mixed layer ocean models, which are thought to closely mimic

[1]These computations were carried out using idealized single-column models of the type described in Chapter 4.5.3 of Pierrehumbert (2010). See Methods section for details.

more complete simulations on longer time scales, once the ocean stops taking up heat.

For the general circulation models listed in Table 8.2 of IPCC, Working Group I, The Physical Science Basis (IPCC, 2007a), the equilibrium ΔT_{2x} has a minimum of 2.1°C, a maximum of 4.4°C, and a median of 3.2°C. Even the least sensitive model has a higher climate sensitivity than the idealized calculations yield for basic clear-sky water vapor and lapse-rate feedback. This is largely because the presence of clouds increases the basic black-body plus water vapor feedback sensitivity to about 1.8°C even if clouds do not change as the climate warms. This form of cloud effect is not conventionally counted as a cloud feedback. It is more robust than feedbacks due to *changing* clouds, because it is based on cloud properties that can be verified against today's climate. This value, too agrees well among models and is considered to be highly certain. Thus, the least sensitive IPCC models correspond very nearly to cloud properties remaining fixed while warming is amplified by water vapor feedbacks alone. The more sensitive models are more sensitive primarily by virtue of having positive cloud feedback. The spread in equilibrium climate sensitivity within the IPCC ensemble of models is primarily due to differences in cloud feedback, and in particular to the feedback of low clouds (Bony et al., 2006). Note that climate sensitivity could only be lower than about 1.8°C if there are negative feedbacks very different from those of any of the models, such as changes in upper tropospheric or lower stratospheric water vapor, or changes in clouds that are opposite to those expected.

It has long been recognized that a symmetric distribution of the uncertainty in the strength of the feedbacks affecting climate sensitivity results in a skewed distribution in the climate sensitivity itself, with a high probability of large values (e.g., Schlesinger, 1986).[2] Roe and Baker (2007) attempt to use this property to argue that it will be extremely difficult to eliminate the significant possibility of very high climate sensitivities. However, there is no a priori reason to expect the uncertainty in the strength of the feedback

[2]This can be understood by noting that the climate sensitivity is proportional to $1/(1-f)$, where f is the strength of the feedbacks, and is positive if the feedbacks are positive. If one starts with the value $f = 0.5$, then increasing f by 0.25, say, increases the sensitivity by 100%, or a a factor of two. Decreasing f by the same amount decreases the sensitivity by only 67%. One can also use Figure 3.1 to understand this result pictorially. Climate sensitivity is proportional to the reciprocal of the slope shown in the figure, with the magnitude of the slope determined by the strength of the feedbacks. A symmetric distribution of slopes does not result in a symmetric distribution of climate sensitivity. A symmetric distribution might include zero slope with a finite probability, resulting in infinite climate sensitivity, which, of course, is ruled out by the observed stability of the climate system.

to be symmetric, especially when one considers observational constraints, such as those that are provided by paleoclimates, that constrain the equilibrium climate response directly rather than by constraining the strengths of feedbacks. Hannart et al. (2009) have highlighted the implications of the arbitrary assumptions regarding uncertainty in Roe and Baker's analysis. Chapter 6 contains a brief discussion of paleoclimatic constraints consistent with the range of equilibrium sensitivities in Table 3.1

Perturbed physics ensembles, notably Stainforth et al. (2005), do show that there are physical mechanisms that can operate in climate models, which yield climate sensitivity well above the top of the IPCC range employed in Table 3.1. Such ensembles do exhibit a distribution of climate sensitivity with a fat tail skewed toward high values, much as one would expect from assuming a symmetric and broad uncertainty distribution in the total feedback strength. In perturbed physics ensembles, it is typical to admit members to the ensemble only if they pass through a "keyhole" requiring that the basic climate of that member is realistic enough to serve as a basis for predicting the future. If the keyhole is made too wide, it can allow unrealistic behaviors to pass through. This is evidently the case for the many of the anomalously high climate sensitivity cases seen in Stainforth et al. (2005), which require a very unrealistic moistening of the upper troposphere and lower stratosphere (Sanderson et al., 2008; Joshi et al., 2010). It cannot unequivocally be ruled out that some unknown future climate state could trigger the onset of such behavior of water vapor, but there is no good basis at present for evaluating the prospects that this might happen. As another example, working with an alternative model, Yokohata et al. (2005) illustrate how simulation of the Pinatubo eruption provides evidence against a version of the model with equilibrium sensitivity as high as 6K.

Our choice to emphasize the CMIP3/AR4 model range is based on the judgment that these models have been analyzed most fully by the research community, thanks in large part to the open archive created by the World Climate Research Program's Couple Model Intercomparison Program and the Department of Energy's Program for Climate Model Diagnosis and Intercomparison (*http://www-pcmdi.llnl.gov/*). The description of this range of equilibrium sensitivity in the CMIP3 models as "likely" is consistent with the conclusions of the review of Knutti and Hegerl (2008), which attempts to take into account observational constraints as well as model results.

In interpreting future climate impacts based on Table 3.1 it is important to keep in mind that these do not necessarily represent worst possible cases, even if defensible physical mechanisms leading to higher climate sensitivity have not yet been identified. Our judgment is that the likelihood of very high equilibrium climate sensitivities cannot be quantified at this time. We recognize the importance to policy makers of quantitative statements concerning high sensitivities but do not attempt to address this issue further in this report.

We consider it to be more difficult to provide useful estimates of uncertainty in equilibrium climate sensitivity than in the transient climate response (Section 3.3), due to the stronger observational constraints on the latter arising from observations on multidecadal-to-century time scales. In Table 3-1, we suggest values for the likely (66%) and very likely (90%) ranges for the transient climate response based on the discussion is Section 3.3

We are cognizant that current climate models have limitations, especially resulting from deficiencies in cloud simulations and in simulations of tropical convection. We do not try to delineate these deficiencies here or relate them to uncertainties in sensitivity. We rely on a general consistency between the range of sensitivities in the AR4 models and various observational constraints. The latter are discussed in Section 3.3 as they relate to the transient climate response and in Section 6.1 on longer time scales.

The equilibrium global mean temperature change results shown in Table 3.1 were computed by logarithmic extrapolation from the equilibrium climate sensitivity values reported for the 17 general circulation models in Table 8.2 of IPCC, Working Group I, The Physical Science Basis (IPCC, 2007a).[3] Over the range of CO_2 covered in the table, logarithmic extrapolation is equivalent to assuming temperature change to be linear in radiative forcing.

On time scales longer than a few years, the oceanic mixed layer and the atmosphere warm as a unit. However, heat loss out of the bottom of the mixed layer keeps the warming below the equilibrium value until the deep ocean has warmed up and equilibrated. It takes a great deal of energy to warm the deep ocean, and consequently it takes many centuries for the temperature to relax to the equilibrium warming. The idealized situation in which CO_2 is held fixed for many centuries through carefully defined and ever decreasing emissions is unrealistic, and this might seem to make the equilibrium sensitivity a concept of mostly academic interest. Nonetheless, if used properly, the equilibrium climate sensitivity yields important informa-

[3]These values were, in turn, obtained from atmosphere/land models coupled to "slab" ocean models. Slab ocean models equilibrate quickly but assume that there are no changes in horizontal ocean heat transports as climate changes. The results are usually considered only rough approximations to the true equilibrium of the coupled system. Danabasogulu and Gent (2009) have recently provided encouraging results on the accuracy of this approximation in one particular model.

tion about the long-term evolution of climate. As discussed in Chapter 2, if the emissions are reduced to zero after some fixed time, the CO_2 peaks at the time of cessation of emissions, and very gradually relaxes back to smaller values over the subsequent millennia. Over the first few centuries, the warming in any given year will fall short of the equilibrium value expected from the CO_2 concentration prevailing in that year, but during the long, slow decline of CO_2, the temperature has time to catch up to the equilibrium curve. This form of approach to equilibrium is illustrated in the simulation shown in Figure 3.2. On time scales longer than about a thousand years, the equilibrium sensitivity applied to the instantaneous CO_2 value provides a good estimate of the warming.

The approach to equilibrium takes long enough that slow feedback processes can intervene and alter the long-term climate evolution. This will be taken up in Chapter 6, where it will be shown that the persistent warming computed on the basis of equilibrium climate sensitivity provides a valuable guide as to whether the human imprint on climate is likely to be

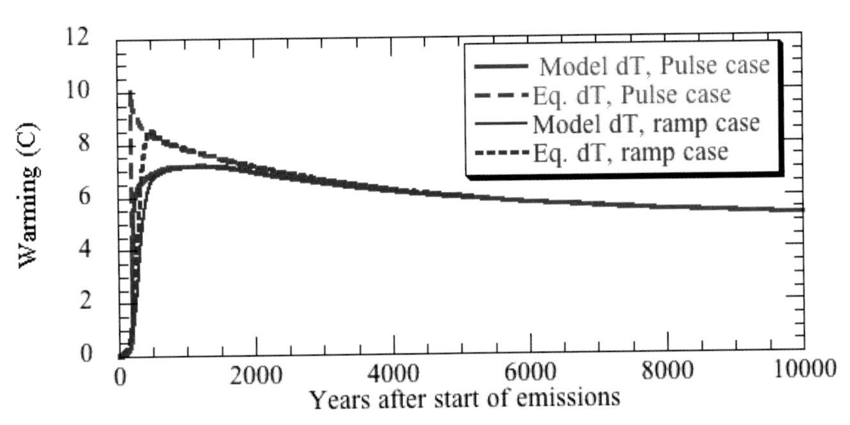

FIGURE 3.2 Comparison of equilibrium warming based on instantaneous CO_2 values with actual modeled temperatures from Eby et al., 2009. The simulation shown is based on cumulative emission of 3840 Gt carbon in the form of CO_2. Results are shown both for a pulse emission and for an exponentially increasing ramp lasting 350 years. "Instantaneous equilibrium warming" at any given moment in time is defined as the warming that would ultimately be reached in equilibrium if the CO_2 prevailing at that moment were held fixed indefinitely. It provides an accurate estimate of the actual warming corresponding to the time varying CO_2 when the CO_2 concentrations are varying sufficiently slowly. In computing the instantaneous equilibrium warming, the climate sensitivity is held fixed at the value appropriate to the model used in the simulation. The equilibrium curves are proportional to the logarithm of the CO_2 time series produced by the carbon cycle model used in this simulation. For this model ΔT_{2X} is approximately 3.5 C.

of sufficient magnitude and duration to trigger significant effects from slow climate system feedbacks.

3.3 TRANSIENT CLIMATE RESPONSE AND SENSITIVITY

The climate system responds differently to perturbations in radiative forcing on different time scales. It is of special importance to carefully distinguish between the "equilibrium climate sensitivity" and the "transient climate response." The concept of equilibrium climate sensitivity and its alternative definitions have been discussed in Section 3.2. The transient climate response is of special relevance for climate change over the 20th and 21st centuries.

The transient climate response, or TCR, is traditionally defined in a model using a particular experiment in which the atmospheric CO_2 concentration is increased at the rate of 1% per year. The increase in global mean temperature in a 20-year period centered at the time of doubling (year 70) is defined as the TCR (Cubasch, 2001). The upper tens of meters of the ocean, the atmosphere, and the land surface are all strongly coupled to each other on time scales greater than a decade, and are expected to warm coherently with a well-defined spatial pattern during a period of increasing radiative forcing. (This pattern is discussed in Section 4.1.) On these time scales, the rate of change of the heat stored in oceanic surface layers, as well as the atmosphere and land surface, can be ignored to first approximation, and we can think of global mean temperature as determined by the energy balance between the radiative forcing F, the change in the radiative flux at the top of the atmosphere U, and the flux of energy D into the deeper layers of the ocean that are far from equilibrium: $F = U + D$.

On long enough times scales, as long as a millennium by some estimates (e.g., Stouffer, 2004), the heat flux into the deep ocean D tends to zero, and the climate response approaches its equilibrium value determined by the balance between F and U. On the shorter times scales at which the changes in the deep ocean are still small, the heat uptake grows in time roughly proportional to the global mean temperature perturbation, $D = \gamma T$ (Gregory and Mitchell, 1997; Raper et al., 2002; Dufresne and Bony, 2008), similar to the more familiar linear approximation to the radiative flux response, $U = \beta T$. The implication is that on these time scales we can use the simple approximation $T = F/(\beta + \gamma)$.

This approximation is useful on a limited range of time scales—longer than a decade but shorter than the centuries required for the heat uptake by the oceans to begin to saturate (Gregory and Forster, 2008; Held et al.,

2010). It is also useful for the canonical 1% peryr experiment, so we can set $TCR = F_{2X}/(\beta + \gamma)$, where F_{2X} equals the forcing due to doubling of CO_2. We can then use the same proportionality constant, $1/(\beta + \gamma)$, to interpret 20th century warming and 21st century projections given only the evolution of the forcing F over time. As an example, Gregory and Forster (2008) show how the same proportionality constant between forcing and global mean temperature holds in a particular GCM for all of the 21st century forcing scenarios utilized by the AR4 and pictured in Figure 3.3 (IPCC, 2007c). The lack of separation of these different scenarios in the first half of the 21st century is not primarily due to some inertia in the physical climate system, but rather to the fact that the net radiative forcings in the various scenarios do not substantially diverge until the latter half of the century.

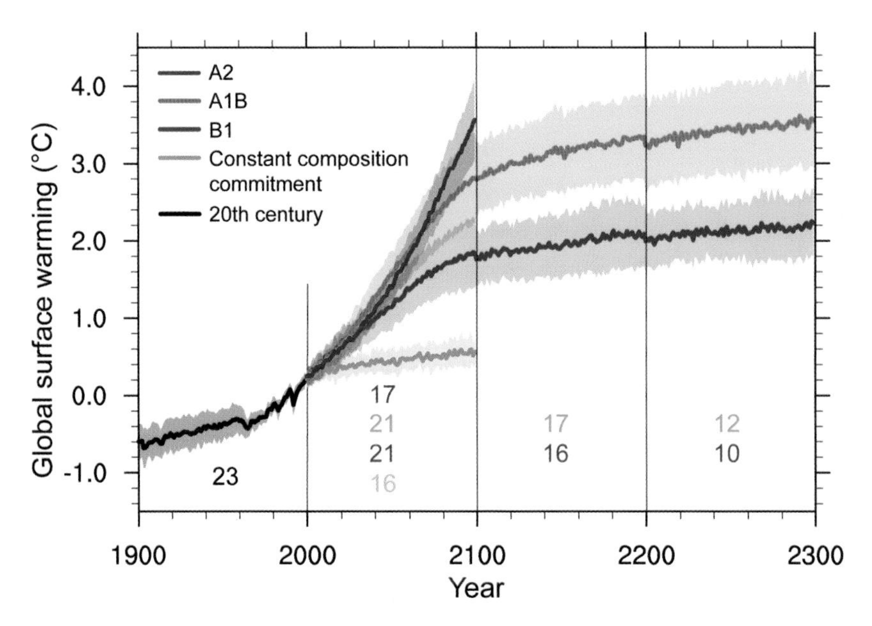

FIGURE 3.3 In stabilization scenarios, such as A1B (green) and B1 (blue) after 2100, or the "constant composition commitment" (yellow) in which the forcing is held fixed at the values in 2000, the warming grows slowly despite the constant forcing. Rescaling the TCR by the forcing will underestimate the surface warming in the stabilization period by an amount that grows with time, as the system slowly makes its transition to its equilibrium response. The average ratio of TCR to the equilibrium sensitivity in the models utilized by the AR4 (using values in Table 8.2 in Chaper 8 of the WG1 report) is 0.55. So the slow growth in the stabilization period, in which the forcing is somehow maintained at a constant level, continues for many centuries beyond that indicated in this figure, until the additional warming in these periods becomes comparable to that in the preceding periods of increasing forcing.

The ratio of TCR to the equilibrium response is smaller than one would expect from the linear analysis described above, using the mean value for these models of the heat uptake efficiency γ (Dufresne and Bony, 2008) and the radiative restoring β implied by the equilibrium climate sensitivity of these models. The explanation is that the strength of the restoring force provided by energy fluxes escaping to space, β, is found to decrease in climate models as the deep oceans equilibrate (Williams et al., 2008; Winton et al., 2010). In the CMIP3 archive the value of the radiative restoring relevant on the time scale of the 20th and 21st centuries is larger than the value relevant for the equilibrium response—by 30% on average, but with considerable model-to-model variations. This weakening of the radiative restoring once the deep oceans equilibrate is likely associated with changes in the horizontal structure of the warming. In the period of increasing forcing, the warming of the subpolar oceans is held back by strong coupling to the deep oceans. After stabilization, the deep oceans slowly warm and polar regions are thereby allowed to warm more rapidly, relative to the average warming of Earth's surface. The radiative restoring is weaker for warming of the subpolar surface than for warming elsewhere because the coupling between the surface and tropospheric layers from which most of the radiation escaping to space is emitted is relatively weak in subpolar latitudes. The result is a reduction in the strength of the radiative restoring per degree warming of Earth's surface as a whole.

The forcing due to well-mixed greenhouse gases (WMGGs) from the mid-19th century till 2010 is estimated to be 2.7 W/m^2 by NOAA's annual greenhouse gas index (*http://www.esrl.noaa.gov/gmd/aggi/*), which is about two-thirds of the forcing due to doubling of CO_2 of about 3.7 W/m^2. Therefore, one can estimate the warming due to the WMGGs since the mid-19th century, for a given value of TCR, as roughly two-thirds of TCR.

The distribution of TCR values generated by the CMIP3 models, as tabulated in Chapter 8 of the WG1/AR4 report (Randall et al., 2007), is displayed as a histogram in Figure 3.4. The mean, or median of this distribution, and the spread of values about this mean, are of interest in describing the values emerging from the best efforts of the major modeling groups around the world. The mean is between 1.8°C and 1.9°C and is somewhat larger than the median, which is close to 1.6°C.

To explain the 20th century warming (0.7-0.8°C) with WMGGs would require Earth to reside at the extreme low end of this distribution. As stated in the SPM of WG1/AR4, "it is likely that increases in greenhouse gases concentrations alone would have caused more warming than observed because volcanic and anthropogenic aerosols have offset some warming that would

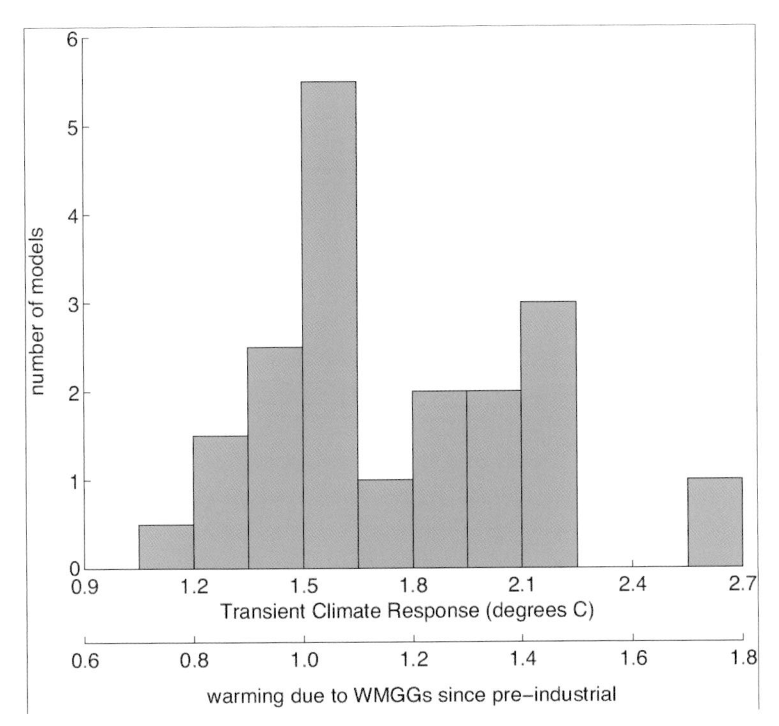

FIGURE 3.4 A histogram of TCR from a set of 19 models from 1% per year experiments described in Ch. 8 of WG1/AR4. The TCR values have also been converted into the warming since the mid-19th century due to well-mixed greenhouse gases by multiplying by 2/3.

otherwise have taken place" (IPCC, 2007c). In particular, a negative forcing of 0.8 W/m^2 due to non-WMGG forcing, or a total forcing of 1.9 W/m^2, would be consistent with the 20th century warming if TCR were roughly 1.5°C—while a negative forcing of 1.2 W/m^2, or total forcing of 1.5 W/m^2, would be consistent with a TCR of about 1.9°C. Total aerosol forcing of this magnitude is consistent, for example, with the estimate of Quaas et al. (2008), but the range of estimates is large.

Best estimates of TCR typically range from 1.5-2°C. For example, Stott et al. (2006) use fingerprinting techniques to adjust the results from GCMs to improve fits to the spatial structure of observations and obtain a best estimate close to 2°C. Knutti and Tomassini (2008) use a simple model to fit the 20th century temperature record and ocean heat uptake, using a Bayesian analysis to determine optimal parameters and uncertainties. Their best estimate for TCR is 1.5-1.6°C, with a sharp cutoff at 1 and with a fairly long tail on the high sensitivity end. The cutoff at 1 is common to most estimates, related to the fact that lower values of TCR requires positive forcing from unknown forcing agents or very large contributions to the century long trend from internal variability unlike any produced by current GCMs.

The temperature record over the past 30 years provides a potentially useful constraint on TCR (Gregory and Forster, 2008; Murphy et al., 2009). The anthropogenic aerosol forcing, although it has probably grown over much of the industrial era, was likely slowing if not fully leveling off over this period, leaving the well-mixed greenhouse gas forcing more dominant. Additionally, solar forcing, which is well constrained by satellite measurement of total solar irradiance, contributed very little to the trend over this period. Removing estimates of volcanic and ENSO signals from the global mean temperature record results in a fairly linear residual (e.g., Lean and Rind, 2009; Thompson et al., 2009). The residual yields a warming of about 0.48 K, or 0.16 K/decade, over the 30-year period. The WMGG forcing over this period is close to 1 W/m^2. Rescaling implies a TCR of 1.8°C, similar to the estimate of Gregory and Forster. Murphy et al. (2009), estimating TCR directly from TOA fluxes and ocean heat uptake over a similar time period, paint a picture consistent with relatively flat aerosol forcing in recent decades and a TCR of around 1.8.

The difficulty in using this relatively short period for estimating TCR is that internal climate variations are capable of modifying this trend substantially, so we cannot assume that the observed trend is entirely forced. Gregory and Forster (2008) estimate uncertainty by using the variability in 30-year trends of global mean temperature from a particular GCM. Their result is 1.3-2.3°C as the 90% confidence interval around their best estimate of 1.8°C. This is about the same range as in the CMIP3 models, but with a very different source of uncertainty. The Gregory and Forster estimate of uncertainty assumes that we have no information as to whether the contribution from internal variability was positive or negative during this period, although the CMIP3 range of values is due to uncertain physics in the models, especially cloud feedbacks.

There is a body of work on multi-decadal variability, especially in the North Atlantic, which suggests that internal variability has contributed positively to the temperature trends over the past 30 years. If correct, this would lower estimates of TCR that are based on comparison to temperature trends in recent decades. The North Atlantic is likely to be the source of much of the multi-decadal variability, and a variety of oceanographic and coupled model studies (Zhang, 2008; Knight, 2009; Latif et al., 2009; Polyakov et al., 2009) indicate that the North Atlantic has been in a warm phase over much of this period. Polyakov et al. estimate that as much as 50% of the trend over this period in the North Atlantic is internal variability. The modeling work by Zhang and Knight also indicate that the influence of the Atlantic, despite its small size, can spread preferentially over Eurasia and contribute to global temperature signals. For example, the model analyzed by Knight generates a 0.1 K global mean warming for an increase of 1 Sverdrup (about 5% of the 20 Sverdrup mean value) in the Atlantic overturning.

Volcanic responses and the response to the 11-year solar cycle can also be used to constrain TCR, as can the autocorrelations of internal fluctuations (rather than watching the volcanic response decay in time to estimate the strength of restoring forces, one can watch internal fluctuations decay). Paeloclimatic evidence constrains equilibrium sensitivity, and with modeling guidance and heat uptake measurements constraining the ratio of TCR to the equilibrium response, one can also use these to constrain TCR. But there are also issues related to the decoupling of transient and equilibrium responses and to issues of "Earth system sensitivity", that come into play when considering paleoclimatic constraints (see Chapter 6). We judge the constraints that directly involve fits to the temperature record over the past century and the last few decades to be the most useful in constraining TCR at this time.

The magnitude of many of the impacts discussed in this report scale with the value of TCR. We estimate TCR by starting with the distribution in the CMIP3 ensemble, but lowering the low end of the distribution slightly to take into account the possibility suggested by some recent studies of internal variability that a portion of the most recent Northern Hemisphere warming is internal. This results in a best estimate of **1.65°C,** likely lying in the range **1.3-2.2°C** (i.e., with 2/3 probability) and very likely lying in the range of **1.1-2.5°C** (i.e., with 90% probability). The estimate in the WG1/AR4 report is a very likely range of 1-3°C. Our reduction of the upper limit to this range is consistent with our critique of very high equilibrium sensitivities in Section 3.2.

3.4 CUMULATIVE CARBON

Introduction: Why Use Cumulative Carbon Emissions?

The temperature response to anthropogenic carbon emissions is determined by: (1) the response of the carbon cycle to emissions; (2) the climate response to elevated CO_2 concentrations; and (3) the feedback between climate change and the carbon cycle. There has been significant attention in recent literature on how carbon sinks (and resultant airborne fraction) are affected by both elevated CO_2 and climate changes (concentration-carbon and climate-carbon feedbacks; see Friedlingstein et al., 2006, Gregory et al., 2009, and Section 2.4). There is also a large body of literature aimed at estimating the temperature response to elevated CO_2, typically defined as equilibrium climate sensitivity or transient climate response (Meehl et al., 2007; see also Sections 3.2 and 3.3). Each stage in the progression from carbon emissions to the resultant climate warming carries large uncertainty due to our incomplete understanding of the magnitude of both physical and biochemical feedbacks in the climate system.

Matthews et al. (2009) proposed a new metric of the temperature response to carbon emissions, the "carbon climate response," which includes the net effects of both carbon cycle and physical climate feedbacks. The carbon-climate response is defined as the globally averaged temperature response to 1 trillion tons of carbon emissions (3.7 trillion tons of CO_2), thus framing the climate response to emissions in the context of cumulative emissions of carbon dioxide over time. In effect, the carbon-climate reponse is a generalization of the concept of climate sensitivity as it pertains to carbon dioxide forcing. By including the carbon cycle response to emissions in addition to the temperature response to CO_2 forcing, the carbon-climate response represents a metric that relates global mean temperature change directly to cumulative carbon emissions. This concept of measuring the climate response to cumulative emissions was also proposed concurrently by three other studies (Allen et al., 2009; Meinshausen et al., 2009; Zickfeld et al., 2009), all of which demonstrated a remarkably consistent temperature response to a given level of cumulative carbon emissions. Although CO_2 radiative forcing decreases logarithmically with increasing CO_2 concentrations, this is balanced by a near-exponential increase in the airborne fraction of emissions due to weakening carbon sinks at higher CO_2 concentrations (Caldeira and Kasting, 1993). As a result, global mean temperature change is almost linearly related to cumulative carbon emission and is independent of the time during which the emissions occur (Matthews et al., 2009).

Estimates of the Temperature Response to Cumulative Emissions

Matthews et al. (2009) estimated the temperature response to cumulative carbon emissions as 1-2.1°C per trillion tons of carbon (1000 GtC) emitted, based both on 21st century simulations by coupled climate-carbon cycle models and on historical observations of CO_2-induced temperature change and anthropogenic CO_2 emissions (Figure 3.5). This study produced a best

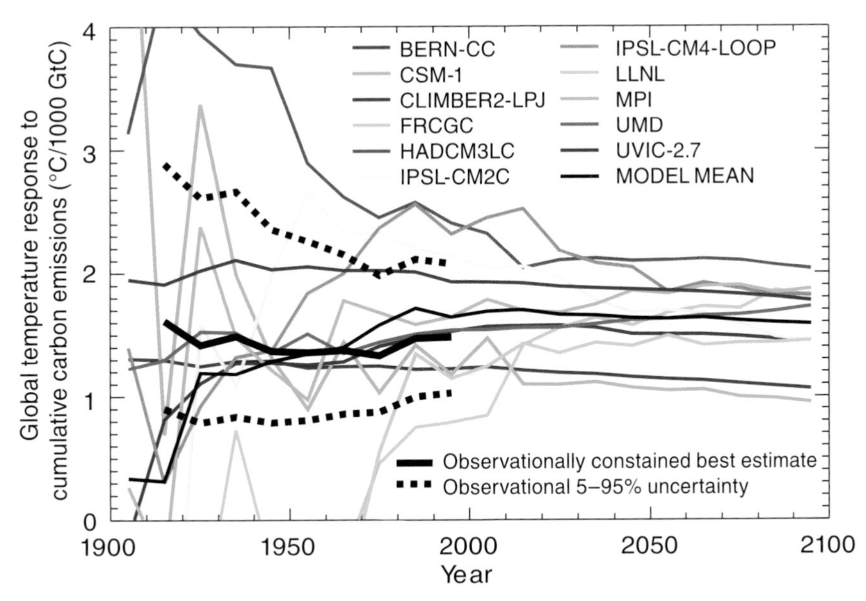

FIGURE 3.5 Temperature response to cumulative carbon emissions from coupled climate-carbon model simulations (thin colored lines) and historical observations of CO_2-induced warming and anthropogenic CO_2 emissions (thick black line solid and dotted lines). (Figure adapted from Figures 3 and 4 of Matthews et al., 2009).

estimate of 1.5°C/1,000 GtC emitted based on observational constraints, and 1.6°C/1,000 GtC based on the model average. Allen et al. (2009), using a simpler climate model but considering a larger range of climate sensitivity, found a most likely peak temperature response of 2°C/1,000 GtC, with a 5-95% confidence range of 1.3-3.9°C/1,000 GtC. Allen et al. also provided an estimate for the *instantaneous* temperature response to cumulative emissions (corresponding to the definition of the carbon-climate response from Matthews et al.) of 1.4-2.5°C/1,000 GtC. Both Matthews et al. (2009) and Allen et al. (2009) concluded that the temperature response to cumulative emissions is remarkably constant over time and over a wide range of CO_2 emissions scenarios. Based on this, they provided best estimates of the allowable emissions for 2°C global temperature increase of 1,000 GtC (Allen et al., 2009) and 1,300 GtC (Matthews et al., 2009).

Zickfeld et al. (2009) also presented an estimate of the cumulative emissions required to meet a 2°C temperature target. The authors considered both climate sensitivity uncertainty and the uncertainty in climate-carbon feedbacks, and thus were able to generate a probabilistic estimate of the emissions associated with various temperature targets. To restrict the probability of exceeding 2°C to 33%, Zickfeld et al. concluded that emissions from 2001 to 2500 must be kept to a median estimate of 590 GtC, with a range of 200 to 950 GtC owing to different estimates of climate sensitivity uncertainty, as well as uncertainty in climate-carbon feedbacks. For an exceedence probability of 50%, the median estimate from this study was approximately 840 GtC (range: 500 to 1,210). Including also historical CO_2 emissions up to 2000 (~460 GtC; Houghton, 2008; Boden et al., 2009), this best estimate for 2°C from Zickfeld et al. corresponds to emissions of approximately 1,050 GtC (1,300 GtC) for an exceedence probability of 33% (50%). Meinshausen et al. (2009) also presented an estimate of the cumulative carbon emissions required to meet a 2°C temperature target. This study used a simpler model and a narrower time window (2000-2050) but considered also the effect of non-CO_2 greenhouse gases and aerosols; Meinshausen et al. estimated that cumulative emission from 2000-2050 must be restricted to 390 GtC to avoid 2°C warming with 50% likelihood.

"Stabilization" Framework Based on Cumulative Carbon

The cumulative carbon framework is well suited to relating instantaneous global temperature change to a given level of cumulative carbon emitted. Based on the above studies, we select 1.75°C global temperature change per 1,000 GtC emitted to be a representative best estimate for the climate response to cumulative carbon emissions. There is large uncertainty, however, in this estimate of the temperature response to carbon emissions owing both to uncertain carbon cycle response to elevated CO_2 and climate changes (Section 2.4) as well as to uncertainty in the physical climate system response to CO_2 forcing (Sections 3.2 and 3.3). We use here a *very likely* uncertainty range on this central estimate of 1 to 2.5°C per 1,000 GtC emitted, based on the lower and upper 5-95% confidence limits given in Matthews et al. (2009) and Allen et al. (2009).

This estimate of the temperature response to cumulative emissions corresponds to approximately 1,150 GtC (4200 billion tons of CO_2) in allowable emissions consistent with 2°C temperature change (Figure 3.6). It is critical to recognize, however, that the uncertainty range on this number is very large, with a possible lower limit of 500 GtC (1,800 Gt CO_2) inferred from Allen et al. (2009) and a possible upper limit of 1,900 GtC (7000 Gt CO_2) from Matthews et al. (2009); this range of CO_2 emissions (500-1,900 GtC) can be considered to be a *very likely* range for emissions consistent with 2°C global warming. We can narrow this range somewhat and apply

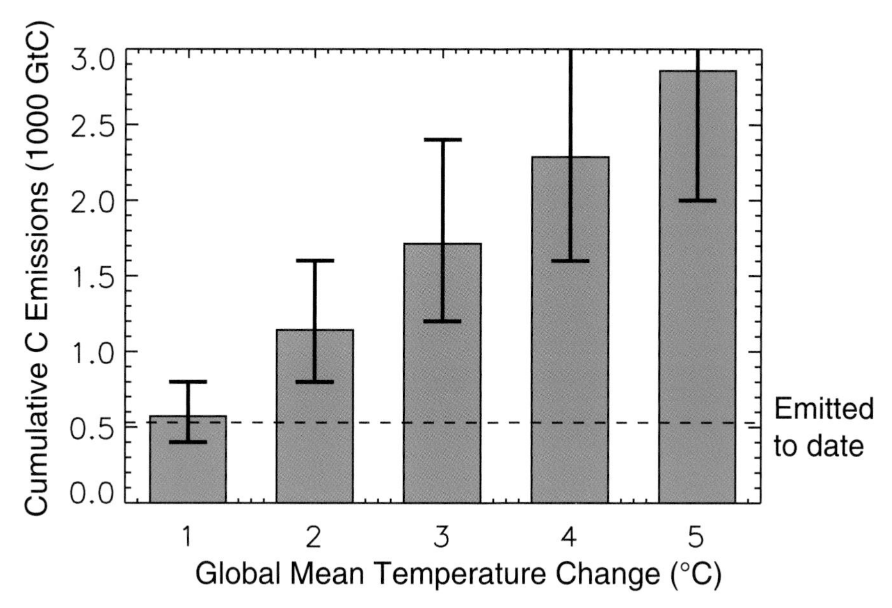

FIGURE 3.6 Cumulative carbon emissions consistent with global mean temperature changes of 1 to 5°C. Best estimates are based on 1.75°C per 1,000 GtC emitted, taken as a representative best estimate from Matthews et al. (2009) and Allen et al. (2009). *Likely* uncertainty ranges of 70-140% of the best estimate are based on Zickfeld et al. (2009) and Matthews et al. (2009). The dashed line shows cumulative emission to the year 2009 (530 GtC).

it to other warming levels based on a combination of the results of Zickfeld et al. (2009) and Matthews et al. (2009). Matthews et al. (2009) presented a 5-95% uncertainty range of 1,000 to 1,900 GtC on emissions for 2°C, based on a central estimate of 1400 GtC; this corresponds to a uncertainty range of approximately 70-140% of the central estimate. Zickfeld et al. (2009) also provided an uncertainty range for the cumulative emissions associated with 2, 3, and 4°C global mean temperature change, and found that the relative uncertainty scaled approximately with the median value. Based on this, we use the relative uncertainty range from Matthews et al. (2009) and apply this to our central estimate of emissions consistent with each temperature target. For 2 degrees, this represents a best estimate of 1,150 GtC, with an uncertainty range of 800 to 1,600 GtC (Figure 3.6), which we adopt here as the *likely* range of emissions for 2°C global warming.

While global mean temperature change is well constrained by cumulative emissions, this cumulative carbon framework is less consistent with the more widely used framework of CO_2 concentration stabilization. A given level of cumulative CO_2 emissions do not result in stable CO_2 concentrations, but rather in CO_2 concentrations that peak at some value and then decrease slowly when emissions fall below the level of persistent natural carbon sinks (see Figure 3.7). For rates of emission reduction of the order of 1-4% per year, and even if CO_2 emissions become close to zero, the decrease in atmospheric concentrations may, however, occur very slowly over

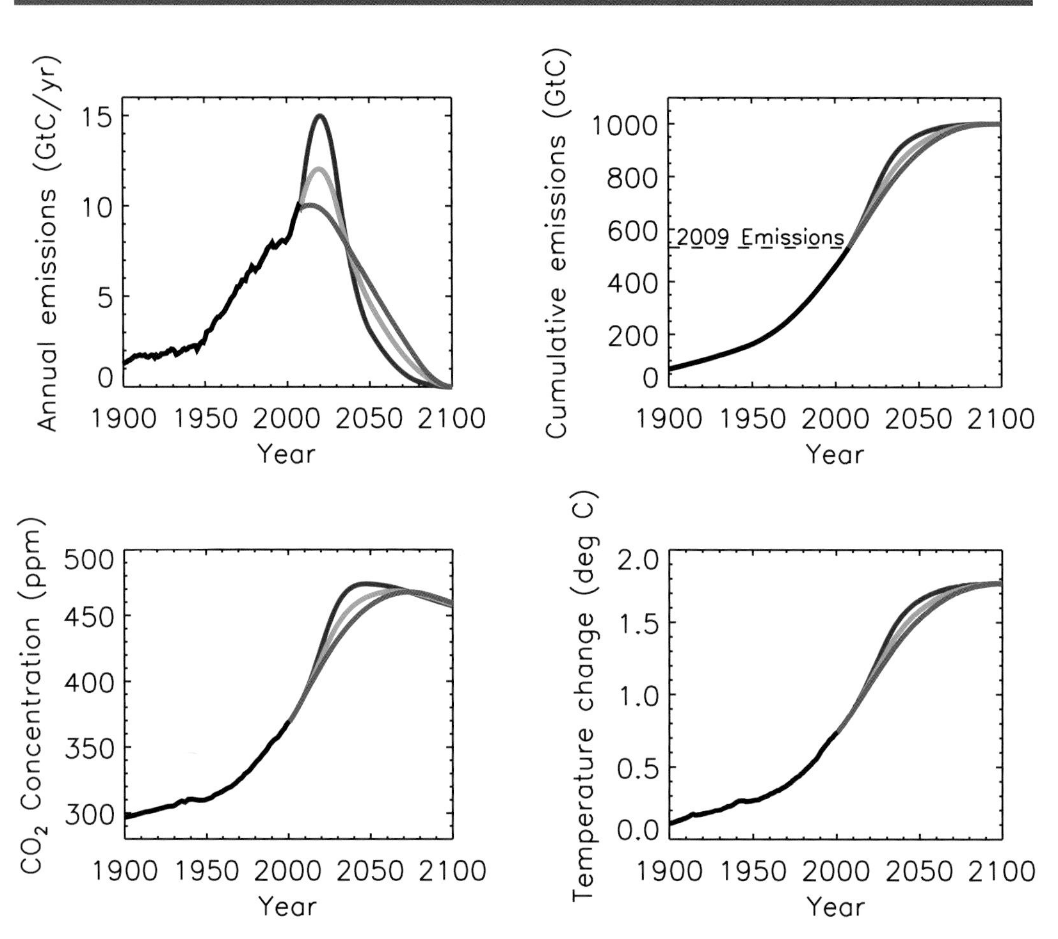

FIGURE 3.7 Illustrative emissions scenarios with cumulative emissions from 1750 to 2100 totaling 1,000 GtC (3,700 GtC). For all scenarios, the year-2100 temperature change and CO_2 concentration do not depend on the shape of the emissions scenario, but rather on the total cumulative emitted. These scenarios were constructed such that total cumulative carbon emissions were the same for each scenario, the rate of emissions decline varied from 1.5 to 4.5% per year relative to the peak emissions, and emissions were constrained to reach zero at the year 2100. CO_2 concentrations and temperature changes shown here were simulated by the UVic ESCM. Source: Weaver et al. (2001) and Eby et al. (2009).

centuries (see Section 2.2). In a framework of cumulative carbon emissions, CO_2 concentrations do not necessarily "stabilize" but rather change over time in response to a given CO_2 emissions scenarios; in this case, it is the total cumulative carbon emitted over time, rather than the atmospheric CO_2 concentration itself, that indicates the level of expected climate warming.

The clear advantage of the cumulative carbon framework is that a given level of cumulative emissions corresponds to a unique temperature change, which remains approximately constant for several centuries after the point of zero emissions (Matthews et al., 2008; Solomon et al., 2009). As can be seen in Figure 3.7, for this particular model, cumulative emissions of 1,000 GtC from 1750 to 2100 result in a year-2100 global temperature change of 1.8°C over pre-industrial temperatures, which corresponds to a year-2100 CO_2 concentration of 460 ppm; both the year-2100 temperature change and the year-2100 CO_2 concentration are independent of the shape of the CO_2 emissions scenario and depend only on the total cumulative carbon emitted. By contrast, the rate of temperature change, as well as the peak CO_2 concentration in these simulations, varied as a results of differences in the rates of increase and decline of emissions in each scenario.

Figure 3.8 illustrates how the concept of CO_2 stabilization can be reconciled with the cumulative emissions framework. This figure shows idealized CO_2 concentration scenarios that reach between 350 and 1,000 ppm at the year 2100, along with the cumulative carbon emissions and temperature changes associated with each scenario.[4] From this analysis, restricting global temperature change to 2°C requires best-guess cumulative emissions of 1,150 billion tons of carbon between the years 1800 and 2100; this corresponds to a "stabilization" CO_2 concentration of between 450 and 550 ppm at the year 2100, with the caveats that CO_2 concentrations changes and warming after the year 2100 would depend on the level of additional post-2100 emissions. In general, at a given time (e.g., the year 2100) both the atmospheric CO_2 concentration and the associated temperature change can be inferred from cumulative carbon emissions to date. If carbon emissions were subsequently eliminated, atmospheric concentrations would slowly decrease over time, whereas temperature would remain elevated for several

[4]The temperature responses to cumulative emissions shown in Figure 3.8 include both the carbon cycle and climate sensitivity to emissions, but do not correspond directly to either the transient climate response or the equilibrium climate sensitivity associated with a given CO_2 concentration. In general, for higher emissions scenarios where forcing is still increasing rapidly at 2100, this temperature change will more closely reflect the transient climate response. For lower emissions scenarios with stable or declining forcing during the latter half of the 21st century, the temperature change at 2100 will more closely reflect the equilibrium climate sensitivity.

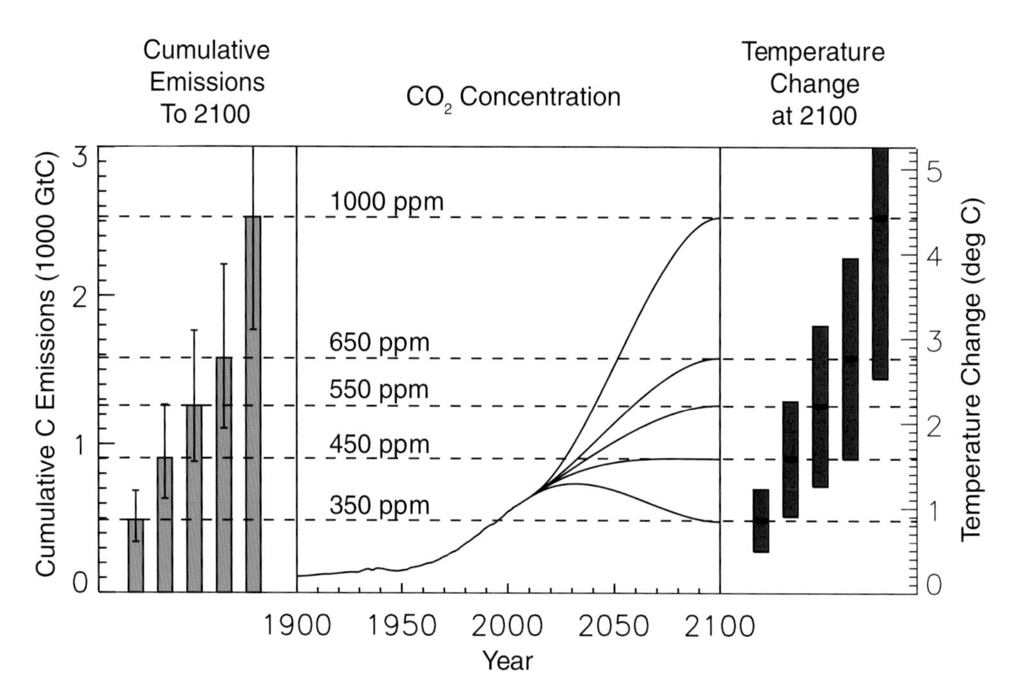

FIGURE 3.8 Idealized CO_2 concentration scenarios reaching between 350 and 1,000 ppm at the year 2100. At the year 2100, the atmospheric CO_2 concentration and global mean temperature change is dependent on cumulative carbon emissions to date, with variation in the rate of emissions over time affecting only the rate of increase of forcing and consequent rate of temperature change. Stabilization of CO_2 concentrations after the year 2100 would require continued low-level CO_2 emissions (leading to increasing cumulative carbon emitted), whereas zero emission after 2100 would result in slowly declining CO_2 concentrations and approximately stable global temperature. Cumulative emissions for each scenario shown here are based on simulations with the UVic ESCM, with uncertainty ranges of 70-140% of the central value based on Matthews et al. (2009) and Zickfeld et al. (2009). Temperature changes are calculated using 1.75°C per 1,000 GtC emitted, with a 1-2.5°C/1,000 GtC uncertainty range based on Matthews et al. (2009) and Allen et al. (2009). These uncertainty ranges reflect both uncertainty in the response of carbon sinks to elevated CO_2 and climate changes, as well as uncertainty in the physical climate system response to change in CO_2 forcing.

centuries. Similarly, should emissions continue at a low level (resulting in increasing cumulative carbon emissions), atmospheric concentrations may remain stable, but global mean temperature would continue to increase over time. Atmospheric CO_2 stabilization is consistent with a small amount of continued CO_2 emissions at a rate equal to the level of persistent natural carbon sinks, whereas atmospheric temperature stabilization is only consistent with near-zero CO_2 emissions (Matthews and Caldeira, 2008; Solomon et al., 2009).

4

Physical Climate Change in the 21st Century

4.1 REGIONAL PATTERNS OF WARMING AND RELATED FACTORS

An approximation on which we will rely in this report is "pattern scaling", in which a robust pattern of climate change is assumed to exist that describes the geographical and seasonal structure of the temperature response that does not depend on climate sensitivity or on details of the forcing or emission scenario. All of the scenario and sensitivity dependence is captured within the time evolving global mean surface temperature, $T_G(t)$. Letting the symbol ξ stand for the two horizontal spatial coordinates and the time of year, the assumption is that

$$T(t,\xi) = T_G(t)\ \tau(\xi)$$

The pattern $\tau(\xi)$ has a spatial-annual mean of unity by definition. We focus on temperature here and discuss pattern scaling for precipitation in Section 4.2.

The validity of this approximation is discussed by Santer et al. (1990), Mitchell et al. (1999), and Mitchell (2003). It has been used extensively for regional temperature (and precipitation) change projections (Dessai et al., 2005; Murphy et al., 2007; Watterson, 2008) and impacts studies, as a substitute to running fully coupled simulations under different scenarios or with different models with a range of climate sensitivities.

The pattern is derived from experiments with fully coupled Global Climate Models (GCMs), with validation from efforts to isolate the well-mixed greenhouse gas signal from the historical temperature record. The value of the method relies on the pattern remaining fairly constant during a simulation, across different concentration pathway scenarios and across different model settings. Regionally and temporally differentiated results under different scenarios or climate sensitivities can be derived by first characterizing the stable geographical pattern of warming (and its spatial variability if some measure of uncertainty is sought) on the basis of available coupled model simulations, then adjusting the profile of transient warming by using a fast and simple energy balance model to predict the global mean surface temperature evolution. There are significant differences between the patterns generated by different models—for example, in the amount of polar amplification in the Arctic per degree of global warming. These differences are averaged over in the ensemble-based estimates of mean patterns, but uncertainty can be characterized by the inter-model spread in the pattern $\tau(\xi)$.

The choice of the pattern in the studies available in the literature are often as simple as the ensemble average (across models and/or across scenarios, for the coupled experiments available) of the spatial change in temperature, normalized by the corresponding change in global average temperature, choosing the end of the simulations (usually last two decades of the 21st century) and a baseline of reference (pre-industrial or current climate). Similar properties and results have been obtained using more sophisticated multivariate procedures that optimize the variance explained by the pattern.

There are limitations to this approach. It can break down if aerosol forcing is significant, not only because aerosols and greenhouse gases can have different spatial footprints, but also because the effects of aerosols themselves are more difficult to characterize in this simple way. For example, Asian and North American aerosol production are likely to have different time histories in the future. Our focus in this report is on the greenhouse gas component of climate change, making the pattern scaling assumption more justifiable.

Simple pattern scaling is regarded as especially useful for summarizing model projections of transient climate change due to well-mixed greenhouse gas increases on a time scale of a few centuries. But it is less accurate for stabilization scenarios, as the temperature changes approach an equilibrium response. From the early work of Manabe and Wetherald (1980) and Mitchell et al., (1999) it has been clear that the pattern of temperature response evolves as the slow component of the warming, associated with equilibration of the deep oceans on multi-century time scales, equilibrates. In particular, on these long time scales the warming of high latitudes in the Southern Hemisphere is much larger relative to the global mean warming than in the earlier periods. Held et al. (2010) emphasize that this slow warming pattern is present, but of small amplitude, during the initial transient adjustment phases of the response as well.

There are some regions of sharp temperature gradients, near the ice edge for example, where pattern scaling will break down. As the climate warms, temperature changes will be large as the ice edge moves across a particular location, but then return to small values with additional warming, because the ice edge is now further poleward.

Given other uncertainties, we find that pattern scaling is justified for current attempts to link stabilization targets and impacts, keeping in mind limitations due to the evolution of the pattern of warming on long, stabilization time scales and the limitations in regions of sharp gradients.

On the basis of CMIP3 simulations, Chapters 10 and 11 of IPCC AR4, WG1 analyzed geographical patterns of warming and measures of their variability across models and across scenarios. The executive summary of Chapter 10 reports that "[g]eographical patterns of projected SAT warming show greatest temperature increases over land (roughly twice the global average temperature increase) and at high northern latitudes, and less warming over the southern oceans and North Atlantic, consistent with observations during the latter part of the 20th century …". Figure 10.8 of the report depicts the patterns of annual average warming across three scenarios (A2, A1B and B1) and three time periods (2011-2030, 2046-2065, and 2080-2099) over which change is computed. Figure 10.9 shows seasonal patterns for DJF and JJA under A1B. Chapter 10 also reports that the spatial correlation of fields of temperature change is as high as 0.994 in the model ensemble mean when considering late 21st century changes between A2 and A1B. A table in the same section (Table 10.5) quantifies the strict agreement between the A1B field, as a standard, and the other scenario patterns using a measure proposed by Watterson (1996) with unity meaning identical fields and zero meaning no similarity. Values of this measure are consistently above 0.8 and increase as the projection time increases (later in the 21st century fields agree better than earlier in the century), with values of 0.9 or larger for the late 21st century. The same table also shows that the agreement deteriorates if considering commitment scenarios. The results are documented as applying to seasonal warming patterns besides annual averages.

On the basis of the previous discussion and results, we compute patterns of standardized warming from the available CMIP3 SRES scenario simulation and produce maps of the ensemble average warming (these—in their non-normalized version—are available from AR4 WG1 Figures 10.8 and 10.9, and individual models' maps are available in supplementary material in Chapter 10 (*http://ipcc-wg1.ucar.edu/wg1/Report/suppl/Ch10/Ch10_indiv-maps.html*, and Chapter 11 (*http://www.ipcc.ch/pdf/assessment-report/ar4/wg1/ar4-wg1-chapter11-supp-material.pdf*) along with measures of the regionally differentiated variability of this pattern across models and scenarios (IPCC, 2007a).

Figure 4.1 shows patterns of warming normalized "per degree C of global annual average warming". Ranges across models or across scenarios are shown in Figures 4.4 and 4.5. We show maps of global geographical patterns and North American patterns for annual average warming and December-January-February and June-July-August average warming.

We highlight here the main features characterizing warming patterns:

The annual and DJF mean patterns present a typical gradient of warming that decreases from north to south, with the higher latitude of the Northern Hemisphere seeing the largest increases, and the land masses warming more than the oceans. The JJA patterns' main characteristics are an enhanced warming of the interior of the continents and the Mediterranean Basin, with a gradient that is generally equator to poles rather than north-south. The largest source of variation resides in the inter-model spread rather than the inter-scenario spread, and it is mainly localized over and at the edge of the ice sheets of the Arctic and Antarctica.

We also present in Figure 4.2 a number of scatter plots depicting the relation between the magnitude of global average warming (by 2080-2099 compared to 1980-1999) and the magnitude of regional warming across models (each model one dot) and scenarios (color-coded) for several of the "Giorgi regions" (Giorgi and Francisco, 2000). We chose four regions that subdivide the North American continent (western, central, and eastern North America—WNA, CNA, and ENA respectively—and Alaska, ALA) plus two regions in other climates for comparison, the Mediterranean Basin (MED) and Southern Australia (SAU). The linearity of the relationship and the fairly tight spread around it is clear. As already noted, the source of variation between different models is larger than from the scenarios: averaging each model across the three scenarios would not reduce the scatter as much as averaging all the models within one scenario, as shown by the larger star-shaped marks.

Finally, in Figure 4.3 we show December-January-February and June-July-August warming patterns as calculated in models including both anthropogenic and natural forcing *for the 20th century*, together with 20th century observations. Figure 4.3 shows that the warming patterns in the models and observations display many common large-scale features, and a comparison with Figure 4.1 demonstrates how 20th century patterns contain already many of the large-scale features that characterize 21st century patterns. Among these, the relatively larger magnitude of the warming in December-January-February than in June-July-August in both observed and modeled patterns; the amplification of the warming in the high latitudes of the Northern Hemisphere characteristic of December-January-February

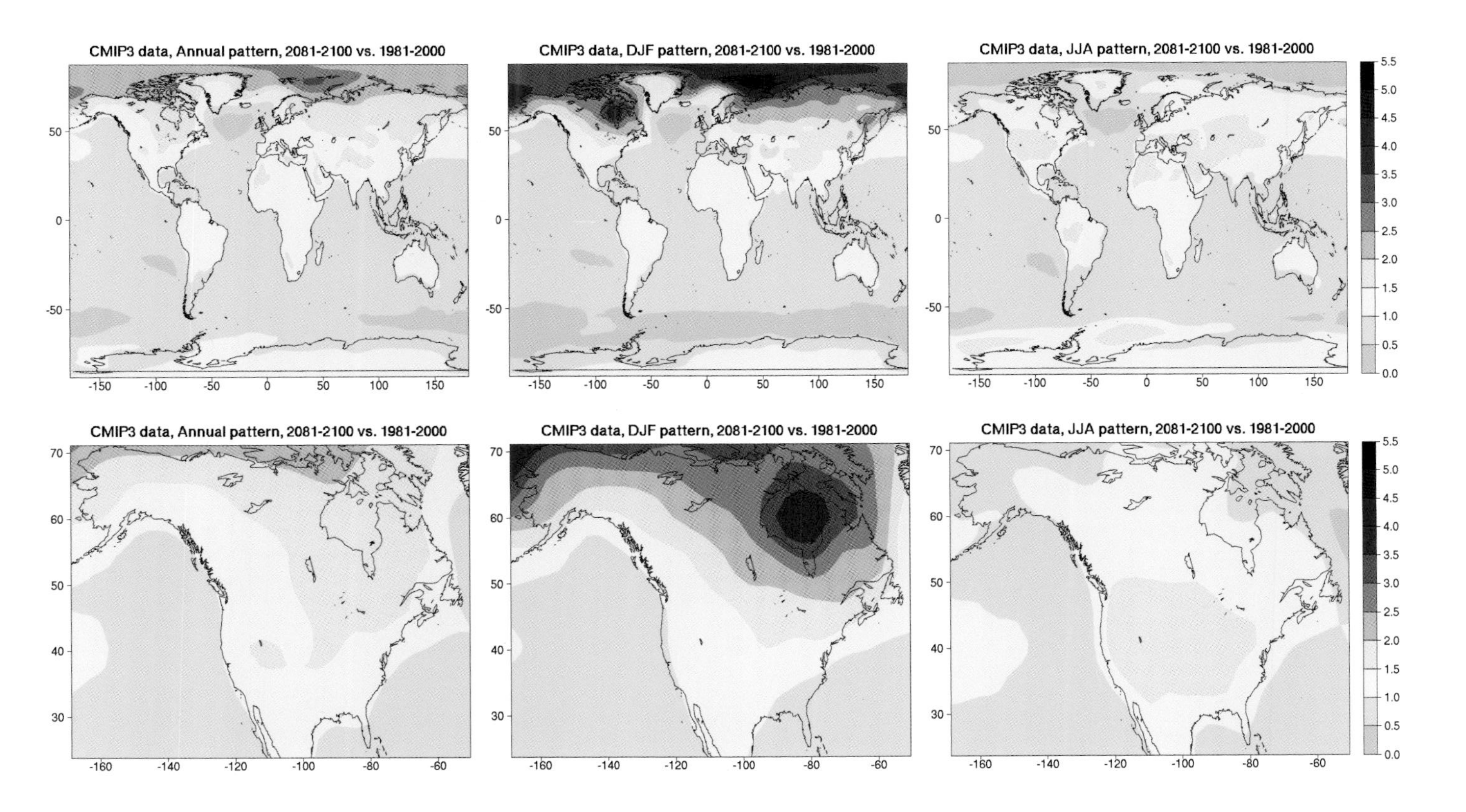

FIGURE 4.1 Geographical patterns of warming per 1°C of global annual average warming, over the whole world (top) and over North America (bottom). From left to right: patterns for average temperature over the whole calendar year or for average temperature in boreal winter (December, January, and February) or summer (June, July, and August). Patterns are obtained by scaling end-of-21st-century changes (compared to end-of-20th-century climatology) by global average annual warming over the same period, using temperature at the surface from the 18 CMIP3 models whose output is available for SRES scenarios A2, A1B, and B1.

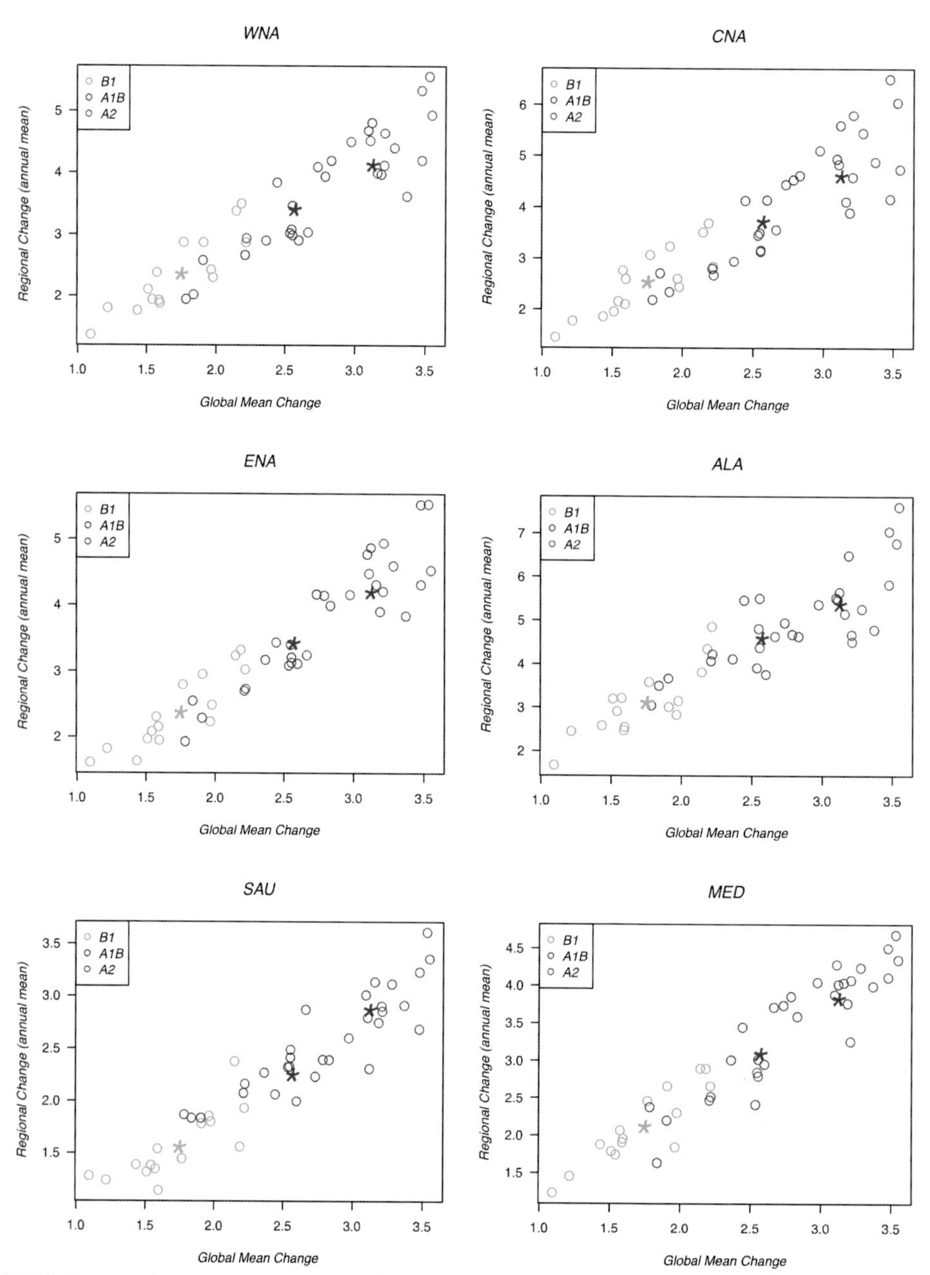

FIGURE 4.2 Scatterplots comparing regional average warming versus global average warming (both averaged over the calendar year) for western North America (WNA), central North America (CNA), eastern North America (ENA), Alaska (ALA), Southern Australia (SAU) and the Mediterranean (MED). Each point indicates results from an individual model under one of the three SRES scenarios (A1B, blue; A2, red; and B1, green). Stars indicate the multi-model ensemble averages for each of the three scenarios.

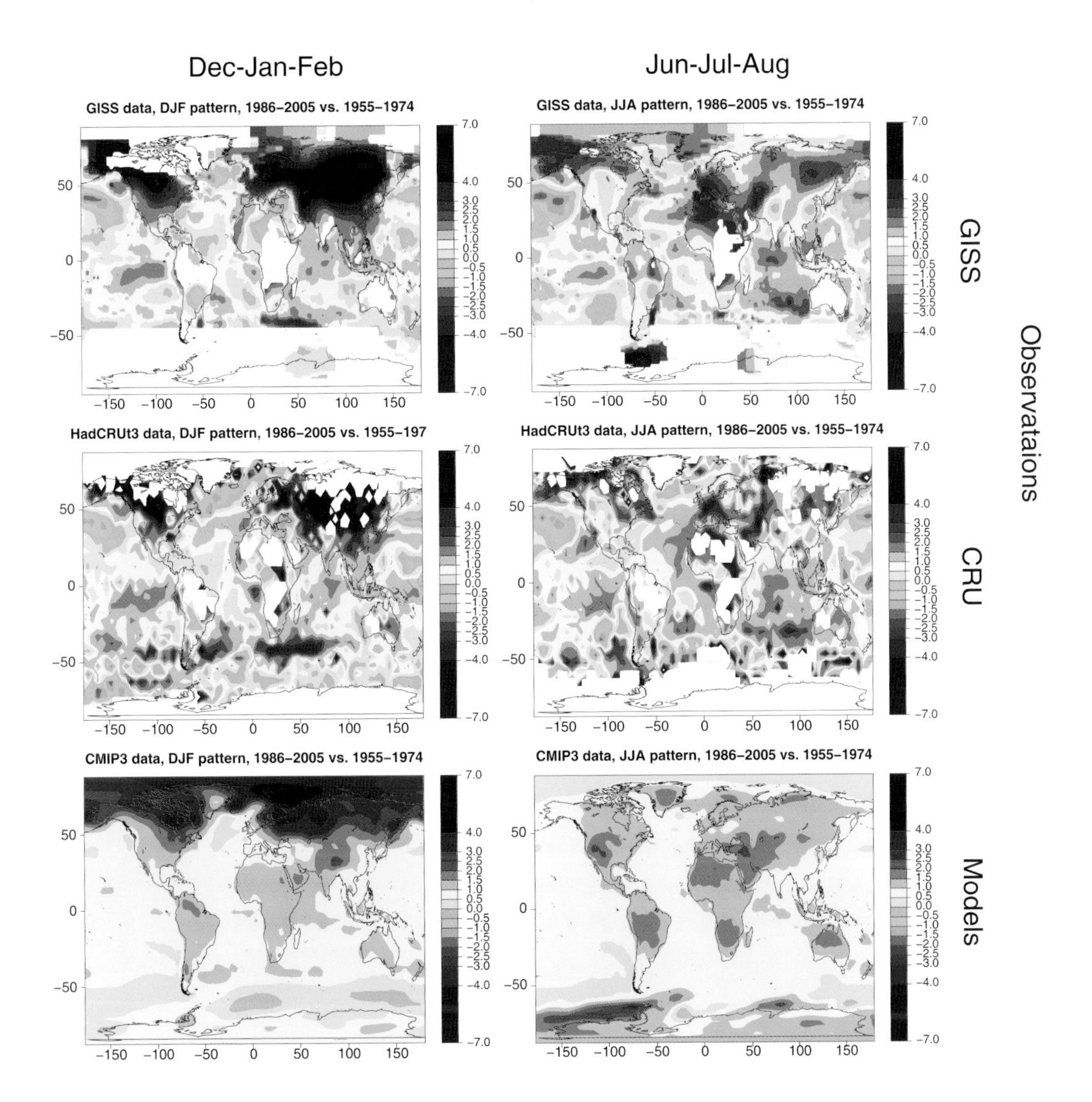

FIGURE 4.3 Relative patterns of warming (normalized to one for the globe),for December-January-February (left) and June-July-August (right) for 1955-2005 (obtained as differences between 20-year average temperatures for 1986-2005 and 1955-1974).The top two panels show results from two instrumental temperature records, NASA GISS and CRU. White indicates regions where data are not available. The bottom panels show results for the Climate Modelling Intercomparison Project (CMIP3) multi-model ensemble. Patterns expected from projections for the 21st century are largely similar to those shown here, as seen in Figure 4.1.

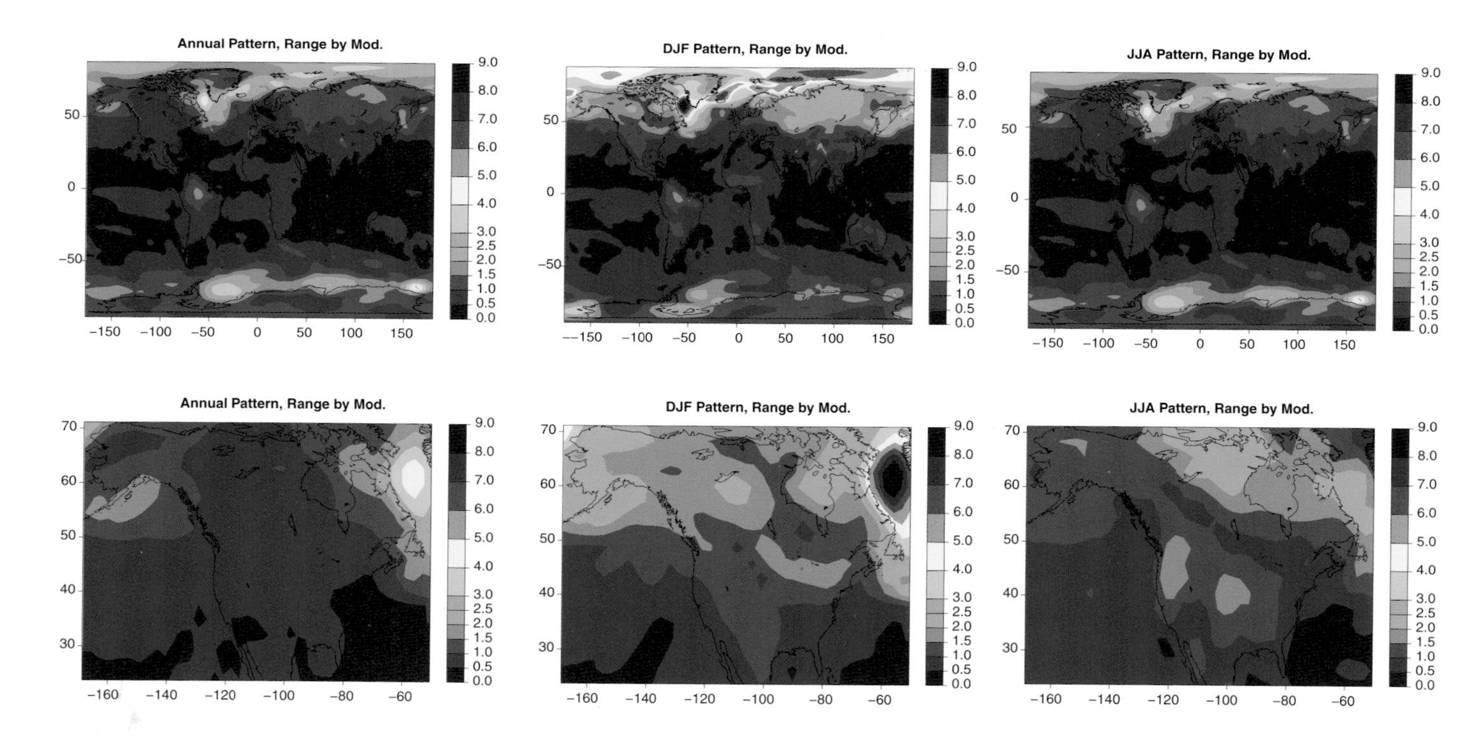

FIGURE 4.4 For the same patterns as in Figure 4.1, the range across the 18 models of the average of each model's patterns obtained under the three different scenarios is shown as a measure of the variability of the patterns produced by different GCMs.

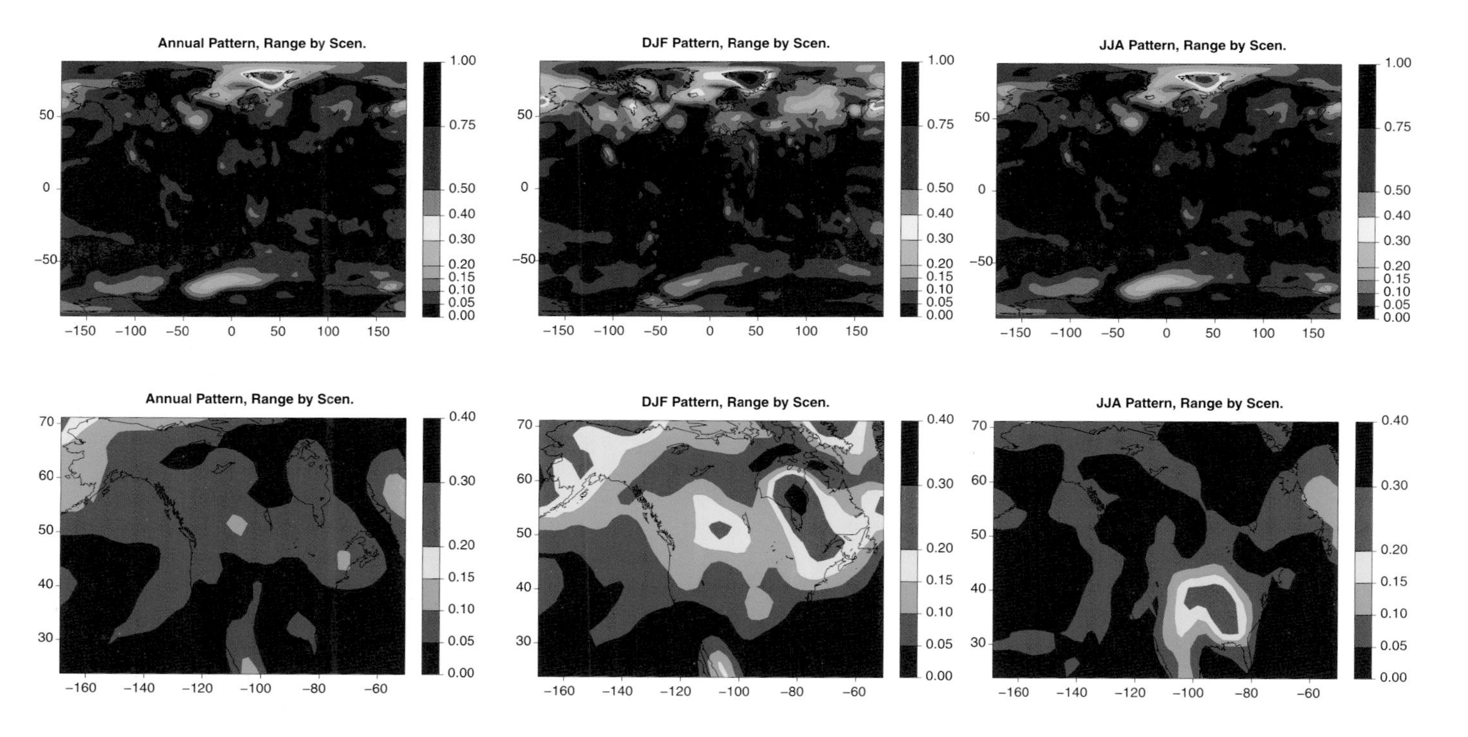

FIGURE 4.5 For the same patterns as in Figure 4.1, the range across SRES A1B, A2, and B1 of the patterns derived as a multi-model ensemble is shown as a measure of the variability of the patterns under different emission scenarios.

warming; and the strong signal of warming in the semi-arid to arid regions especially in the Mediterranean basin and nearby portions of the Asian continent in June-July-August. Areas of disagreement between models and observations remain, particularly in the form of more homogeneous patterns of warming at the subcontinental scales in the modeled patterns than in observations, with one example being the area of cooling in the southeast of the United States appearing in the HadCRU data and not represented in the models.

4.2 PRECIPITATION RESPONSE

As described in Section 4.1, it is reasonable to assume that the local temperature response to an increase in a well-mixed greenhouse gas is proportional to the global mean temperature response, with a well-defined spatial and seasonal pattern. There are also good reasons to assume that the local precipitation response scales with the global mean surface temperature response, although the uncertainties are greater, both with regard to the spatial and seasonal structure of the pattern and with regard to the limitations of this pattern scaling assumption.

Using the CMIP3 archive and computing the precipitation response, measured as a percentage change, divided by the global mean warming and then averaging over models and scenarios, just as for the temperatures in Figure 4.1, one obtains the pattern shown in Figure 4.6, both globally for the annual average and the summer and winter seasons, and focusing on North America (which is not found in the IPCC AR4 report [IPCC, 2007a]). The patterns are very similar to those shown for a particular scenario and time frame in the Summary for Policy Makers of the AR4/WG1 report (IPCC, 2007e).

There is a general increase in precipitation in subpolar and polar latitudes, and a decrease in the subtropics, and an increase once again in many equatorial regions. The boundary between the subtropical decrease and subpolar increase cuts through the continental United States, but with the boundary moving north in the summer and south in the winter. As a result, this ensemble mean projection is for an increase in precipitation in much of the continental United States in the winter and a reduction in the summer. In contrast, Canada is more robustly wetter and Mexico drier, being located closer to the centers of the subpolar region of increasing precipitation and the subtropical region of decreasing precipitation, respectively. The Mediterranean/Middle East and southern Australia are other robust regions of drying in these projections.

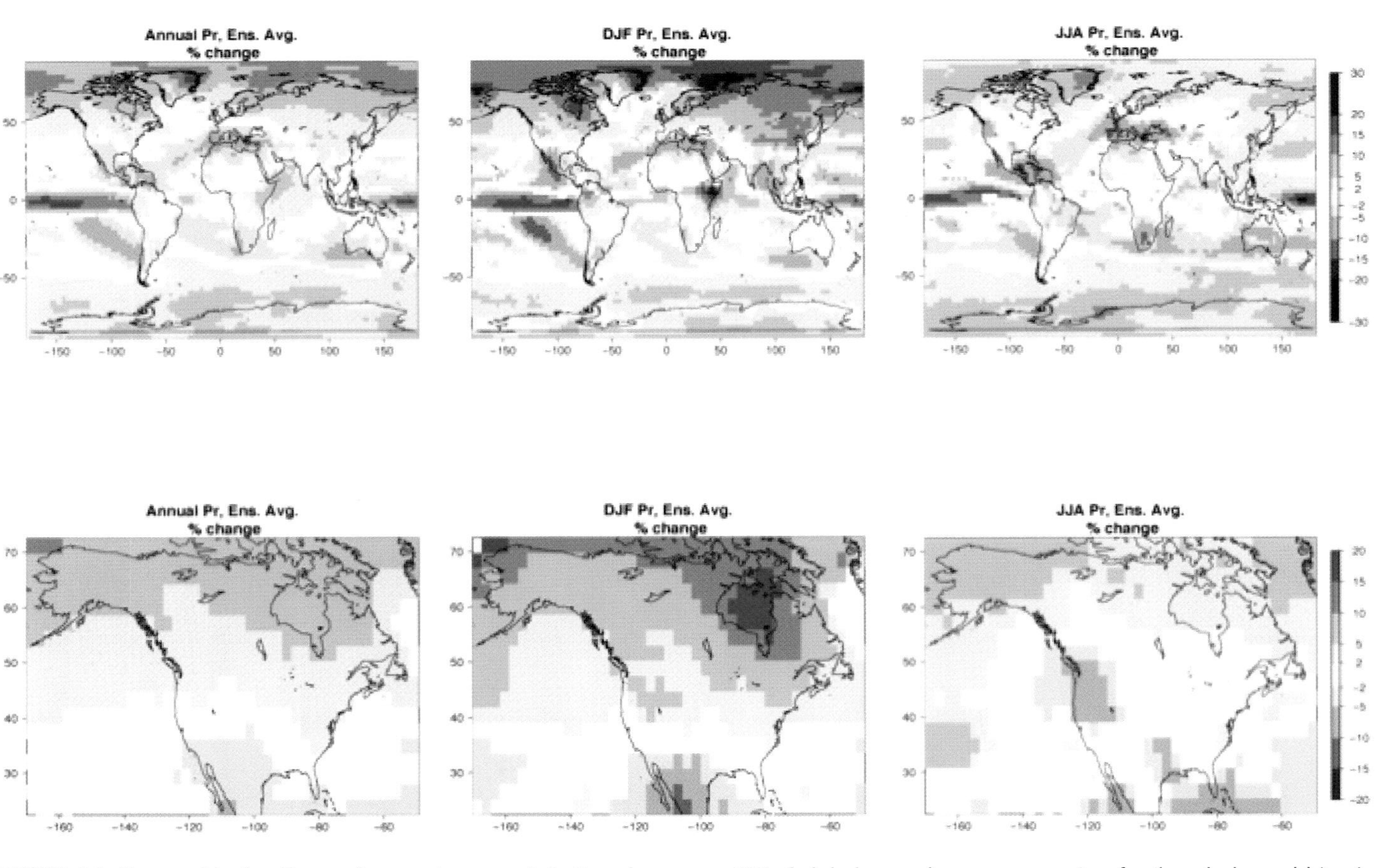

FIGURE 4.6 Geographical patterns of percentage precipitation change per 1°C of global annual average warming, for the whole world (top) and over North America (bottom). From left to right: patterns of changes in precipitation averaged over the whole calendar year or over boreal winter (December, January, and February) or summer (June, July, and August). Patterns are obtained by scaling end-of-21st-century percentage changes in precipitation (compared to end-of-20th-century climatology) by global average annual *warming* over the same period for the 18 CMIP3 models whose output is available for all three SRES scenarios A2, A1B, and B1. White regions are where less than two-thirds of the models agree over the sign of the change.

Focusing on fixed seasons provides one snapshot into these projected patterns. An alternative procedure, emphasizing the extent to which the "dry get drier", is to plot the precipitation changes in the driest 3 months at each location, irrespective of calendar date (Solomon et al., 2009). This alternative version is provided in the section, "Overview of Climate Changes and Illustrative Impacts." Local details differ, but the large-scale pattern is unchanged.

Our confidence in this model-generated pattern is enhanced by an understanding of the simple underlying mechanism controlling its structure. The starting point on which this understanding is based is the increase in the saturation pressure for water vapor with increasing temperature. Most of the water vapor in the atmosphere resides in the lowest 2-3 Km, where the saturation vapor pressure increases at roughly 7% per 1°C of warming. The ratio of water vapor in the atmosphere to this saturation value is referred to as the relative humidity. Especially near the surface, observational trends (e.g, Trenberth et al., 2005) and simple physical arguments (Held and Soden, 2006) consistently show that relative humidity cannot be expected to change substantially on average, and certainly not enough to counterbalance the increase in water vapor expected from the increase in saturation vapor pressure. As a result, there is high confidence in the projection that the total water vapor in the atmosphere will increase at roughly 7% per 1°C of warming.

If one averages around latitude circles, the hydrological cycle on Earth can be pictured schematically as in Figure 4.7. Evapotranspiration is largest in lower latitudes and decreases steadily as one moves poleward. Precipitation is larger than evapotranspiration in subpolar latitudes, and near the equator, and less than evapotranspiration in the subtropics. Fluxes of water vapor in the atmosphere are responsible for these local excesses and deficits. The atmosphere is converging water into the regions of excess and diverging water from the regions of deficit. As temperature increases and the amount of water vapor increases, the divergence and convergence of the water vapor flux increase proportionally, increasing the magnitudes of these excesses and deficits This simple picture also explains the magnitude of the increases and decreases in precipitation expected: because water vapor increases by roughly 7% per 1°C, the atmospheric fluxes increase by the same percentage, so that the pattern of precipitation minus evaporation is also amplified by about the same factor.

The final element in the picture is that the changes in the global mean hydrological cycle, the globally averaged precipitation and evapotranspiration, increase more slowly than do the increase in these water vapor fluxes.

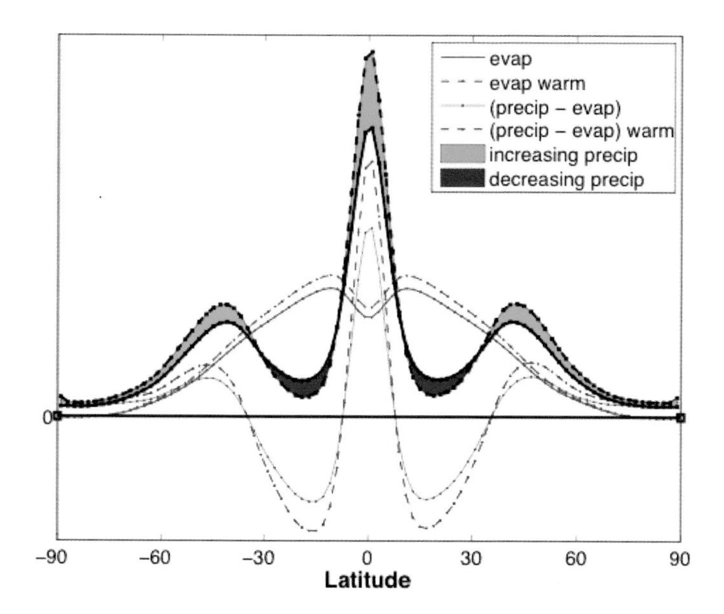

FIGURE 4.7 A schematic of the precipitation P (black) and evapotranspiration E (red) and P-E (blue) averaged around latitude circles. Solid lines indicate the unperturbed climate, and dotted lines a warmer climate. E increases slightly but is overwhelmed, on average, by the increase in magnitude of P-E, which is controlled by the magnitude of the horizontal water vapor fluxes.

The increase in precipitation in subpolar latitudes, for example, is due to an increase in the flux of vapor into these latitudes and not primarily to an increase in evaporation.

This picture is conservative, in that it assumes that changes in the atmospheric circulation are small so that the fluxes of vapor are dominated by the increases in the vapor itself and not in the winds carrying the vapor. This assumption has to be modified to understand model projections in more detail (see Section 4.4).

In the tropics the circulation is primarily driven by the heat released where water vapor condenses, so one cannot assume that the tropical winds will remain the same as water vapor increases. Partly due to the complexity of the interactions between winds and rainfall in the tropics, model projections for changes in rainfall are not as robust as in other areas. Some models predict dramatic drying of the Amazon, with important consequences for the forest and the global carbon cycle, but other models do not. Some models predict dramatic dying in parts of sub-Saharan Africa, but others do not. Patterns of precipitation changes in middle and higher latitudes are more robust on large scales but remain uncertain on the regional scales of importance for impact and adaptation strategies. For example, the changes in precipitation in the western United States in wintertime are sensitive to the El Niño phenomenon in the equatorial Pacific, the response of which is itself uncertain. And especially over semi-arid land surfaces, arguments based on the assumption that relative humidities do not change can break down. Given these complexities, we consider the CMIP3 ensemble as providing our best estimates of the pattern of precipitation change accompanying global warming, with consistency or inconsistency across this ensemble provided some indication of uncertainties.

4.3 HURRICANES

Atlantic Hurricanes

Atlantic hurricane activity has increased markedly in the past 20 years, concurrent with increases in ocean temperatures in the tropical North Atlantic region in which hurricanes are formed. But it is well established that hurricane frequency is dependent on other aspects of the large-scale climate as well as the local ocean temperatures. Many of the issues regarding detection, attribution, and projection of tropical cyclone trends in the Atlantic and throughout the tropics have recently been assessed by a WMO expert team (Knutson et al., 2010). We highlight a few of these issues here, basing the discussion in large part on this recent assessment, and refer the reader to Knutson et al. (2010) for a more detailed discussion.

Several detailed studies of the tropical storm record in the Atlantic have converged on a consensus view that the recent trend is more likely due to internal multi-decadal variability than a part of a century-long trend associated with greenhouse gas increases. While the raw data for Atlantic storms suggest a significant century-long trend, three separate lines of analysis cast doubts on the reality of this trend: (1) the frequency of landfalling storms, for which the completeness of the data is less of an issue than for basin-wide statistics, shows no significant long-term trend (Landsea, 2007); (2) estimates based on historical ship tracks indicate enough storms were likely missed in the early part of the century to account for most of the long-term trend in storm frequency (Chang et al., 2007; Vecchi and Knutson, 2008); and (3) the long-term trend in the raw data is primarily from short-lived storms (<3 days), which also suggests data artifacts are dominating the long-term trend (Landsea et al., 2010). The tentative picture that emerges is of Atlantic cyclone frequency being strongly modulated by internal multi-decadal variability, with levels of activity as high in the mid-20th century and the late 19th century as in the most recent decade. Additional clarification of tropical cyclone records outside of the Atlantic are needed to provide a clearer global picture of trends and variability.

Recent studies with relatively high-resolution global atmospheric models and with dynamical and statistical downscaling techniques have supported this picture. Several of these studies have successfully simulated the trend in Atlantic storm frequency in recent decades when provided with boundary conditions, especially ocean temperatures, as observed over this period (Knutson et al., 2008; LaRow et al., 2008; Zhao et al., 2009). When these or similar models are used to simulate the tropical storm statistics consistent with climate projections for the late 21st century, they consistently predict responses that cannot be obtained by extrapolating recent Atlantic frequency trends (e.g., Sugi et al., 2002; Oouchi et al., 2006; Bengtsson et al., 2007; Knutson et al., 2008; Zhao et al., 2009). Averaged over the globe, models typically predict that the frequency of tropical cyclones will stay roughly unchanged or be reduced by 0-10% per degree C global warming. Within the Atlantic basin, there is more variance in the results, with projections either increasing or decreasing frequencies by as much as 25% per degree C global warming.

New high-resolution studies are appearing regularly, so this picture is likely to evolve, but results to date suggest that the recent increase in Atlantic activity is likely due to an increase in tropical Atlantic ocean temperatures relative to the rest of the tropics and not to the absolute increase in temperatures (i.e., Vecchi et al., 2008). Consistently, the spread in projections for the future may have less to do with the details of the downscaling approach and more to do with the spatial pattern of tropical ocean warming projected by different global climate models, with those models that project larger (smaller) warming in the Atlantic than in other ocean basins resulting in an increase (decrease) in Atlantic storm activity (Zhao et al., 2009).

Insights into likely changes in storm intensity are primarily guided by the theory of the maximum intensity that tropical cyclones can attain in a given environment (Emanuel, 1987), by idealized modeling studies (Knutson and Tuleya, 2004), and a small number of attempts at dynamical and statistical downscaling (Emanuel et al., 2008). A consensus picture emerges in which storms are expected to become more intense on average, roughly by 1-4% per degree global warming, as measured by maximum wind speeds (Knutson et al., 2010), or by 3-12% per degree for the cube of this wind speed, often taken as a rough measure of the destructive potential of storm winds (Knutson et al., 2010). Whether the frequency of the most intense (category 3-5) cyclones should be expected to increase is less certain, in that it is also affected by the potential for reduction in total cyclone numbers.

An increase in rainfall within tropical cyclones is projected by arguments based on the increase in atmospheric water vapor. Typical values in models are on the order of an 8% increase per degree warming for the rainfall within 100 km of the storm center (Knutson et al., 2010).

4.4 CIRCULATION AND RELATED FACTORS

As discussed in Section 4.2, increasing greenhouse gas concentration is expected to cause an increase in atmospheric moisture of about 7% per degree C global temperature increase. In general, this will lead to greater rainfall in the tropical monsoon regions and drier conditions in the subtropical regions. Wet regions over the tropical oceans should also see increases in precipitation, although the positioning of these centers may shift due to coupled atmosphere-ocean interactions. The increase in atmospheric moisture is also expected to increase precipitation in the mid-latitude storm track regions by roughly 5-10% per degree C global temperature increase; however, it is not clear that storminess will increase.

Arguments based on first principles also indicate that increasing greenhouse gas concentrations should cause the mid-latitude jets and attendant storm tracks to shift poleward (e.g., Hartmann et al., 2000). Since in the atmosphere the temperature increase due to increasing greenhouse gases is confined to the troposphere and the tropopause is higher in the tropics than in high latitudes, increased greenhouse gases causes the poleward pressure gradient at the tropopause in mid-latitudes to increase. This, in turn, causes changes in the wave-mean flow interactions that shift the jets and their associated storm tracks poleward. This poleward shift in the jets and stormtracks should persist as long as the greenhouse gas concentrations remain elevated. A poleward shift in the jets and stormtracks due to global warming is a robust feature of the climate models (Miller et al., 2006; Hegerl et al., 2007; Meehl et al., 2007), with a zonal mean poleward shift of about 100 km per 3°C global temperature increase.

As a result of the poleward shift in the storm tracks, precipitation will likely increase on the poleward—and decrease on the equatorward side—of the location the climatological jets in the present climate. As noted in Section 4.2, this is indeed the pattern of precipitation change that is found across western North America in summertime when greenhouse gas concentrations are enhanced.

The same argument for a poleward shift in the storm tracks in the Northern Hemisphere in a warmer climate also applies in the Southern Hemisphere. Because the jet and storm tracks are intimately related and in the southern hemisphere they are circumpolar and vary weakly with longitude, this pattern of climate change is called the Southern Annular Mode (SAM; named after the leading pattern of circulation variability in the Southern Hemisphere due to wave-mean flow interactions—largely in the troposphere). There is ample evidence that the SAM-like changes in the mid- and high-latitude circulation in the Southern Hemisphere that were observed in the latter third of the 20th century (a poleward shift in the jet) was in part due to the human-induced depletion of stratospheric ozone (see, e.g., Miller et al., 2006 and references therein). Hence, a poleward shift in the jet in the Southern Hemisphere that would otherwise be expected because of increasing greenhouse gases will be somewhat tempered until about mid-21st century—when the stratospheric ozone is expected to be nearly recovered.

Many of the AR4 climate models forced by increasing greenhouse gases show that the Meridional Overturning Circulation (MOC) in the Atlantic Ocean will slow down over the 21st century in direct response to increased buoyancy in the sinking regions of the North Atlantic (Meehl et al., 2007). The increased buoyancy is due to both warming and freshening, with the former being more important than the latter. The slowdown in the MOC will have several impacts including a somewhat muted change in surface temperature in the North Atlantic and in maritime western Europe compared to other oceanic regions and modifications in the gradients in tropical sea surface temperature, which will further influence hurricane activity in the tropical Atlantic. The poleward movement of the jet in the Southern Hemisphere will cause the Antarctic Circumpolar Current to narrow and shift southward, impacting the sea-ice extent, the temperature of the water in contact with the floating ice sheets, and the exchange of carbon between the atmosphere and ocean.

There is much interest in knowing, as the planet warms: "How will the El Niño/Southern Oscillation (ENSO) and its teleconnected climate impacts change?" Unfortunately, this question cannot as yet be answered with confidence because almost all of the present climate models have poor representations of ENSO—as measured by the incongruities between the observed and simulated spatial and temporal characteristics of the observed ENSO and those simulated by the climate model. These nature-model inconsistencies likely result from the rather large biases in the climatology of the atmosphere and ocean in the tropical Pacific in current generation of global climate models (e.g., the double Intertropical Convergence Zone (ITCZ) common to many of the models). Whether or not the spatial, temporal and amplitude characteristics of ENSO change in a warmer world, however, the associated far-field impacts of ENSO will be different for several reasons. For example, the pattern and amplitude of the mid-latitude wintertime climate anomalies associated with ENSO will change as the mid-latitude jets shift poleward; places that experience drought during the warm (Australia) or cold (central United States) phase of ENSO and that are projected to dry with increasing global average temperature will experience enhanced drought conditions associated with identical ENSO cycles.

4.5 TEMPERATURE EXTREMES

The literature on temperature (as well as that on precipitation) extremes clearly suggests that any method trying to link changes in extremes of maximum and/or minimum temperature, and their consequent effects on the number of very hot or very cold days, and on the duration of hot and cold spells, to global or even local average temperature changes would fall short. As a recent indication of this, a paper by Ballester et al. (2010) shows how an accurate scaling of temperature extremes would have to involve not only average temperature change under future scenarios, but also change in temperature variability and in the skewness of its distribution. No paper has addressed explicitly the quantitative differences across multiple scenarios (or stabilization targets) of temperature extremes that we could straightforwardly utilize to describe expected changes under different atmospheric concentrations.

We chose here to adopt the perspective offered by Battisti and Naylor (2009)—BN09 from now on—on the expected changes in the likelihood of experiencing extremely hot or unprecedented *average summer temperatures*. This way of characterizing warming from the perspective of the tail of the distribution of *average temperature* has the strength of utilizing model output (seasonal averages of surface temperature, or TAS) whose reliability has been more extensively corroborated than other parameters representing more traditional definitions of extreme or rare events (e.g., daily temperatures exceeding high thresholds). We adapt BN09's analysis to our report's focus on different stabilization targets and make it consistent with its reliance on pattern scaling in order to infer geographically detailed projections of temperature changes as a linear function of changes in global average temperature (see Methods, Section 4.5).

Maps similar to those in Figure 3 of BN09 will result from each shift in mean temperature, except they will correspond to different atmospheric concentrations of CO_2 (and their implied expected global average warming) rather than to different future periods. The main difference in our approach compared to the analysis in BN09 is the choice of shifting the distribution uniformly to the right rather than trying to build a new distribution of average temperature anomalies on the basis of model output. This is a choice dictated by the use of pattern scaling, which lacks the availability of an ensemble of fully coupled model runs for each CO_2 target.

It could be argued that our analysis represents a conservative estimate of the expected changes in extreme seasonal temperatures, because we are assuming no change in the climatological distributions of temperatures besides a shift of their central locations. In particular, no change in the variability of seasonal average temperature is taken into consideration. Some studies (see for example Scherrer et al., 2005; Fischer and Schar, 2009 for the European region; Giorgi and Bi, 2005; Kitoh and Mukano, 2009 for global patterns) have shown that future variability of summer temperature is projected to increase, in association with drying.

In Figures 4.8 and 4.9 we show the resulting likelihoods of exceeding the 95th percentile, or the warmest anomaly of current average JJA and DJF temperatures (1971-2000 of 20C3M simulations) for the three levels of global warming. The patterns become redder (higher likelihood) as we look down each column (larger global average warming implies greater chances of exceeding the thresholds) and bluer (smaller likelihood) as we look across rows (higher thresholds make exceedances rarer).

4.6 PRECIPITATION EXTREMES

A general increase in atmospheric water vapor is predicted by essentially all climate models as temperatures increase, and an upward trend in column integrated water vapor has been observed in many regions (Trenberth et al., 2005). The consequences of this increase for the distribution of mean precipitation has been discussed in Section 4.2. As outlined below, models and simple theories also suggest that this increase in water vapor will increase the intensity of heavy rainfall events. Increasing trends in extreme precipitation have been documented in many regions, including much of North America (USCCSP, 2008c), but evidence for associated increases in floods is not compelling to date (Lins and Slack, 1999, 2005; WWAP, 2008).

As articulated by Trenberth (1999) and many others, precipitation in storms is related mostly to atmospheric moisture convergence rather than

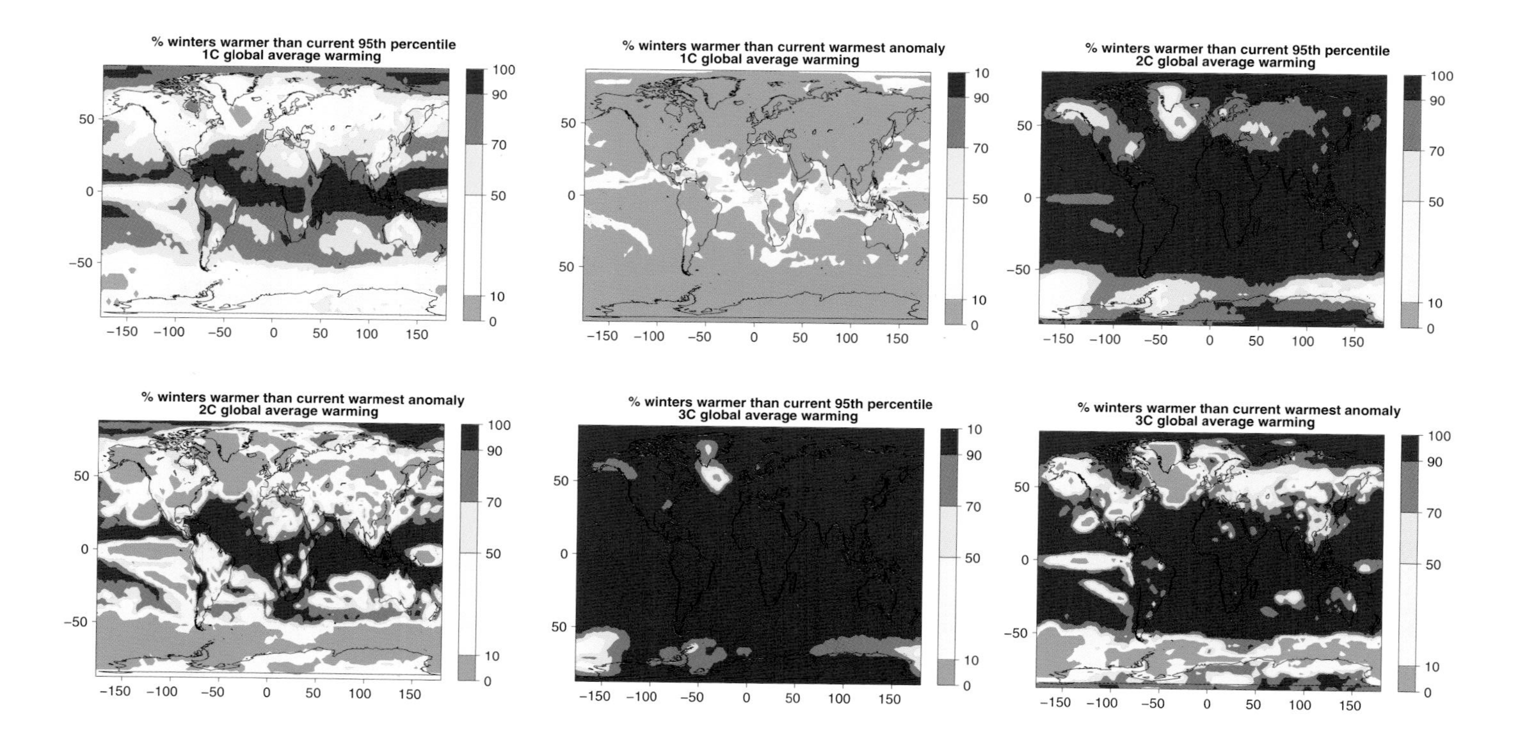

FIGURE 4.8 Chances that December-January-February average temperatures will be warmer than the 5th percentile of the climatological distribution (1971-2000)—left column—and warmer than the warmest (boreal) winter in the climatological distribution—right column—for different degrees of global annual average warming (1°C, 2°C, and 3°C ,respectively, along the three rows). Global average warming is calculated with respect to 1971-2000 averages. To obtain warming with respect to pre-industrial add 0.8°C.

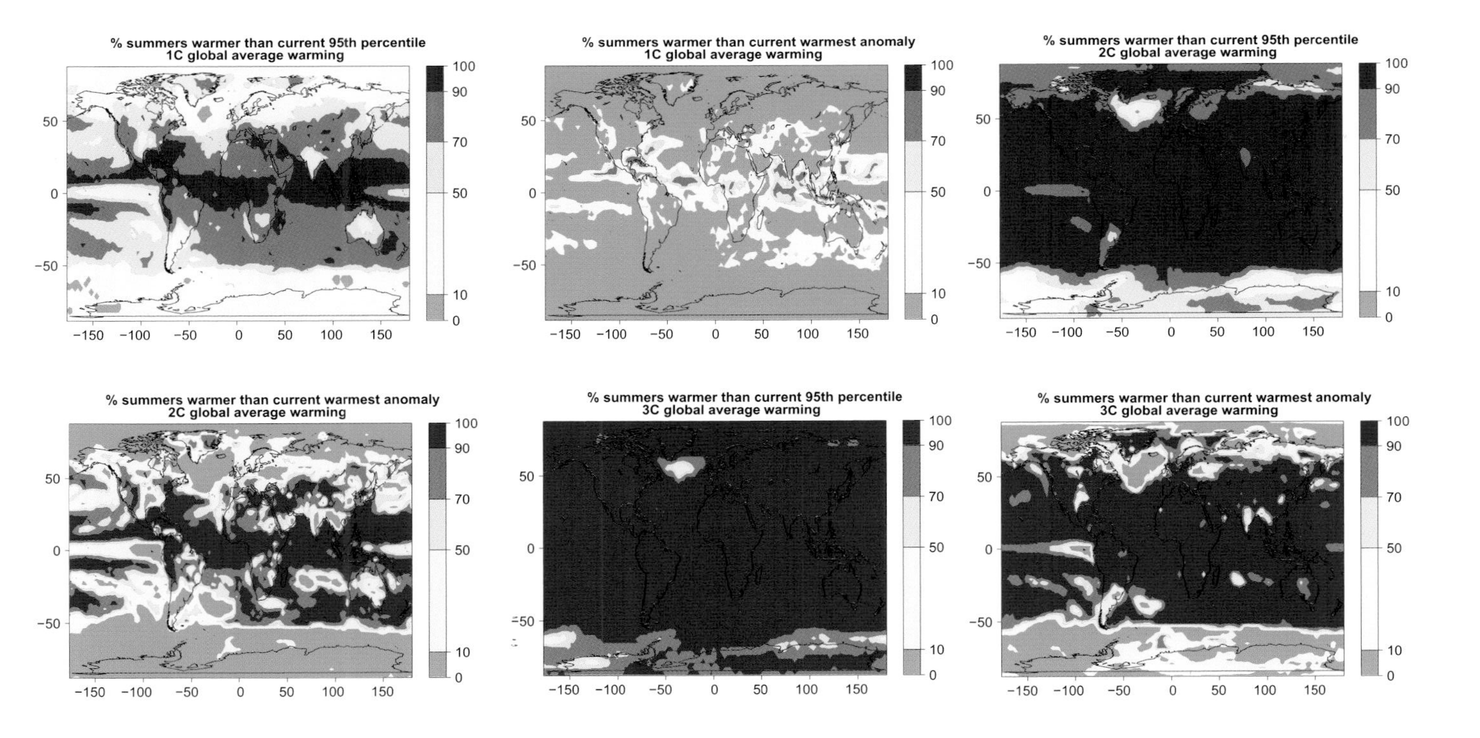

FIGURE 4.9 Chances that June-July-August average temperatures will be warmer than the 5th percentile of the climatological distribution (1971-2000)—left column—and warmer than the warmest (boreal) summer in the climatological distribution—right column—for different degrees of global annual average warming (1°C, 2°C, and 3°C, respectively, along the three rows). Global average warming is calculated with respect to 1971-2000 averages. To obtain warming with respect to pre-industrial add 0.8°C.

evapotranspiration during the storm; therefore, a storm with a given structure (with given vertical motion in particular), will generate more precipitation in a warmer climate, roughly following the increase in saturation vapor pressure, to the extent that the circulation within the storm does not change significantly. Additionally, as discussed in Section 4.2, evapotranspiration on average increases more slowly than does the water-holding capacity of the atmosphere; therefore, if precipitation intensity increases, then the frequency of precipitation must decrease to preserve a global balance of precipitation and evapotranspiration.

That climate models behave in this fashion, to first approximation, is documented in a number of papers (Emori and Brown, 2005; Pall et al., 2007; Allan and Soden, 2008; Sugiyama et al., 2010). Although there are some changes in the intensity of vertical motion, increases in extreme precipitation are dominated by the changes in atmospheric moisture. Even in the subtropical semi-arid zones in which mean precipitation is projected to decrease, the highest percentile daily precipitation events increase in frequency, as shown by Pall et al. (2007).

O'Gorman and Schneider (2009) have analyzed an idealized atmospheric model and point to the importance of the vertical gradient of the saturation mixing ratio in the lower troposphere rather than the water vapor mixing ratio itself as the key quantity whose increase controls intensity changes. The percentage change in this gradient is somewhat smaller than that of the mixing ratio but is of the same magnitude for typical atmospheric conditions, roughly 5% per degree warming. Limitations of this simple thermodynamic perspective are most likely to occur in the tropics, where latent heating is essential in driving circulations and where guidance from climate models is more suspect than in extratropical latitudes.

Allan and Soden (2008) compare precipitation estimates over the tropical oceans and GCM simulations to investigate the dependence of extreme precipitation intensities on surface temperature. They find the greatest rate of increase at the highest (most intense) precipitation percentiles in both models and observations. Furthermore, the inferred increases from the satellite precipitation data were considerably larger—in some cases over double—those predicted by the GCM simulations for the most extreme precipitation rates investigated.

The most extreme storms in mid-latitudes are also likely to be affected in structure by the increased latent heating, possibly causing changes in extreme precipitation larger than suggested by the thermodynamic arguments. For example, Lenderink and Van Meijgaard (2008) analyzed a 99-year hourly precipitation record from the Netherlands along with a regional climate model simulation and found that the most extreme precipitation rates increased at a rate nearly double that inferred from the roughly 7 percent per degree C Clausius-Clapyron rate of increase in atmospheric moisture content for constant relative humidity. The regional climate model generally produced changes in precipitation extremes that roughly matched those inferred from observations, i.e., approximately 14% increase per degree C warming.

Kharin et al. (2007) compare 20-year return period precipitation amounts simulated by IPCC AR4 model runs globally for mid- and late 21st century relative to model simulations for 1981-2000. They find that the models' ability to reproduce current climate precipitation extremes in the extratropics is plausible, but that there are large differences in the simulations in the tropics, which are compounded by observational issues. On a global basis, the average rate of increase in 20-year return period precipitation was slightly less than the Clausius-Clapyron rate of increase in atmospheric water holding capacity (see Figure 4.10).

Considering the combined literature on the physical basis for changes in extreme precipitation in a warming climate, along with model results and observational studies, there is a strong basis for concluding that precipitation extremes should increase with temperature in most parts of the globe. Model results are roughly consistent with simple physical arguments, although there are major issues with model parameterizations that complicate model-based interpretations in the tropics. Attempts to estimate changes in precipitation extremes from observations are complicated by large natural variability and spatial differences, but nonetheless observed changes over the past century are mostly consistent in sign with the expectation of increases in a warming climate (see also USCCSP, 2008c; and Section 4.2). We conclude the following: *Extreme precipitation is likely to increase as the atmospheric moisture content increases in a warming climate, with changes likely to be greater in the tropics than in the extratropics.* Typical magnitudes are 3-10% per degree C warming, with potentially larger values in the tropics, and in the most extreme events globally. However, despite general agreement in the likely direction of future changes, our current understanding of precipitation extremes is not sufficient to infer the likely magnitudes of future changes for the large return periods used for infrastructure design. Although these changes in precipitation extremes could lead to changes in flood frequency, the linkage between precipitation changes and flooding will be modulated by interactions between precipitation characteristics and river basin hydrology, the nature of which are not yet well known.

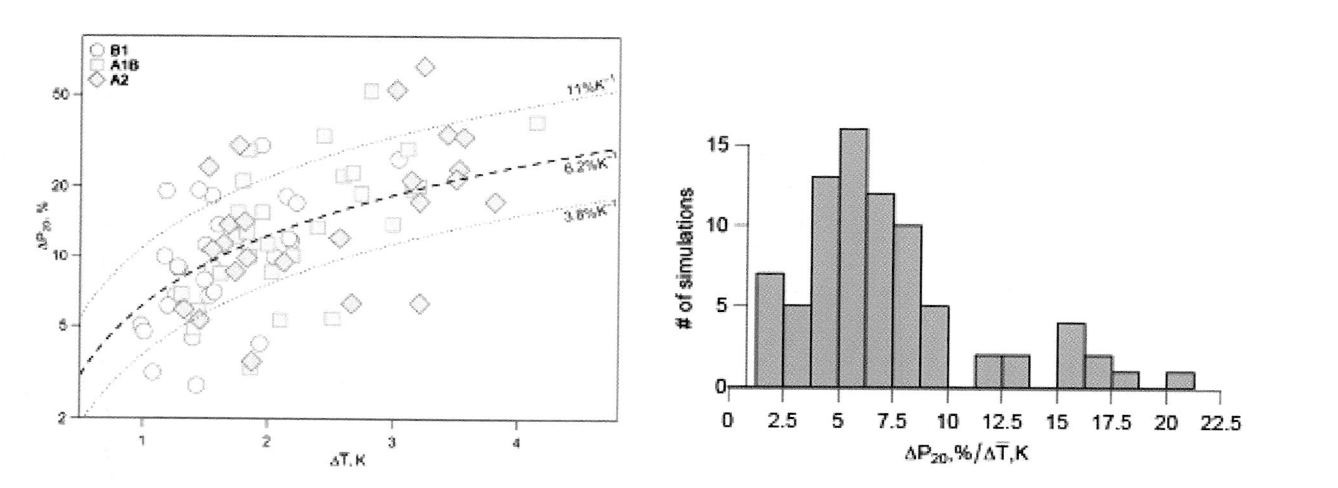

FIGURE 4.10 Relative changes in 20-year return values averaged over the global land area of annual 24-h precipitation maxima ($\Delta P20$) as a function of globally averaged changes in mean surface temperature for B1, A1B, and A2 global emissions scenarios, with results pooled from 14 GCM runs and for 2046-2065 and 2081-2100 relative to 1981-2000. In the left panel, the pooled results are shown along with the median slope of 6.2% per °C and the 15th and 85th percentiles (dashed and dotted lines, respectively). The right panel shows the results as a histogram. Source: Replotted from Kharin et al. (2007, Figure 16).

4.7 SEA ICE, SNOW, AND RELATED FACTORS

Current State of Sea Ice

Sea ice plays a critically important role in our climate system. It controls the rate of heat exchange between the polar ocean and atmosphere, reflects much of the solar energy incident on it, helps to maintain the equator-pole temperature gradient, influences ocean circulation, and is of great ecological importance. Typically, sea-ice area is about 14 to 16 million Km^2 in late winter in the Arctic and 17 to 20 million Km^2 in the Antarctic Southern Ocean. In late summer, on average, only about 3 to 4 million Km^2 remain in the Southern Ocean while in the Arctic there are approximately 7 million Km^2 (NSIDC; *http://nsidc.org/sotc/sea_ice.html*). Sea-ice extent, defined here as *the area of ocean with at least 15% sea-ice concentration*, is negatively correlated to global average surface temperature so that as globally averaged surface temperature increases, sea-ice extent decreases (e.g., Gregory et al., 2002). This sea-ice response to the increasing global surface temperatures occurs directly through thermally driven flux exchanges with the atmosphere and their impact on sea-ice extent and thickness, and indirectly, through the additional impact of increasing temperatures on dynamic mechanisms such as ENSO (e.g., Timmermann et al., 1999) and SAM (e.g., Arblaster and Meehl, 2006). Satellite observations show that there has been a decrease in globally averaged sea-ice extent since 1979. This decrease has occurred in the Arctic, while the Antarctic sea ice has increased slightly (Figure 4.11).

Since the 1950s Arctic sea-ice extent has exhibited a statistically significant decrease (Vinnikov et al., 1999), and the rate of decrease has been faster in summer (–7.8% per decade) than in winter (–1.8% per decade) (Stroeve et al., 2007). Satellite observations beginning in 1978 show that the annual average Arctic sea-ice extent has shrunk by 2.7% per decade (IPCC, 2007a) with larger decreases at the end of summer (9.1% per decade) than at the end of winter (2.9% per decade) (Stroeve et al., 2007). By some measures, from 1979 to 2006, September (late summer) sea-ice extent decreased by almost 25% or about 100,000 km^2 per year (Serreze et al., 2007). Comparison of observed Arctic sea-ice decline to IPCC AR4 projections show that the observed rate of ice loss is faster than that predicted by any of the IPCC AR4 models (Stroeve et al., 2007).

The decrease in sea-ice extent has been accompanied by thinning perennial and seasonal ice (Kwok and Rothrock, 2009), a decrease in multiyear ice (Maslanik et al., 2007), and record minima in September sea-ice cover (Serreze et al., 2007; Stroeve et al., 2007; Stroeve et al., 2008). Spring

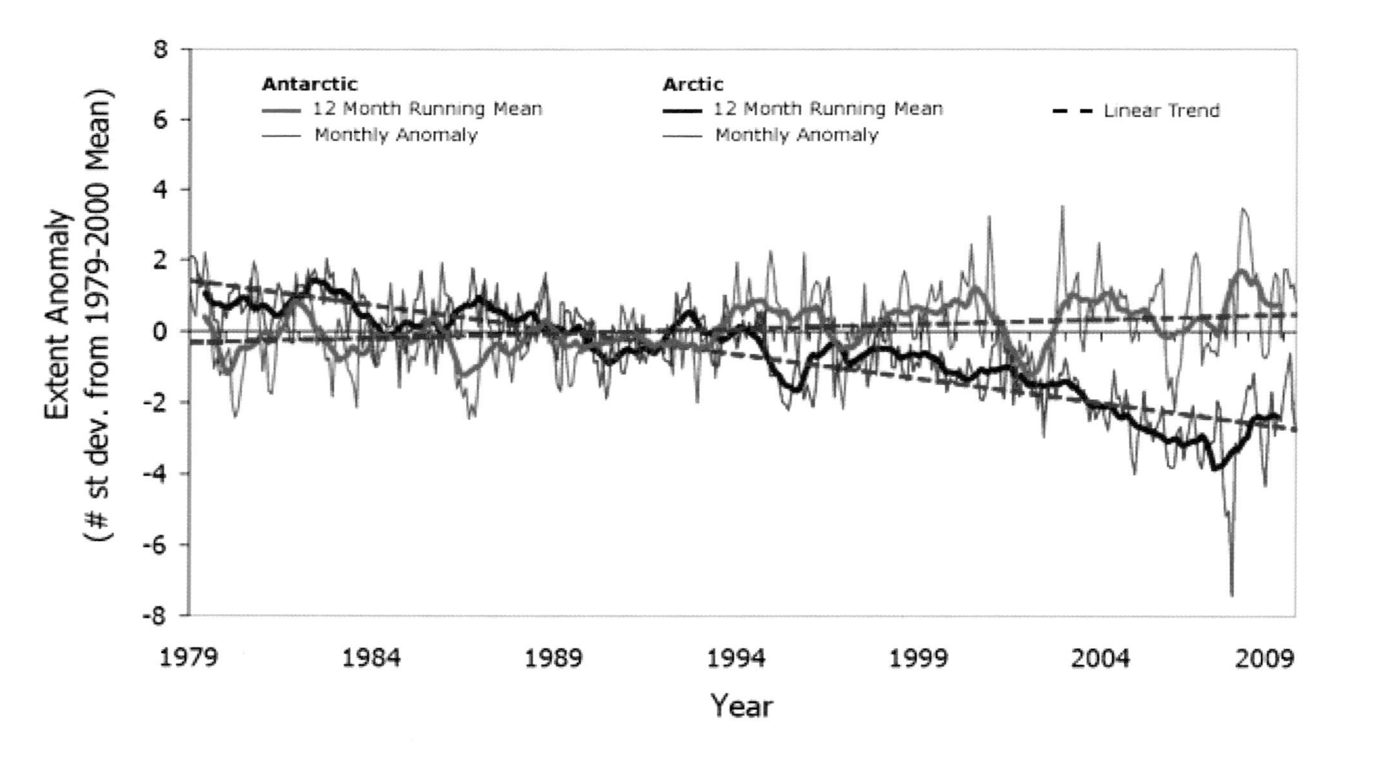

FIGURE 4.11 Arctic and Antarctic sea ice extent anomalies, 1979-2009: Although Arctic sea ice extent underwent a strong decline from 1979 to 2009, Antarctic sea-ice underwent a slight increase. Source: Image provided by National Snow and Ice Data Center, University of Colorado, Boulder.

melt seasons begin earlier and last longer (Serreze et al. (2007). The reduction in sea ice has occurred concurrently with a general increase in surface air temperatures over the Arctic Ocean (Comiso, 2003).

Part of this negative trend in Arctic sea ice has been attributed to natural variability in the atmospheric circulation, which influenced the ice circulation (Rigor and Wallace, 2004), as well as in ocean heat transport from the Atlantic (Polyakov et al., 2005) and the Pacific (Shimada et al., 2006) Oceans. However, natural variability does not account for all of the trend. Part of it has been attributed to increased atmospheric greenhouse gas (GHG) loading. Stroeve et al. (2007) suggest that some 33-38% of the overall observed trend is due to GHG forcing and that for the more recent period 1979-2006 this value may be as high as 47-57%.

Although the whole Arctic has experienced a trend of decreasing sea ice extent, trends in Antarctic sea ice extent have a strong regional signal and have been of opposing sign as demonstrated in Figure 4.12, taken from Liu et al. (2004).

In the Bellingshausen/Amundsen and western Weddell Seas there have

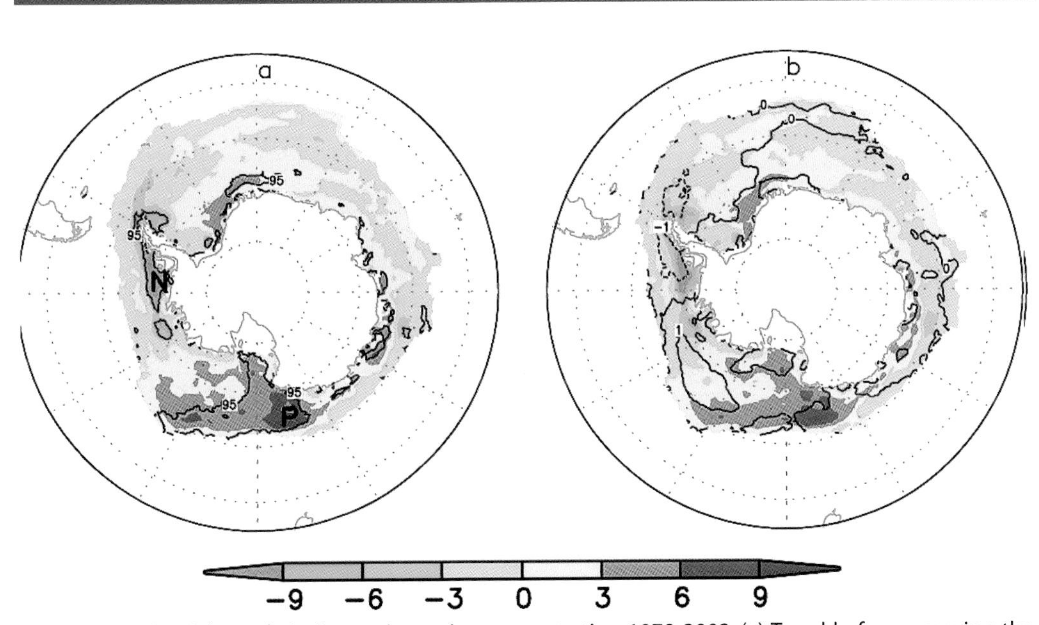

FIGURE 4.12 Spatial trends in Antarctic sea-ice concentration 1979-2002. (a) Trend before removing the influence of AAO and ENSO; (b) Trend after removing the influence of ENSO and the AAO. Source: Liu et al., (2004).

been marked decreases in sea-ice extent, while the Ross Sea has exhibited increasing sea ice extent (Figure 4.12). Due to the opposing regional trends, Antarctic sea-ice extent (Figure 4.11) shows a slight, but statistically significant, positive trend of 0.96±0.61% to 1.0±0.4% per decade, from 1979 to present (Cavalieri and Parkinson, 2008; Turner et al., 2009). This positive trend is led by autumn increases in the Ross Sea followed by the western Pacific Ocean and the Weddell Sea. Recently, the negative trends in the Bellingshausen/Amundsen Seas have decreased, while that in the Indian Ocean has become positive (Cavalieri and Parkinson, 2008). These regional trends in the Antarctic sea ice have been attributed to the influence of atmospheric circulation mechanisms, such asthe SAM (e.g., Thompson and Wallace, 2000), ENSO (e.g., Carleton et al., 1998), and zonal wave three (Raphael, 2007), and stronger cyclonic flow over the Amundsen and Ross Seas associated with stratospheric ozone depletion (Turner et al., 2009). As in the Arctic, there have also been changes in seasonality of the sea ice. The sea-ice maximum appears to occur slightly later in the year, while the sea-ice minimum tends to occur earlier. However, neither trend is statistically significant.

Predictions for Arctic Sea Ice over the 21st Century

The consistent conclusion from all of the model studies is that Arctic sea-ice extent will continue to decrease during the 21st century if emissions of GHGs remain unchecked. This is clearly illustrated in Figure 4.13, which shows the September results from IPCC models using 21st century forcings based on the SRES A1B scenario, a medium-high emissions scenario. In this scenario CO_2 concentration is projected to reach 720 ppm by 2100. Although much smaller, the late winter trends are of similar sign. Even with reductions in GHG emissions, Arctic sea-ice extent is predicted to decrease, and the rate of decrease changes with the GHG emission scenario as well as individual model physics. Washington et al. (2009) conducted an analysis of what would occur if CO_2 were to be stabilized at 450 ppm at the end of the 21st century. Their results for late summer/early fall (Figure 4.14) show that sea-ice extent for an unmitigated CO_2 scenario is reduced by 76% compared to present-day values, whereas for 450 ppm it is reduced only by 24%, preserving 4 million Km^2 of sea ice.

Although the predicted year varies, by the end of the 21st century most IPCC models predict an essentially ice-free Arctic (less than 1 million Km^2) in late summer. Estimates range from 2037 to 2100 (e.g., Arzel et al., 2006; Boe et al., 2009; Wang and Overland, 2009). The timing depends in part

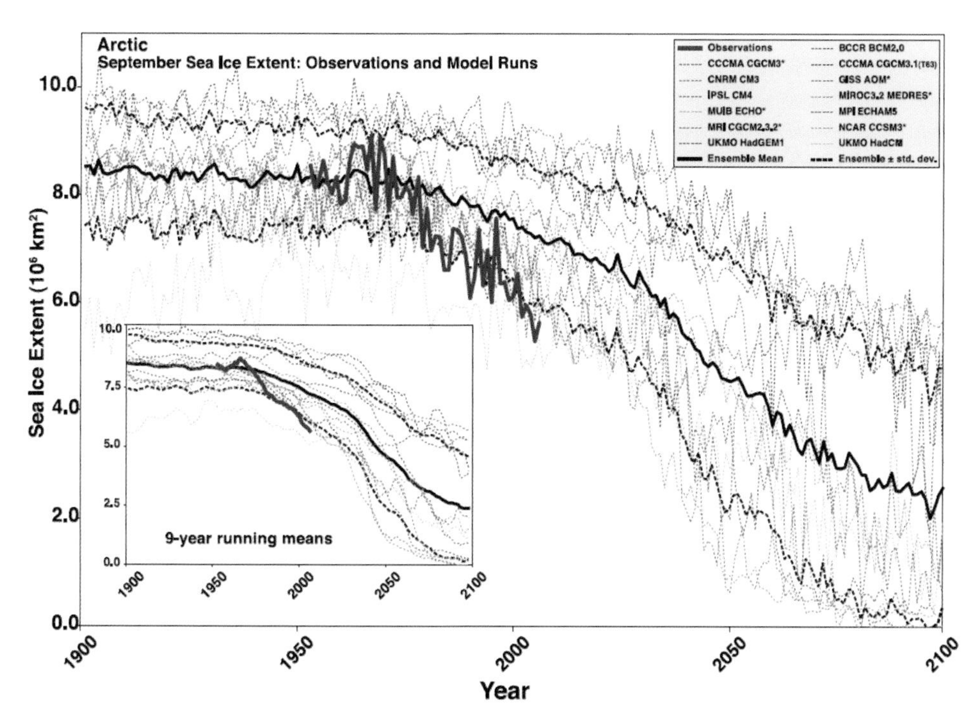

FIGURE 4.13 Arctic September sea-ice extent (× 10^6 km^2) from observations (thick red line) and 13 IPCC AR4 climate models, together with the multi-model ensemble mean (solid black line) and standard deviation (dotted black line). Models with more than one ensemble member are indicated with an asterisk. Inset shows 9-year running means. Source: Stroeve et al. (2007).

---

on the emission scenario used by the models but may also depend on the natural variability simulated by the individual models. Figure 4.15 taken from Wang and Overland (2009) illustrates this using six IPCC models that simulate the observed mean minimum and seasonality of sea ice very well. Two SRES scenarios are represented: the A1B, which reaches CO_2 concentration of 720 ppm by the end of the 21st century, and A2, which reaches 850 ppm at the same time.

The predicted reduction in sea-ice extent is expected to be accompanied by reduction in sea-ice thickness in summer and winter as more areas are replaced by first year ice. Ice in the Northern Hemisphere is expected to thin dramatically as the projected reduction in sea-ice volume is about twice that of the sea-ice extent reduction. Using the same six models in Figure 4.15, Wang and Overland (2009) compare the ice thickness at the time when the

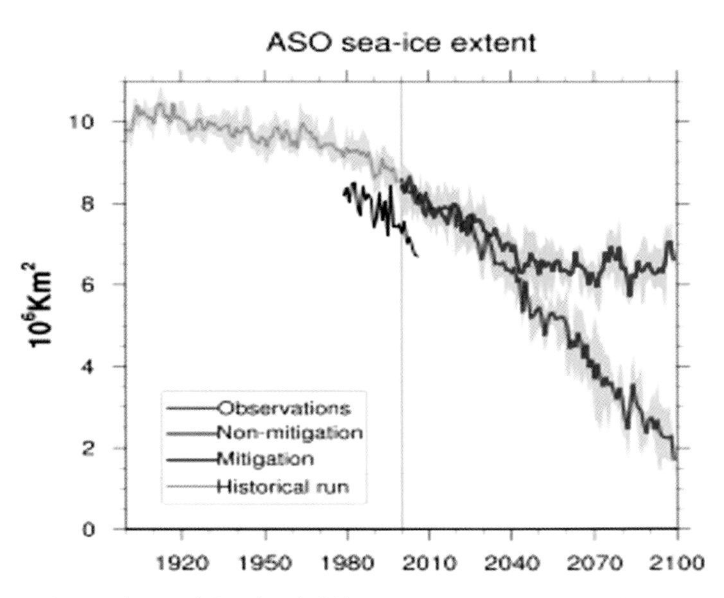

FIGURE 4.14 Time series of August, September and October (ASO) season average Arctic sea ice extent. Dark solid lines are the model ensemble means and the shaded areas show the range of the ensemble members. Observed sea ice is in black. SOURCE: Washington et al., 2009, their Figure 3.

simulated sea-ice extent reaches present-day values (4.6 million km^2) in September and March compared to when the sea-ice extent reached nearly ice-free conditions (less than 1.0 million km^2) near the end of the 21st century (see Figure 4.16). When sea-ice extent reaches present-day observed values in March much of the central Arctic is covered by sea ice less than 2.5 m thick. By the end of the century this sea ice is less than 2.0 m thick. In September over the same period sea ice moves from being less than 1.2 m thick to nearly ice-free conditions.

The mechanisms involved in reducing sea-ice cover are all positively correlated with temperature increase, giving rise to a linear relationship between annual Arctic sea-ice area reduction and global-averaged surface air temperature. According to one set of estimates, if GHG emissions continue to increase, corresponding temperature increases of 1°C, 2°C, 3°C, and 4°C are associated with Arctic sea-ice area reductions of 13%, 25%, 36% and 50% respectively (e.g., Gregory et al., 2002: Figure 4). Greater reductions are expected for summer compared to winter. For summer these values are on the order of 24% per degree warming resulting in an ice-free summer

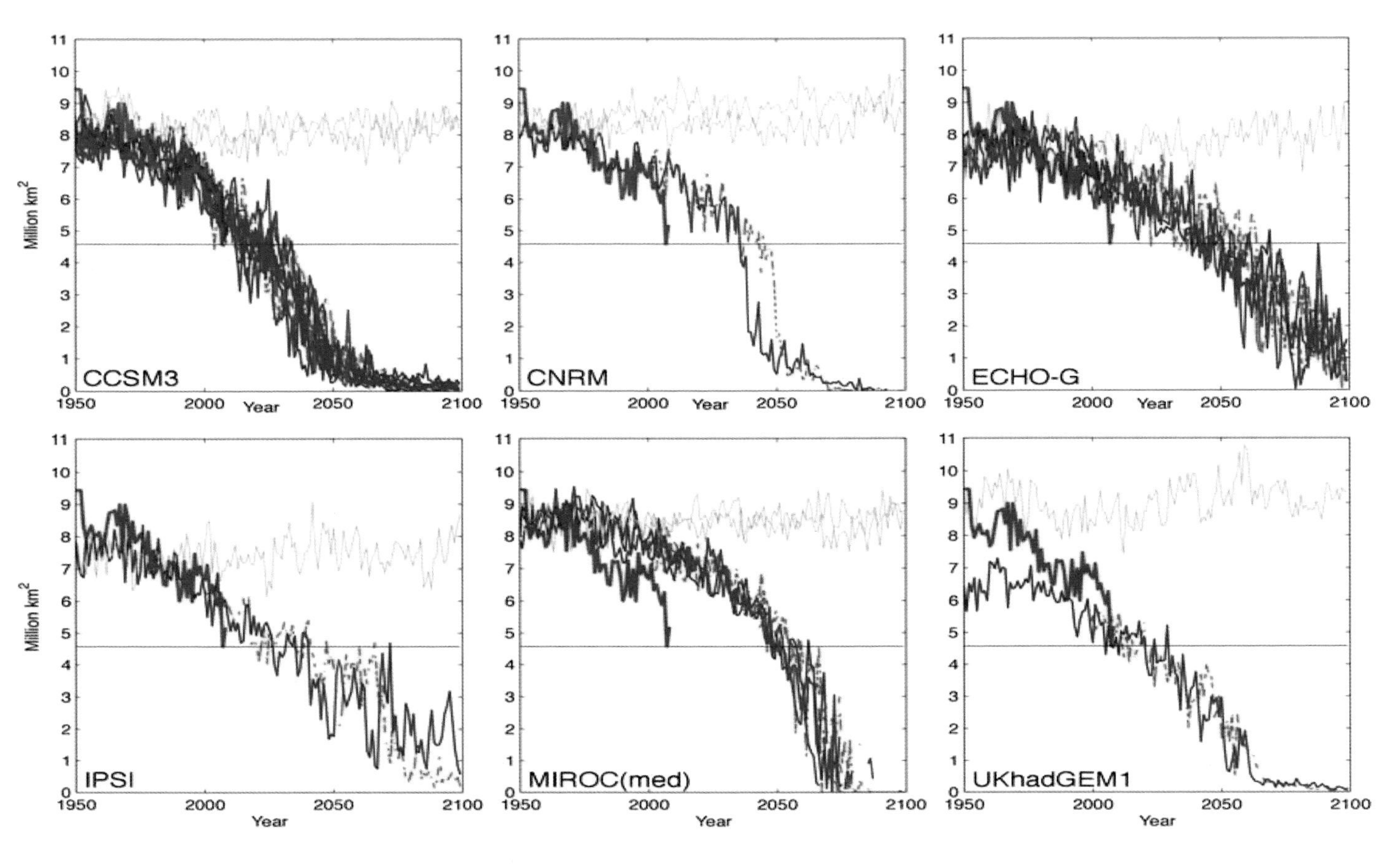

FIGURE 4.15 September sea-ice extent as projected by the six models that simulated the mean minimum and seasonality with less than 20% error of the observations. The colored thin line represents each ensemble member from the same model under A1B (blue solid) and A2 (magenta dashed) emission scenarios, and the thick red line is based on HadISST analysis. Grey lines in each panel indicate the time series from the control runs (without anthropogenic forcing) of the same model in any given 150-year period. The horizontal black line shows the ice extent at 4.6 million km^2 value, which is the minimum sea-ice extent reached in September 2007 according to HadISST analysis. All six models show rapid decline in the ice extent and reach ice-free summer (<1 million km^2) before the end-of-the-21st-century. Source: Wang and Overland (2009): their Figure 1).

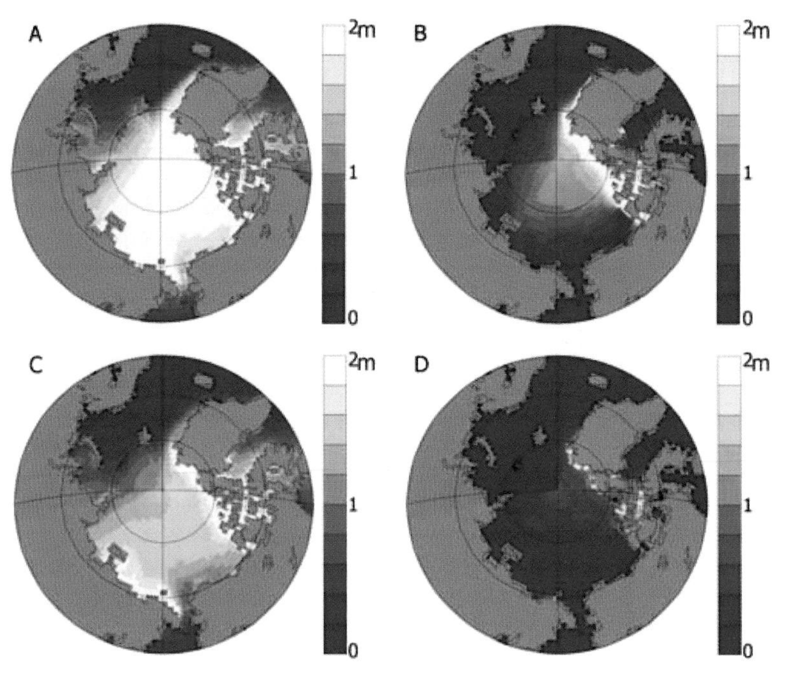

FIGURE 4.16 Mean sea-ice thickness for (left) March and (right) September based on ensemble members from six models under A1B emissions scenario. (a and b) Year when the September ice extent reached 4.6 million km^2 by these models and (c and d) year when the Arctic reached nearly sea-ice conditions (less than 1.0 million km^2) in September. Source: Wang and Overland, (2009: Figure 3).

by the end of the 21st century for a global warming of 2-4°C above late 20th century values or 3-5°C above pre-industrial values This linear relationship between sea-ice loss and global-averaged surface air temperature has implications for sea-ice recovery. One set of simulations using the A1B scenario suggests that Arctic sea ice may recover if GHG emissions were reduced. This response is linear for the annual sea-ice extent but nonlinear for September (Holland et al., 2006, 2010).

Predictions for Antarctic Sea Ice over the 21st Century

Compared to the Arctic very few studies examine the predicted sea-ice changes for the Antarctic. However, many of the characteristics of projected change are like those for the Arctic. Over the 21st century Antarctic sea-ice cover is projected to decrease. Note that late 20th century and early 21st century observations suggest a slight but significant increase in Antarctic sea-ice extent, contrary to that simulated by the models. This increase is associated with stratospheric ozone depletion and is expected to reverse as ozone returns to normal levels. The predicted decrease will occur more slowly than in the Arctic, particularly in the Ross Sea where temperatures are expected to remain cooler. This is attributed to the fact that the Southern Ocean stores much of its heat increase at depths below 1 km, while in the Arctic Ocean and subpolar seas the heat remains in the upper 1 km (Gregory, 2000; Bitz et al., 2007). In addition, horizontal heat transport poleward of about 60°N increases in many models (Holland and Bitz, 2003). These differences in the depth where heat is accumulating in the high-latitude oceans influence the relative rates of sea ice decay in the Arctic and Antarctic so that ice loss is faster in the Arctic (IPCC, 2007a).

A comparison of the last 20 years of the 21st century under the SRES A1B scenario with the last 20 years of the 20th century shows a decrease in the sea-ice concentration in both summer (JFM) and winter (JAS) by the end of the 21st century (Meehl et al., 2007; Sen Gupta et al., 2009). (See Figure 4.17.)

IPCC models predict a loss in sea-ice cover in summer and winter ranging from 10-50% in winter, and 33% to total loss in summer by the end of 2100. This is associated with a global warming ranging from 1.7-4.4°C above late 20th century values. The volume of sea ice is also reduced, and

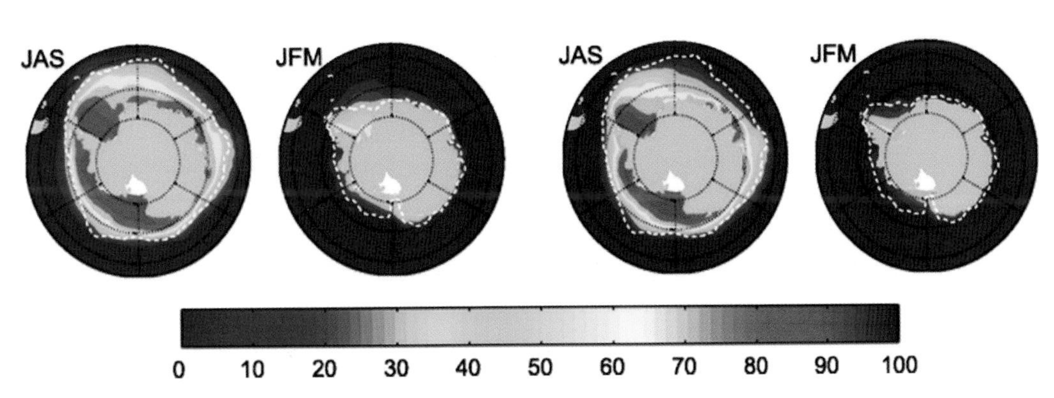

FIGURE 4.17 Multi-model mean sea-ice concentration (%) for January to March (JFM) and June to September (JAS), in the Antarctic for the periods (a) 1980 to 2000 and (b) 2080 to 2100 for the SRES A1B scenario. The dashed white line indicates the present-day 15% average sea-ice concentration limit. Source: Modified from Flato et al. (2004); IPCC (2007a, their Figure 10.14).

this decrease in volume is larger than the decrease in sea-ice concentration, indicating thinning of the sea ice. Substantial sea-ice loss occurs at all longitudes especially in the west Antarctic. In summer the greatest loss is east of the peninsula. Ice loss at these levels suggests a reduction in the seasonal cycle of sea-ice concentration and volume and means a reduction of brine rejection at the margins of Antarctica. This has implications for the future formation of Antarctic deep water and therefore for global ocean heat transport (Sen Gupta et al., 2009).

Snow Cover

Snow cover depends on both temperature and precipitation and is strongly negatively correlated with air temperature. Due to its high albedo, much less solar energy is absorbed by snow-covered areas as contrasted with snow-free areas, and surface temperatures tend to be lower, especially during periods of snow melt. Snow also plays an important role in storing moisture that is released when temperatures rise above freezing. It is a significant variable in the water balance of snow-covered regions and plays an important role in water management, especially in the western United States, where much of the region's runoff originates in snow that accumulates in mountainous areas in winter. Along with snowfall amounts and snow-covered area, snow water equivalent (SWE) and snow cover duration (SCD) are variables that can be used to characterize snow cover.

Current State of Snow Cover

In the Northern Hemisphere the snow-cover area (SCA) in March has exhibited a negative trend since about 1970, and this decrease has been associated with increased winter temperatures. This correspondence between March SCA and winter temperature appears to have strengthened toward the end of the 20th century (McCabe and Wolock, 2010). Since 1950, SWE has displayed a negative trend in the Pacific Northwest, particularly at lower elevation (Mote, 2003). A shift in the Pacific Decadal Oscillation (PDO) may have contributed to this trend but Mote et al. (2005), in an analysis expanded to include all of the western United States, argued that the PDO alone could not explain the trend in SWE. These conclusions are supported by Hamlet et al. (2005) who attributed the downward trend in SWE to increases in temperature rather than a decrease in precipitation. Figure 4.18 illustrates the observed trends in SWE as well as the trends simulated by the Variable Infiltration Capacity (VIC) hydrologic model. Although some positive trends

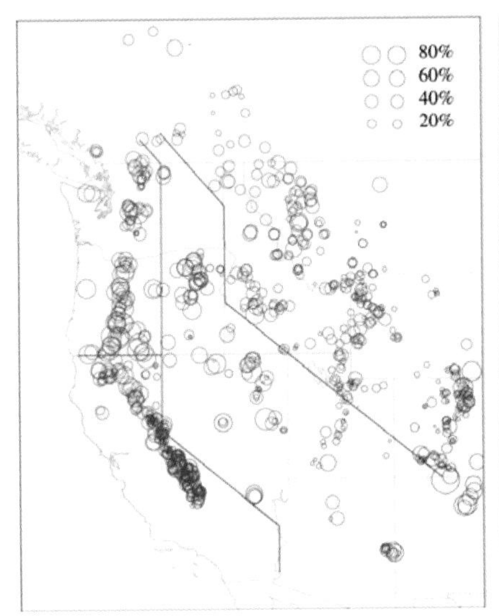

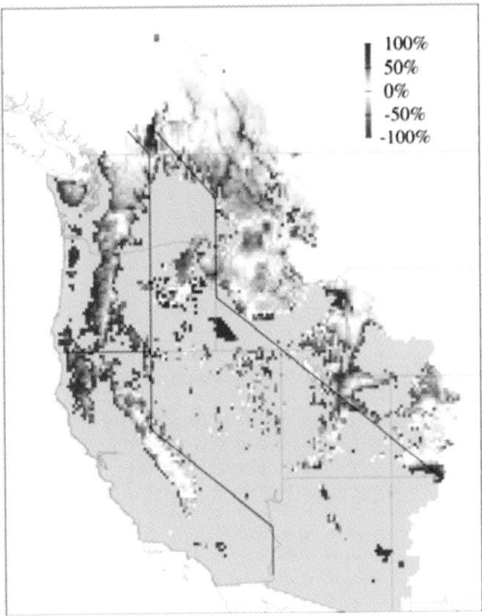

FIGURE 4.18 Linear trends in 1 Apr SWE relative to the starting value for the linear fit (i.e., the 1950 value for the best-fit line): (a) at 824 snow course locations in the western United States and Canada for the period 1950-1957, with negative trends shown by red circles and positive by blue circles; (b) from the simulation by the VIC hydrologic model (domain shown in gray) for the period 1950-1997. Source: Mote et al. (2005: Figure 1).

occurred in, for example, the Southwest, negative trends dominated the region with the largest reductions occurring in western Washington, western Oregon, and northern California.

Accompanying that trend in SWE is a shift in snow melt timing to earlier than average for the contiguous western United States, Alaska, and Canada. Over the period 1966-2007 satellite observations show snow cover reduction over western North America and maritime regions, eastern North America, Scandinavia, and the Pacific coast of Russia (Brown and Mote, 2009).

Predicted Changes in Snow Cover

Snow cover formation and melt are closely related to temperature; therefore snow cover is expected to decrease as temperatures increase. Model results project widespread reductions in snow cover over the 21st century. The Arctic Climate Impact Assessment (ACIA) model mean projects a reduction in Northern Hemisphere snow cover of 15% by the end of the 21st century under the B2 scenario. Individual, projected reductions range from 9-17%, and are greatest in spring and late summer/early winter, resulting in a shorter snow cover season. The timing of snow accumulation and snow melt is also projected to change. Accumulation is projected to begin later and melt to begin earlier. Fractional snow coverage is also predicted to decrease (Hosaka et al., 2005).

Regionally, the changes are a response to both increased temperature and increased precipitation and are complicated by the competing effects of warming and increased snowfall in those regions that remain below freezing. Taken across models, snow amount and snow coverage decrease in the Northern Hemisphere. Although a number of assessments have been performed of the implications of changes in snow cover on run-off, especially in the western United States (e.g., Hamlet and Lettenmaier, 1999; Christensen et al., 2004; Payne et al., 2004; Christensen and Lettenmaier, 2007), they have mostly been based on off-line simulations wherein hydrology models have been forced with downscaled output from GCMs, rather than as sensitivity analyses, for instance, per degree of global warming. One study that has adopted a sensitivity analysis approach is Casola et al. (2009), which estimated that for each increase in (local, not global) temperature of 1°C the snowpack should decrease by about 20%. Precipitation changes (precipitation is expected to increase slightly as the climate warms) complicate the snow cover response, which varies with latitude and elevation. When the expected global rate of increase in precipitation was considered, Casola et al. (2009) estimated that the sensitivity of Cascades snowpack would be reduced to 16% per degree C of local warming. Snow amount is projected to increase in other regions, for example, in Siberia. Such increases are thought to be due to increases in precipitation (snowfall) during autumn and winter (Hosaka et al., 2005; Meleshko et al., 2005).

The snow cover variables that are most sensitive to a warming climate are snow cover duration (SCD) and snow water equivalent (SWE). This sensitivity varies with climate regime and elevation, where the largest changes are projected to occur over lower elevations of regions with maritime winter climate, that is, moist climates with snow season temperatures in the range of –5°C to 5°C, and the lowest sensitivities are in mid-winter in cold continental climates. In general the models predict an enhanced early response of snow cover to warming over the western cordillera of North America and maritime regions. Over continental interiors, snow cover changes are slower. However, during the 21st century, annual mean SCD is expected to decrease. Although few models show any significant decrease in SCD during the 20th century, by 2020 the majority of models show significant decrease in SCD over eastern and western North America, Scandinavia, and Kazakhstan. By 2080 the majority of models show significant decreases everywhere. The regions that exhibit the least consensus tend to be continental and that might be due to a slower snow cover change signal. Like SCD, SWE decreases into the 21st century; however, model consensus on the statistical significance of this decrease does not become clear until 2050 when significant decreases occurred over the mid-latitude coastal regions of western Europe and North America (Brown and Mote, 2009).

Permafrost

Permafrost is defined as soil that remains at or below freezing temperatures for two or more years successively. Classified as continuous, discontinuous, and sporadic, permafrost zones occupy about 24% of the Northern Hemisphere exposed land or about 26 million km^2 but permafrost underlies only 13-18% of the exposed land (Nelson et al., 1997; Zhang et al., 1999). The active layer of permafrost is the upper layer of soil that thaws in summer and refreezes in winter (Sazonava et al., 2004). In this layer almost all of the below-ground biological processes take place. When it is deep, thawing increases soil moisture storage. Permafrost degradation occurs when temperatures increase in the active layer, and as a result the depth of thaw increases in successive summers and becomes greater than the depth of refreezing.

Although there have been exceptions to the trend (e.g., Brown et al., 2000), in general, observations indicate that the temperature of the permafrost has increased over the 20th century. For example in northern Alaska permafrost temperatures increased by 2 to 7°C over the 20th century (Lachenbruch and Marshall, 1986; Nelson, 2003). Although the size varies, this warming trend is repeated in northwestern Canada (e.g., Smith et al., 2003), northwestern Siberia (e.g., Pavlov and Moskalenko, 2002), and Scandinavia (Isaksen et al., 2000) among other regions. Along with the higher temperatures, observations show that the active layer is deepening and the permafrost extent is decreasing (Jorgenson et al., 2001; Serreze et al., 2002; Zhang et al., 2005). Since 1900, the maximum area of seasonally frozen ground has decreased by about 7%, led by decreases in spring of up to 15% (Lemke et al., 2007).

Predicted Changes in Permafrost

Climate models predict increases in the depth of thaw over much of the permafrost region. By 2050 seasonal thaw depths are projected to increase by more than 50% in the permafrost regions to the far north including Siberia and northern Canada, while in the southern extents increases of 30-40% are predicted (Stendel and Christensen, 2002; ACIA, 2005; Sasonova et al., 2004). In eastern Siberia permafrost degradation is projected to begin as early as 2050. The increases in thaw depth are associated with warming at the high northern latitudes (e.g., Lawrence and Slater, 2005; Yamaguchi et al., 2005; Kitabata et al., 2006).

The total area covered by continuous, near-surface permafrost (less than 4 m deep) is also projected to decrease and shrink poleward during the 21st century. This is demonstrated in Figure 1 of Stendel and Christiansen (2002), which shows model results for the A2 scenario. Predicted median values for this change are 18, 29, and 41% by 2030, 2050, and 2080 respectively (ACIA, 2005). The size of the decrease varies by model and by warming scenario. For example, Washington et al. (2009) show a range of permafrost loss with the largest losses occurring in their SRES A2 (high emission scenario) and the least in their low emission (450 ppm CO_2) scenario. (See Figure 4.19.)

Estimates of near-surface permafrost degradation rates to warming forced by the A1B GHG emissions scenario is on the order of 81,000 km^2 per year (Lawrence et al., 2008a). Rates of permafrost degradation may be influenced by rapid Arctic sea ice loss. One climate simulation of such loss predicted warming rates in the western Arctic of 3.5 times greater than the current 21st century climate change trends. This warming signal penetrated inland up to 1,500 km and, although most apparent in autumn, is there year-round. This warming leads to substantial ground heat storage, which in turn degrades the permafrost (Lawrence et al., 2008b).

4.8 SEA LEVEL RISE

The coastal zone has changed profoundly during the 20th century, primarily due to growing populations and increasing urbanization. In 1990, 23 percent of the world's population (or 1.2 billion people) lived both within a 100 km distance and 100 m elevation of the coast at densities about three times higher than the global average. By 2010, 20 out of 30 mega-cities are on the coast with many low-lying locations threatened by sea level rise. With coastal development continuing at a rapid pace, society is becoming in-

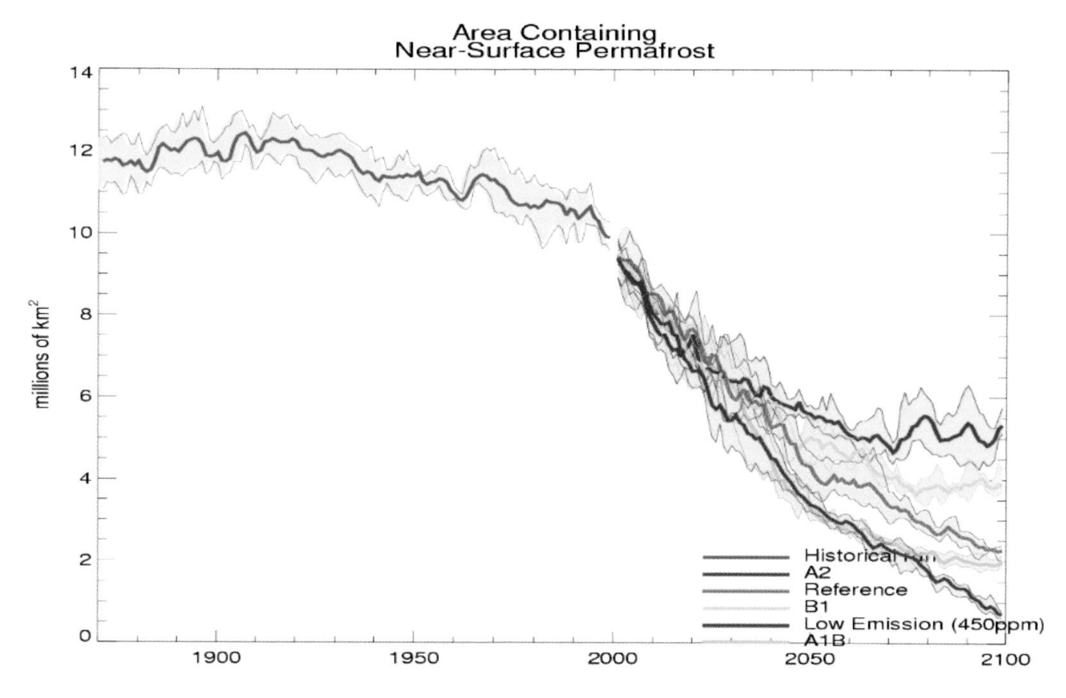

FIGURE 4.19 Time series of model simulations of Northern Hemisphere permafrost up to 2000 and predicted permafrost for several different scenarios after 2000. Dark lines are the model ensemble means and shaded areas show the range of ensemble members. Source: Washington et al. (2009: supplementary figures).

creasingly vulnerable to sea level rise and variability—as Hurricane Katrina recently demonstrated in New Orleans. Rising sea levels will contribute to increased storm surges and flooding, even if hurricane intensities do not increase in response to the warming of the oceans. Rising sea levels will also contribute to the erosion of the world's sandy beaches, most of which have been retreating over the past century. Low-lying islands are also vulnerable to sea level rise. An increase in global temperature will cause sea level rise and will change the amount and pattern of precipitation, which is important for the stability of ice sheets. The current increase in precipitation over Greenland does not offset the ice loss by melt and fast-flowing glaciers.

Mean, Local, Eustatic, and Steric Sea Level

Mean sea level (MSL) is a measure of the average height of the ocean's surface (such as the halfway point between the mean high tide and the mean low tide). Local mean sea level (LMSL) is defined as the height of the sea with respect to a land benchmark, averaged over a period of time (such as a month or a year) long enough that fluctuations caused by waves and tides are smoothed out. One must adjust perceived changes in LMSL to account for vertical movements of the land, which can be of the same order (mm y^{-1}) as sea level changes. Some land movements occur because of isostatic adjustment of the mantle to the melting of ice sheets at the end of the last ice age. The weight of the ice sheet depresses the underlying land, and when the ice melts away the land slowly rebounds, whereas some coasts are sinking as a result of isostatic adjustment due to collapsing forebulges in the near-field ocean of previously glaciated regions (Mitrovica and Milne, 2002). Atmospheric pressure, ocean currents and local ocean temperature changes also can affect LMSL. The term eustatic refers to global changes in the sea level brought about by the alteration to the volume of the world ocean (e.g., melting of ice sheets). The term steric refers to global changes in sea level due to thermal expansion and salinity variations. The term isostatic refers to changes in the level of the land masses due to thermal buoyancy or tectonic effects and implies no real change in the volume of water in the oceans.

Global Sea Level

Since 1870, global sea level has risen by about 0.2 m (Bindoff et al., 2007). Since 1993, sea level has been accurately measured globally from satellites. Before that time, the data come from tide gauges at coastal stations around the world. Satellite and tide-gauge measurements show that the rate of sea level rise has accelerated (Figure 4.20). Satellite measurements show sea level is rising at 3.1±0.4 mm y^{-1} since these records began in 1993 through 2003 (Figure 4.21). This rate has decreased for the most recent time period (2003-2008) to 2.5±0.4 mm y^{-1} due to a reduction of ocean thermal expansion from 1.6±0.3 mm y^{-1} to 0.37±0.1 mm y^{-1}, whereas contributions from glaciers, ice caps, and ice sheets increased from 1.2±0.41 mm y^{-1} to 2.05±0.35 mm y^{-1}, respectively (Cazenave et al., 2008). Statistical analysis reveals that the rate of rise is closely correlated with temperature. Sea level rise is an inevitable consequence of global warming for two main reasons: ocean water expands as it heats up, and additional water flows into the oceans from the ice that melts on land.

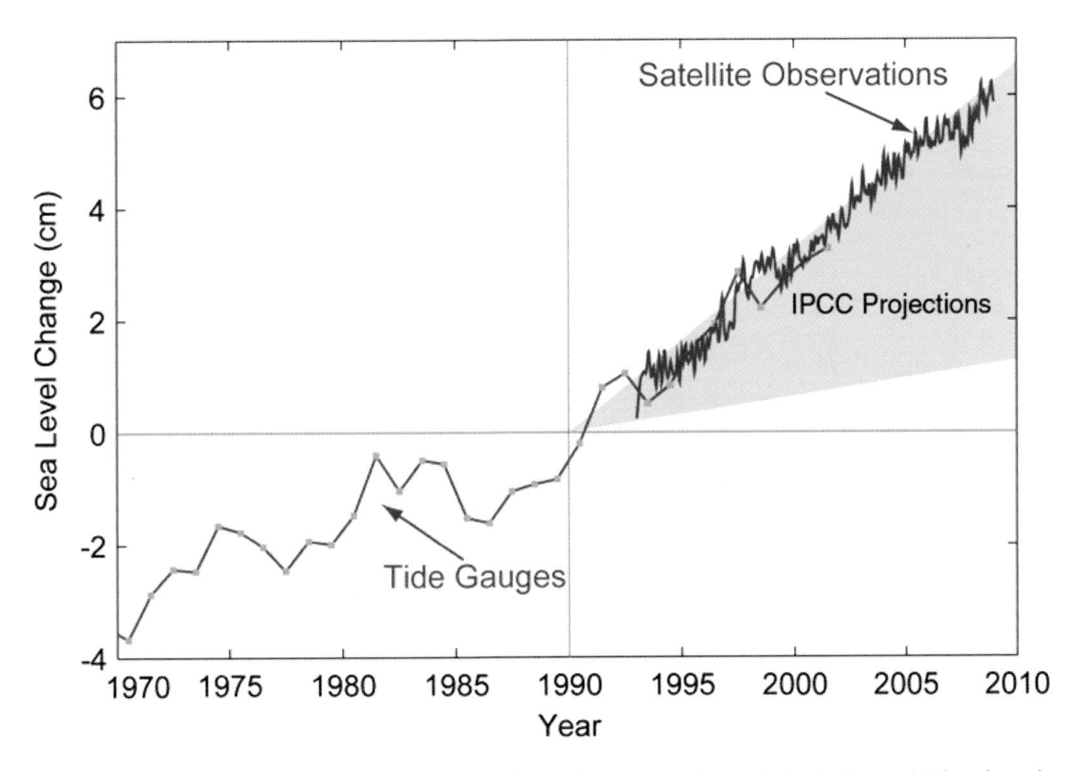

FIGURE 4.20 Sea level change during 1970-2009. The tide gauge data are indicated in red (Church and White, 2006) and satellite data in blue (Cazenave et al., 2008). The grey band shows the projections of the IPCC Third Assessment report for comparison.

Polar Precipitation

Each year, about 8 mm of sea level equivalent accumulates as snow on Greenland and Antarctica. Small changes in precipitation in polar regions will have major affects on the mass balance of the Greenland and Antarctic ice sheets. Increasing temperatures tend to increase evaporation, which leads to more precipitation. Greenland ice sheet precipitation—downscaled from ECMWF operational analyses and re-analyses (Hanna et al., 2005)—follows a significantly increasing trend for 1958-2006. Additional precipitation will occur due to warmer air temperatures mainly in the form of snow accumulation, therefore largely (i.e., ~80%) off sets rising Greenland run-off over the same period (Hanna et al., 2008). The increase in precipitation is confirmed by satellite altimetry data (Johannessen et al., 2005) with a significant growth of 2-5 cm y^{-1} of the Greenland ice sheet interior from 1992-2004.

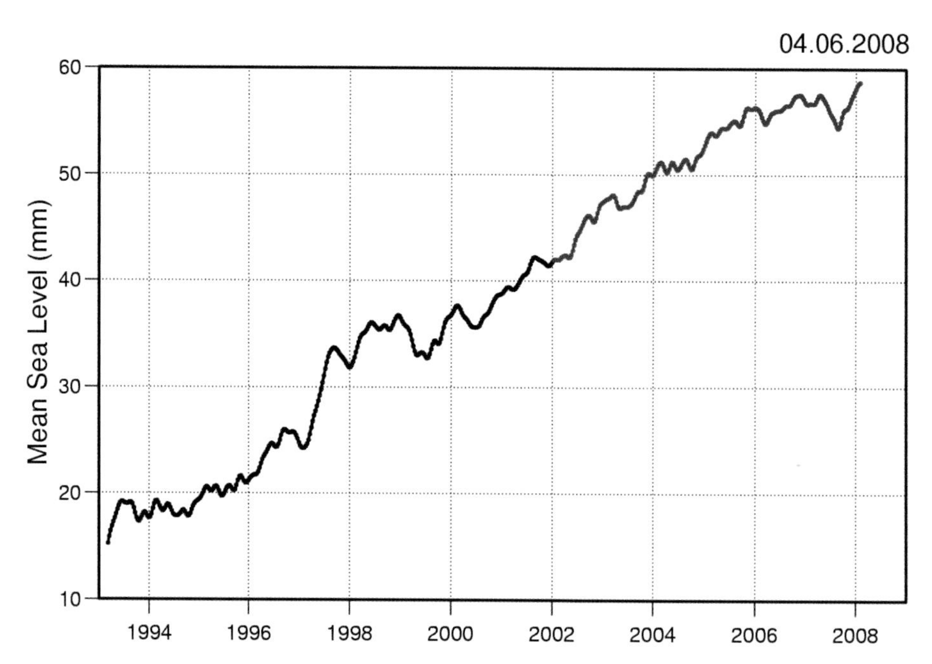

FIGURE 4.21 Sea level curve from Topex/Poseidon and Jason-1 satellite altimetry over 1993-2008 (data averaged over 65°N and 65°S; 3-month smoothing applied to the raw 10-day data). Black: Topex/Poseidon; red: Topex/Poseidon plus Jason-1; blue: Jason-1. Source: Cazenave et al. (2008).

The lack of a 20th century trend in net precipitation in recent comprehensive analyses over Antarctica is particularly problematic in the context of 21st century model projections. Almost all climate models simulate a continuing robust precipitation increase over Antarctica in the coming century. During the first half of the 21st century, models using the A1B scenario predict anything from a maximum upward trend of 0.71 mm y^{-1} to a maximum downward trend of 0.13 mm y^{-1}. Some models project stronger net precipitation increases in the first half of the 21st century, while other models project stronger increases after the middle of the century.

Subsidence

Subsidence is the motion of a surface as it shifts downward relative to a datum such as sea level. Subsidence can be caused by mining, extraction of natural gas, groundwater-related subsidence, isostatic subsidence (change in sedimentation rate), just to name a few, and can locally account for several millimeters per year. Some coastal regions are particularly vulnerable to erosion and inundation due to the rapid deterioration of coastal barriers combined with relatively high rates of land subsidence. Subsidence is very important for local sea level assessments and can outweigh the global sea level rate of change on short time scales.

Thermal Expansion

Changes in the climate systems' energy budget are revealed in ocean temperatures and the associated thermal expansion contribution to sea level rise. Like air and other fluids, water expands as its temperature increases (i.e., its density decreases as temperature increases). As climate change increases ocean temperatures, initially at the surface and over centuries at depth, the water will expand, contributing to sea level rise from thermal expansion. Thermal expansion is likely to have contributed to about 0.025 m of sea level rise during the second half of the 20th century (Meehl et al., 2007), and this rate has increased to about three times during the early 21st century.

The method of calculating average ocean heat content and the associated sea level change using only surface temperatures opens up the possibility of expanding our knowledge into the past and future. Projected rise in global thermosteric sea level for the hypothetical situation where all warming of the sea surface has stopped and the present average temperature is maintained into the future shows that we are committed to a further 0.05 m sea level rise over the next two centuries (Marčelja, 2009).

Current estimates of thermal expansion account for approximately half of the change observed in global mean sea level rise during the first decade of the satellite altimeter record, but only about a quarter of the change during the previous half century (Figure 4.22). Because this contribution to sea level rise depends mainly on the temperature of the ocean, projecting the increase in ocean temperatures provides an estimate of future growth. Over the 21st century, the IPCC AR4 projected that thermal expansion will lead to sea level rise of about 0.23±0.09 m for the A1B scenario.

Glaciers

Terrestrial glaciers and the Greenland and Antarctic ice sheets have the potential to raise global sea level many meters. Terrestrial glaciers are shrinking all over the world. During the past decade, they have been melting at about twice the rate of the past several decades (Figure 4.22).

Glaciers and mountain ice caps, including those peripheral to Green-

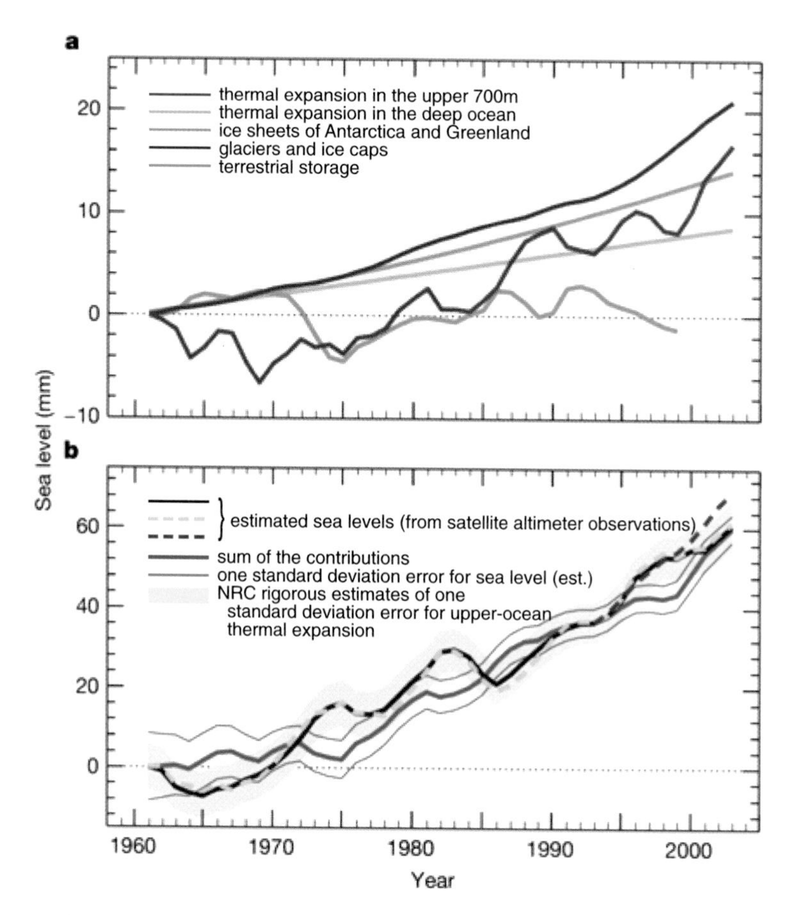

FIGURE 4.22 (a) Total observed sea level rise and its components. The components are thermal expansion in the upper 700 m (red), thermal expansion in the deep ocean (orange), the ice sheets of Antarctica and Greenland (cyan), glaciers and ice caps (dark blue), and terrestrial storage (green). (b), The estimated sea levels are indicated by the black line, the yellow dotted line, and the red dotted line (from satellite altimeter observations). The sum of the contributions is shown by the blue line. Estimates of one standard deviation error for the sea level are indicated by the grey shading. For the sum of components, we include our rigorous estimates of one standard deviation error for upper-ocean thermal expansion; these are shown by the thin blue lines. All time series were smoothed with a 3-year running average and are relative to 1961. Source: Domingues et al. (2008).

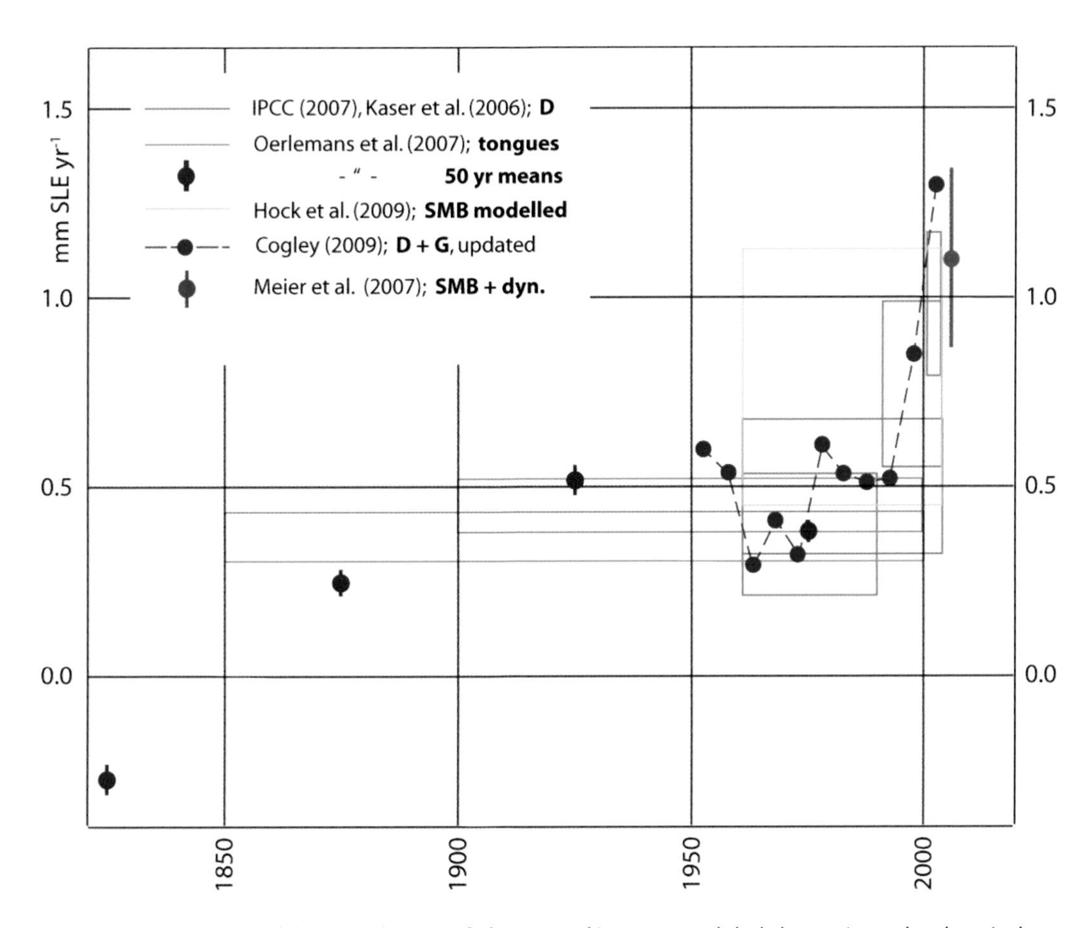

FIGURE 4.23 Estimates of the contribution of glaciers and ice caps to global change in sea level equivalent (SLE), in millimeters SLE per year. Source: Allison et al. (2009b).

land and Antarctica, can potentially contribute a total of approximately 0.7 m to global sea level, and they provide a source of freshwater in many mountain regions worldwide (Bahr et al., 2009). For 1961-2003, glaciers and ice caps contributed 0.5±0.2 mm y^{-1} to global sea-level rise (Figure 4.22), increasing to 0.8±0.17 mm y^{-1} for 1993-2003 (Allison et al., 2009b). This new assessment (Figure 4.23) shows ice loss from glaciers and ice caps slightly higher than those reported in IPCC AR4, contributing now about 1.2±0.2 mm y^{-1} to global sea level rise. Glaciers and ice caps are not in balance with the present climate; glaciers need to decrease in volume by 27% on average, and ice caps need to decrease by 26% to attain equilibrium

(Bahr et al., 2009). A future equilibrium with the current climate implies a change in sea level of 0.089±0.015 m due to glaciers and 0.095±0.029 m due to ice caps, with a total change in sea level of 0.184±0.033 m.

The ratio of accumulation area (AAR) at the end of the melt season and total glacier area dropped from roughly 0.54 in 1961 to 0.44 in 2007, and it is expected to drop further to 0.31 by 2050. This is a conservative estimate—observations indicate a faster than linear decrease in global ice mass balance over the past 40 years (Kaser et al., 2006). Although the actual decrease in AAR may be faster than linear, this conservative estimate represents a 30% decrease from the current value. As a rough approximation, we can assume that the AAR of every glacier decreases by the same percentage, giving an estimate of the fractional volume change for each glacier. In that case, the minimal sea level rise from glaciers and ice caps will be 0.373±0.021 m over the next 100 years (Bahr et al., 2009).

Greenland and Antarctic Ice Sheets

On the polar ice sheets, there is observational evidence of accelerating flow from outlet glaciers both in Greenland and in west Antarctica. Both inland snow accumulation and marginal ice melting have increased over the Greenland ice sheet, but there is little evidence for any significant accumulation trend over the Antarctic ice sheet. Antarctica and Greenland maintain the largest ice reservoirs on land. For 1993-2003, the estimated contributions for the Greenland and Antarctic ice sheets are 0.21±0.07 mm y^{-1} and 0.21±0.35 mm y^{-1}, respectively (Bindoff et al., 2007). There is little information to constrain ice sheet contributions for previous decades, but it is thought that the Greenland contribution has increased significantly in recent years (Lemke et al., 2007). Since IPCC AR4, there have been a number of new studies on ice sheet mass budget that have considerably enhanced our understanding of ice sheet vulnerabilities (Figure 4.24) (Allison et al., 2009a). Recent observations have shown that changes in the rate of ice discharge into the sea can occur far more rapidly than previously suspected (e.g., Rignot, 2006).

The pattern of ice sheet change in Greenland is one of near-coastal thinning, primarily along fast-moving outlet glaciers. Accelerated flow and discharge from some major outlet glaciers (also called dynamic thinning) is responsible for much of the loss (Rignot and Kanagaratnam, 2006; Howat et al., 2007). Pritchard et al. (2009) used high-resolution satellite laser altimetry to show that dynamic thinning of fast-flowing coastal glaciers is now widespread at all latitudes in Greenland. Figure 4.24 shows estimates

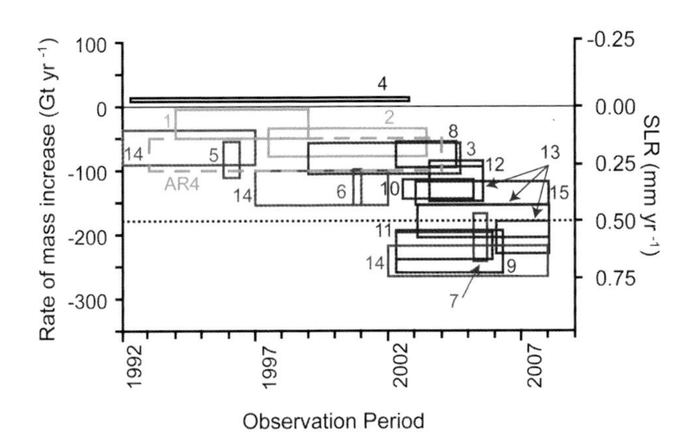

FIGURE 4.24 Estimates of the net mass budget of the Greenland Ice Sheet since 1992 (Allison et al., 2009b). The horizontal dimension of the boxes shows the time period for which the estimate was made, and the vertical dimension shows the upper and lower limits of the estimate. The colors represent the different methods that were used: black is satellite radar altimetry, orange is aircraft laser altimetry, purple is aircraft/satellite laser altimetry, red is the flux component method, and blue is satellite gravity.

of the mass balance of the Greenland ice sheet that have been made since the early 1990s. The horizontal dimension of the boxes shows the time period for which the estimate was made, and the vertical dimension shows the upper and lower limits of the estimate. The colors represent the different methods that were used: black is satellite radar altimetry, orange is aircraft laser altimetry, purple is aircraft/satellite laser altimetry, red is the flux component method, and blue is satellite gravity. The dashed green box represents the estimated Greenland balance of the IPCC AR4 assessment. These data indicate that mass loss from the Greenland ice sheet may be increasing, although it is also clear that the various estimates are frequently not in agreement. Greenland lost roughly 150 Gt y^{-1} since 2000, increasing to 180±50 Gt y^{-1} (0.5±0.14 mm y^{-1} SLR) for the time period 2003-2007. More than 50% of the current ice loss is caused by increase in ice discharge and ocean interaction of tidewater glaciers, the remaining part can be explained by the increase in surface melt due to warmer summer temperatures (Hanna et al., 2008). The interior of the ice sheet is expected to be less vulnerable to future changes than the edge regions. Current discharge rates may represent a transient instability, and whether they will increase or decrease in the

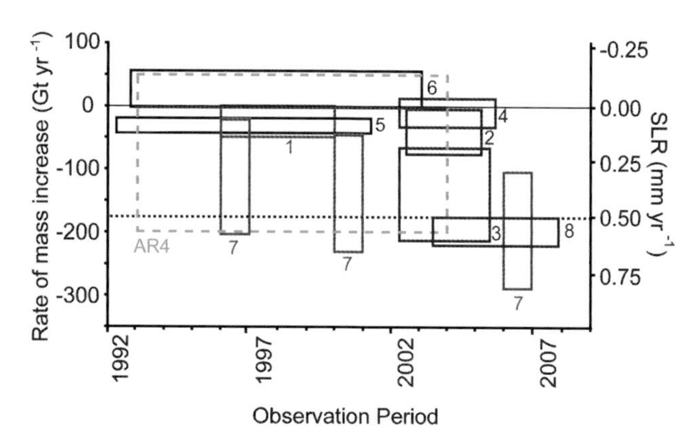

FIGURE 4.25 Estimates of the net mass balance of the Antarctic Ice Sheet since 1992 (Allison et al., 2009b). The horizontal dimension of the boxes shows the time period for which the estimate was made, and the vertical dimension shows the upper and lower limits of the estimate. The colors represent the different methods that were used: black is satellite radar altimetry, orange is aircraft laser altimetry, red is the flux component method, and blue is satellite gravity.

future is unknown. Currently there is no dynamic ice sheet model that can predict the response of the Greenland ice sheet for a warmer climate. We can constrain a possible upper bound of SLR contribution from Greenland assuming a doubling in ice discharge and a continued increase in surface melt using the AR4 A1B scenario (Pfeffer et al., 2008). The total contribution to SLR would be about 0.16 m by 2100, with 0.09 m contribution from ice dynamics, and 0.07 m from surface melt (Pfeffer et al., 2008).

The Antarctic ice sheet shows a pattern of near balance for East Antarctica and mass loss from West Antarctica and the Antarctic Peninsula since 2003 (Cazenave et al., 2009). However, the uncertainties of these measurements are large (Figure 4.25) and there is no strong evidence for increasing Antarctic loss over the period shown.

However, there is a region in West Antarctica that shows increasing ice loss in recent years (Figure 4.25, most of this signal comes from West Antarctica). The Amundsen Coast Basin, including Pine Island ice and Thwaites Glacier, is not confined by large ice shelves, and these marine-based ice masses, which are potentially unstable, contain about 1.5 m sea level equivalent. The average ice velocity in this region is 2 km y^{-1}, which is considerably higher than the average velocity of all Antarctic ice streams (0.65 km y^{-1}) (Pfeffer et al., 2008). If this ice discharge continues to increase, SLR contribution from West Antarctica cannot be ignored. Pfeffer et al. (2008) assume in their sea level rise sensitivity study for Antarctica a doubling of outlet glacier velocities in Pine Island and Thwaites Glacier within the first decade, and ice loss acceleration from the Antarctic Peninsula at the same rate than melt increase at present-day rates of surface mass balance change, resulting in a sea level rise of 0.12 m by 2100, just from the increase in ice discharge.

Summary of Sea Level Change

Greenland and Antarctic ice sheets together would contribute up to 0.285 m sea level under the AR4 A1B warming scenario, assuming a doubling in ice discharge for the Greenland outlet glaciers, and the Amundsen Coast Basin in Antarctica. Such an increase in ice discharge has already been observed for several regions in Greenland. Glaciers and ice caps are expected to contribute 0.37±0.02 m sea level rise under the same warming scenario, and thermal expansion is expected to contribute 0.23±0.09 m by 2100.

Thus, the sea level rise by 2100 is expected to be at least 0.60±0.11 m from thermal expansion and ice loss from glaciers and small ice caps only. Assuming additional ice loss from Greenland at the rate, the total global sea level rise would be about 0.65±0.12 m by 2100. Doubling in ice discharge for both Greenland and Antarctica, the sea level increase could be as high as about 0.88 ±0.12m by 2100. The estimated range in sea level rise in 2100 is therefore from about 0.5 to 1 m.

The dynamic response of ice sheets to global warming is the largest unknown in the projections of sea level rise over the next century. Vermeer and Rahmstorf (2009) made a semi-empirical projection linking sea level to temperatures from past observations; their statistical projection for a temperature scenario A1B (IPCC AR4: 2.3-4.3°C increase for 2100) predicts a SLR of 0.97-1.56 m above 1990 by 2100. This is consistent with what happened during warming in the last interglacial time period (LIG) and cannot be ruled out. The LIG warming was caused by perturbations of Earth's orbit (Overpeck et al., 2006) and arrived much more gradually than is projected for human-induced warming, so a faster sea level rise in the future than in the LIG would not be surprising. On the other hand, changes in the LIG were linked to ice at low elevation, which could behave differently from that at high elevation in the interior of Greenland.

4.9 OCEAN ACIDIFICATION

The oceanic uptake of excess atmospheric carbon dioxide alters the chemistry of seawater, which may impact a wide range of marine organisms from plankton to coral reefs (Doney et al., 2009a,b; NRC, 2010) (see also Section 6.3). Ocean acidification is in fact a series of interlinked and well-known changes in acid-base chemistry and carbonate chemistry due to the net flux of CO_2 into surface waters (Figure 4.26). The chemical shifts include increases in the partial pressure of carbon dioxide (pCO_2), the concentration of aqueous CO_2, and the hydrogen ion (H^+) concentration and decreases in pH ($pH = -\log10[H^+]$). The increase in hydrogen ion concentration acts to lower the concentration of carbonate ions (CO_3^{2-}) through the reaction $H^+ + CO_3^{2-} => HCO_3^-$, even though the total amount of dissolved inorganic carbon (DIC) goes up ($DIC = [CO_2] + [HCO_3^-] + [CO_3^{2-}]$). Declining CO_3^{2-} in turn lowers calcium carbonate ($CaCO_3$) mineral saturation state, $\Omega = [Ca^{2+}][CO_3^{2-}]/K_{sp}$, where K_{sp} is the thermodynamic solubility product that varies with temperature, pressure, and mineral form. Ocean surface waters

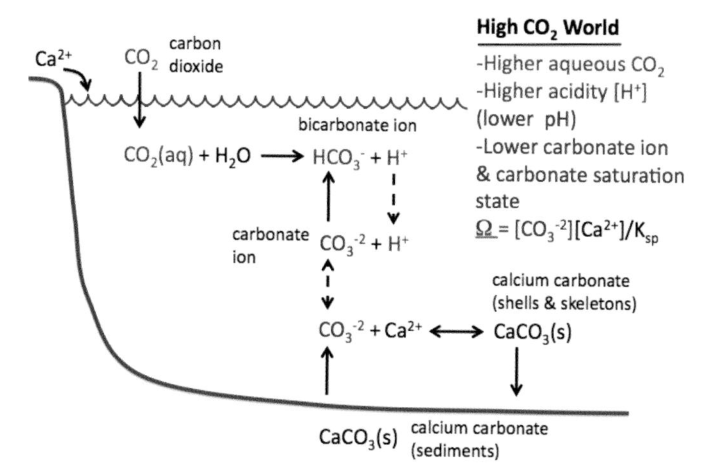

FIGURE 4.26 Schematic indicating the effects on seawater carbonate chemistry due to the uptake of excess carbon dioxide (CO_2) from the atmosphere. Ocean acidification causes increases in some chemical species (red) and decreases in other species (blue). Ocean acidification also causes a reduction in pH ($pH = -\log_{10}[H^+]$) and the saturation states, Ω, of calcium carbonate minerals in shells and skeletons of planktonic and benthic organisms and in carbonate sediments. On millennial and longer time scales, ocean pH perturbations are buffered by external inputs of alkalinity, denoted by calcium ions (Ca^{2+}) and changes in the net burial rate of carbonate sediments.

are currently supersaturated ($\Omega > 1$) for the two major forms used by marine organisms, aragonite (corals, many mollusks) and calcite (coccolithophores, foramaniferia, and some mollusks,). Because of pressure effects and higher metabolic CO_2 from organic matter respiration, Ω decreases with depth often becoming undersaturated ($\Omega < 1$), at which point unprotected shells and skeletons begin to dissolve.

The controls on seawater pH and saturation state vary with time-scale. In the surface ocean, the pH of seawater varies substantially over annual to interannual time scales due to the net biological formation of organic matter (lowers CO_2 and raises pH and Ω) and $CaCO_3$ shells and skeletons (the reverse). Upwelling of CO_2-rich water from below and variations in temperature, salinity, and alkalinity (a measure of the acid buffering capacity of seawater) also influence surface water carbonate chemistry. The saturation state of polar waters is lower in large parts because of colder temperatures. In coastal waters, pH and Ω exhibit large natural spatial and temporal variations due to the interplay of river runoff, strong biological productivity, and in, some locations, coastal upwelling (Salisbury et al., 2008). Over decadal to century time scales, ocean carbon chemistry is modulated by net CO_2 uptake from the atmosphere and trends in ocean circulation and biological productivity, which tend to redistribute dissolved inorganic carbon and alkalinity within the ocean water-column. On even longer time scales of many centuries to millennia, the weathering of calcium carbonate rocks on land adds alkalinity, a measure of the acid buffering capacity, in the form of calcium ions (Ca^{2+}) and carbonate ions (CO_3^{2-}), and alkalinity is removed by the burial, on continental shelves and margins, of biologically formed carbonate sediments made of the shells and skeletons of some plankton, corals, and other calcifying organisms (Figure 4.26). Carbonate sediment burial rates are sensitive to seawater chemistry, and on millennial time scales longer, efficient damping feedbacks act to stabilize mean ocean alkalinity and pH.

At the small number of available open-ocean time-series sites, significant secular trends in surface ocean carbonate chemistry are well documented for the past two decades (Figure 4.27). The time-series records document clearly an increase in surface water pCO_2 and DIC and a decline in pH that is consistent with the rate of change in atmospheric CO_2 (Dore et al., 2009). The WOCE/JGOFS Global CO_2 Survey completed in the 1990s provided a global estimate of ocean anthropogenic CO_2 distributions and a baseline for assessing changes in ocean chemistry with time (Sabine et al., 2004). Decadal resurveys of a subset of the WOCE/JGOFS ocean transects also exhibit decreasing pH through time over the upper thermocline and

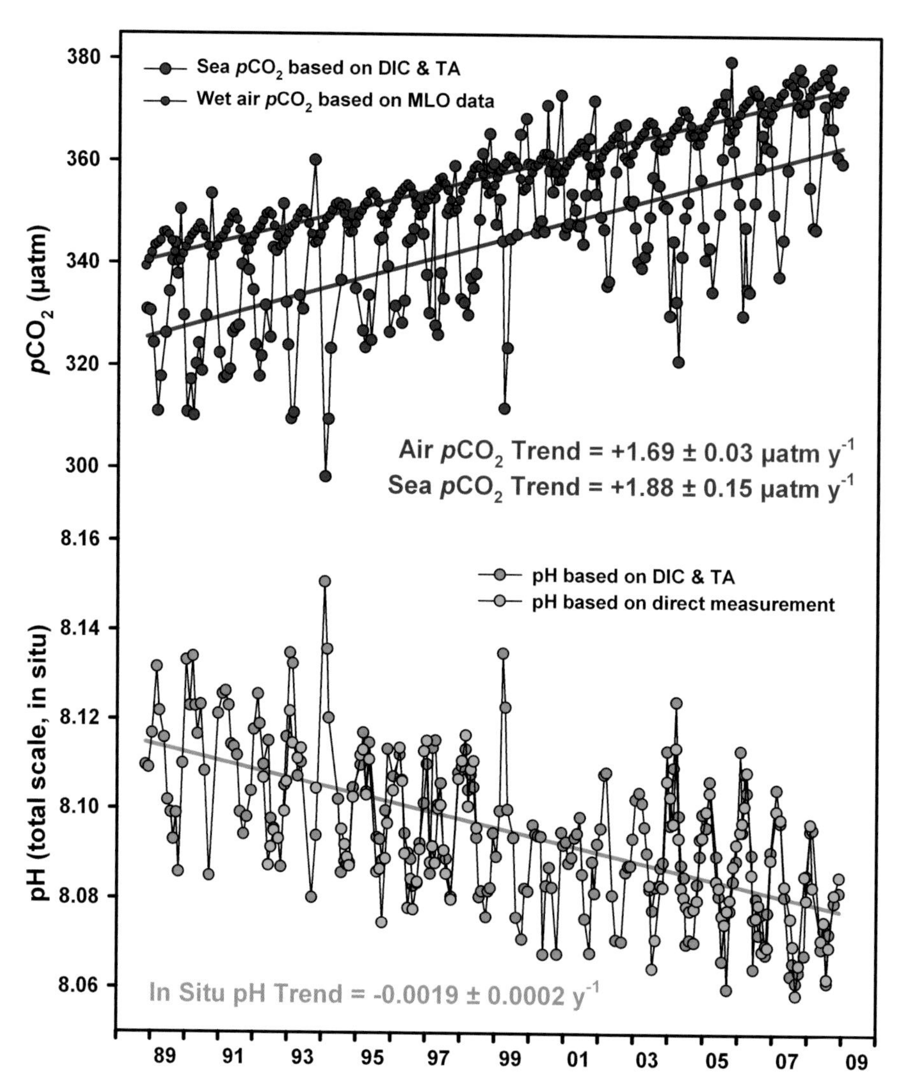

FIGURE 4.27 Time-series surface seawater carbonate system at Station ALOHA in the subtropical North Pacific north of Hawai'i, 1988-2008. The upper panel displays the partial pressure of CO_2 (pCO_2) in seawater calculated from dissolved inorganic carbon (DIC) and total alkalinity (TA) (blue symbols) and in water-saturated air at *in-situ* seawater temperature (red symbols). Atmospheric CO_2 data is from the Mauna Loa Observatory, Hawai'i. The lower panel displays *in-situ* surface pH based on direct measurements (green symbols) or as calculated from dissolved inorganic carbon and total alkalinity (orange symbols). Linear regressions (colored lines) and regression equations are reported for each variable. Source: Adapted from Dore et al. (2009).

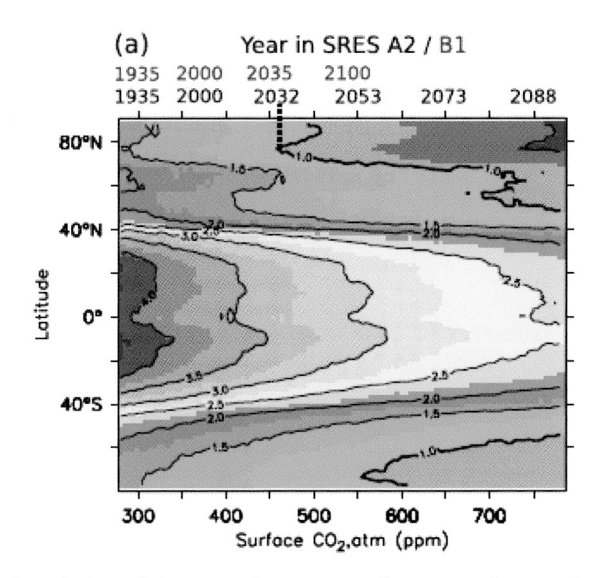

FIGURE 4.28 Projected evolution of the annual-mean, zonally averaged aragonite saturation, Ω, plotted as a function of the annual-mean atmospheric CO_2 mixing ratio at the ocean surface. The corresponding years for the SRES A2 and B1 scenarios are given at the top. The largest decreases in aragonite saturation values occur in the tropics. Arctic and Southern Ocean surface waters transition from supersaturation to undersaturation in the annual-mean beginning at approximately 460 ppm and 550 ppm CO_2, respectively. Undersaturated conditions occur for individual months at even lower atmospheric CO_2 levels, beginning at approximately 410 ppm for the Arctic and 490 ppm for Southern Ocean. Source: Adapted from Steinacher et al. (2009).

across ocean basins (Byrne et al., 2010). Based on ice-core CO_2 data and the WOCE/JGOFS Survey, surface ocean pH has already dropped on average by about 0.1 pH units from pre-industrial levels (pH is measured on a logarithmic scale and a 0.1 pH drop is equivalent to a 26% increase in hydrogen ion concentration) (Orr et al., 2005). The patterns of ocean acidification in subsurface waters depend on ocean circulation patterns; thermocline waters in subtropical convergence regions and deep-waters in polar regions where cold surface waters sink into the interior ocean are affected more than other parts of the subsurface.

Future acidification of surface waters can be predicted for a given atmospheric carbon dioxide level (see Figure 4.28). An additional decline of 0.15 pH unit would occur if atmospheric carbon dioxide increases from current levels to 550 ppm, and larger pH changes would occur, approximately proportionally, for higher CO_2 concentrations. Polar surface waters may become under-saturated with respect to aragonite, a key calcium carbon mineral that can affect the ability of organisms to build their shells, for atmospheric CO_2 levels of 400-450 ppm for the Arctic and 550-600 ppm for the Antarctic (Orr et al., 2005; Steinacher et al., 2010). For tropical surface waters, large reductions in calcium carbonate saturation state are expected to occur, but waters are expected to remain super-saturated for projected atmospheric CO_2 during the 21st century for current scenario projections. Calcium saturation horizons ($\Omega = 1$) have been observed to move upward, that is, shoaled (Feely et al., 2004; Orr et al., 2005), and there is evidence that water undersaturated for aragonite is already upwelling onto the continental shelf off the U.S. west coast due to a combination of strong wind-induced upwelling and the penetration of anthropogenic CO_2 into off-shore source waters (Feely et al., 2008). For most of the surface ocean, climate change feedbacks are weak, and warming and altered ocean circulation have a limited effect on changing pH and Ω that are determined primarily by atmospheric CO_2. An exception is in the Arctic, where sea-ice retreat and changes in surface freshwater balance amplify atmospheric CO_2-driven pH and Ω declines (Steinacher et al., 2010).

5

Impacts in the Next Few Decades and Coming Centuries

5.1 FOOD PRODUCTION, PRICES, AND HUNGER

Even in the most highly mechanized agricultural systems, food production is very dependent on weather. Concern about the potential impacts of climate change on food production, and associated effects on food prices and hunger, have existed since the earliest days of climate change research. Although there is still much to learn, several important findings have emerged from more than three decades of research.

It is clear, for example, that higher CO_2 levels are beneficial for many crop and forage yields, for two reasons. In species with a C_3 photosynthetic pathway, including rice and wheat, higher CO_2 directly stimulates photosynthetic rates, although this mechanism does not affect C_4 crops like maize. Secondly, higher CO_2 allows leaf pores, called stomata, to shrink, which results in reduced water stress for all crops. The net effect on yields for C_3 crops has been measured as an average increase of 14% for 580 ppm relative to 370 ppm (Ainsworth et al., 2008). For C_4 species such as maize and sorghum, very few experiments have been conducted but the observed effect is much smaller and often statistically insignificant (Leakey, 2009).

Rivaling the direct CO_2 effects are the impacts of climate changes caused by CO_2, in particular changes in air temperature and available soil moisture. Many mechanisms of temperature response have been identified, with the relative importance of different mechanisms varying by location, season, and crop. Among the most critical responses are that crops develop more quickly under warmer temperatures, leading to shorter growing periods and lower yields, and that higher temperatures drive faster evaporation of water from soils and transpiration of water from crops. Exposure to extremely high temperatures (e.g., $> 35°C$) can also cause damage in photosynthetic, reproductive, and other cells, and recent evidence suggests that even short exposures to high temperatures can be crucial for final yield (Schlenker and Roberts, 2009; Wassmann et al., 2009).

A wide variety of approaches have been used in an attempt to quantify yield losses for different climate scenarios. Some models represent individual processes in detail, while others rely on statistical models that, in theory, should capture all relevant processes that have influenced historical variations in crop production. Figure 5.1 shows model estimates of the combined effect of warming and CO_2 on yields for different levels of global temperature rise. It is noteworthy that although yields respond nonlinearly to temperature on a daily time scale, with extremely hot days or cold nights weighing heavily in final yields, the simulated response to seasonal warming is fairly linear at broad scales (Lobell and Field, 2007; Schlenker and Roberts, 2009). Several major crops and regions reveal consistently negative temperature sensitivities, with between 5-10% yield loss per degree warming estimated both by process-based and statistical approaches. Most of the nonlinearity in Figure 5.1 reflects the fact that CO_2 benefits for yield saturate at higher CO_2 levels.

For C_3 crops, the negative effects of warming are often balanced by positive CO_2 effects up to 2-3°C local warming in temperate regions, after which negative warming effects dominate. Because temperate land areas will warm faster than the global average (see Section 4.2), this corresponds to roughly 1.25-2°C in global average temperature. For C_4 crops, even modest amounts of warming are detrimental in major growing regions given the small response to CO_2 (see Box 5.1 for discussion of maize in the United States).

The expected impacts illustrated in Figure 5.1 are useful as a measure of the likely direction and magnitude of average yield changes, but fall short of a complete risk analysis, which would, for instance, estimate the chance of exceeding critical thresholds. The existing literature identifies several prominent sources of uncertainty, including those related to the magnitude of local warming per degree global temperature increase, the sensitivity of crop yields to temperature, the CO_2 levels corresponding to each temperature level (see Section 3.2), and the magnitude of CO_2 fertilization. The impacts of rainfall changes can also be important at local and regional scales, although at broad scales the modeled impacts are most often dictated by temperature and CO_2 because simulated rainfall changes are relatively small (Lobell and Burke, 2008).

In addition, although the studies summarized in Figure 5.1 consider several of the main processes that determine yield response to weather, several other processes have not been adequately quantified. These include responses of weeds, insects, and pathogens; changes in water resources available for irrigation; effects of changes in surface ozone levels; effects of

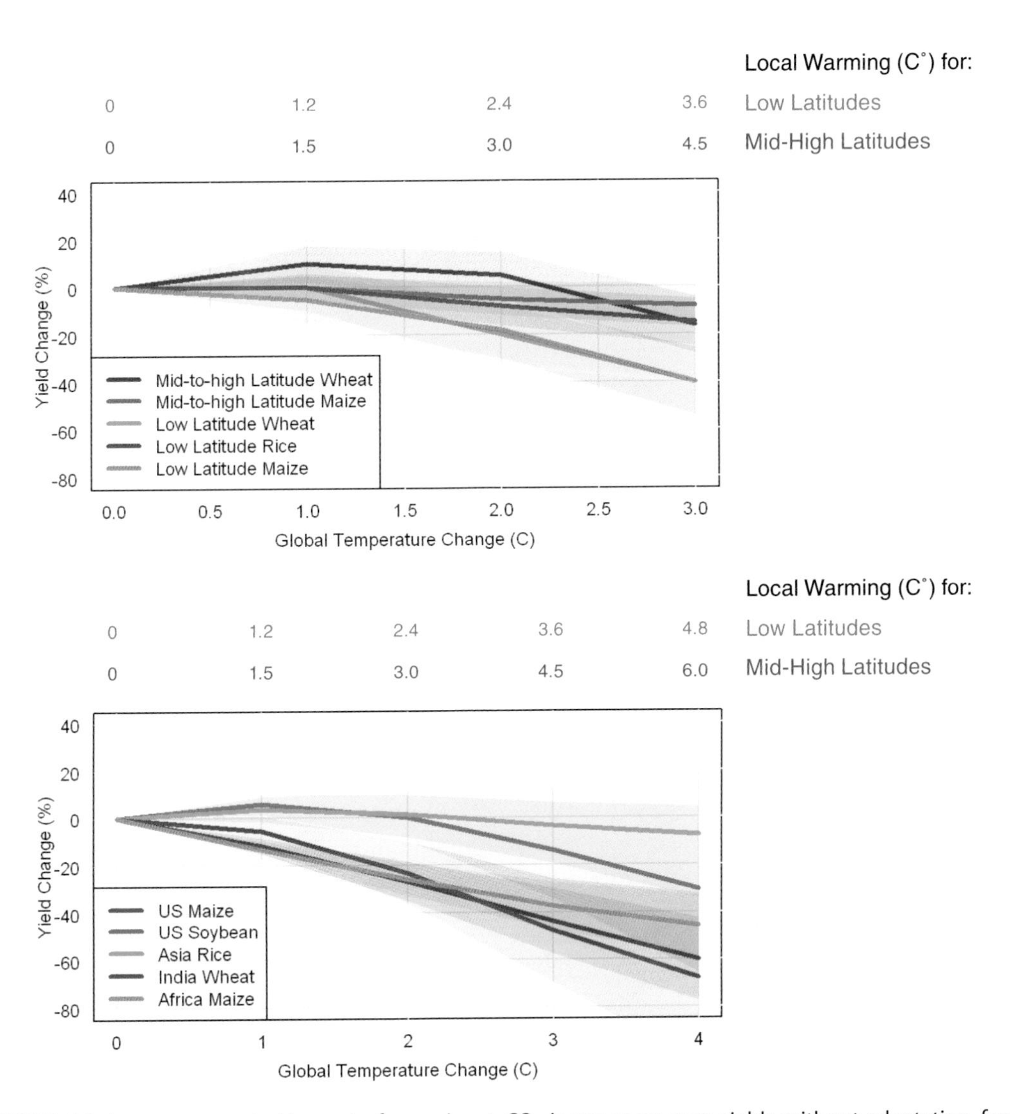

FIGURE 5.1 Average expected impact of warming + CO_2 increase on crop yields, without adaptation, for broad regions summarized in IPCC AR4 (left) and for selected crops and regions with detailed studies (right). Shaded area shows likely range (67%). Impacts are averages for current growing areas within each region and may be higher or lower for individual locations within regions. Temperature and CO_2 changes for the IPCC summary (left) are relative to late 20th century, while changes estimated for regions (right) were computed relative to pre-industrial. Estimates were derived from various sources (Matthews et al., 1995; Lal et al., 1998; Easterling et al., 2007; Schlenker and Roberts, 2009; Schlenker and Lobell, 2010) (see methods in Appendix for details).

BOX 5.1 HOW WILL MAIZE YIELDS IN THE UNITED STATES RESPOND TO CLIMATE CHANGES?

Nearly 40% of global maize (or corn) production occurs in the United States, much of which is exported to other nations. The future yield of U.S. maize is therefore important for nearly all aspects of domestic and international agriculture. Higher temperatures speed development of maize, increase soil evaporation rates, and above 35°C can compromise pollen viability, all of which reduce final yields. High temperatures and low soil moisture during the flowering stage are especially harmful as they can inhibit successful formation of kernels. In northern states, warmer years generally improve yields as they extend the frost-free growing season and bring temperature closer to optimum levels for photosynthesis. The majority of production, however, occurs in areas where yields are favored by cooler than normal years, so that warming associated with climate change would lower average national yields. The most robust studies, based on analysis of thousands of weather station and harvest statistics for rainfed maize (>80% of U.S. production), suggest a roughly 7% yield loss per °C of local warming, which is in line with previous estimates (USCCSP, 2008b). Given the rate of local warming in the Corn Belt relative to global average, this implies an 11% yield loss per °C of global warming (Figure 5.1).

Whether these losses are realized will depend in large part on the effectiveness of adaptation strategies, which include shifts in sowing dates, switches to longer maturing varieties, and

---

increased flood frequencies; and responses to extremely high temperatures. Moreover, most crop modeling studies have not considered changes in sustained droughts, which are likely to increase in many regions (Wang, 2005; Sheffield and Wood, 2008), or potential changes in year-to-year variability of yields. The net effect of these and other factors remains an elusive goal, but these are likely to push yields in a negative direction. For example, recent observations have shown that kudzu (*Pueraria lobata*), an invasive weed favored by high CO_2 and warm winters, has expanded over the past few decades into the Midwest Corn Belt (Ziska et al., 2010).

Adaptation responses by growers are also poorly understood and could, in contrast, reduce yield losses. For example, temperate growers are likely to shift to earlier planting and longer maturing varieties as climate warms, and models suggest this response could entirely offset losses in certain situations. More commonly, however, these adaptations will at best be able to offset 2°C of local warming (Easterling et al., 2007), and they will be less effective in tropical regions where soil moisture, rather than cold temperatures, limits the length of the growing season. Very few studies have considered the evidence for ongoing adaptations to existing climate trends and quantified the benefits of these adaptations.

development of new seeds that can better withstand water and heat stress and better utilize elevated CO_2. A wide range of maize varieties are currently sown throughout the country, customized to local factors such as latitude, growing season length, and soil, and new varieties are continually developed by private seed companies. These companies have historically focused on biotic stresses, but are now releasing the first varieties explicitly targeted for drought resistance. Heat tolerance has not received much investment outside of drought-related traits, likely because of limited economic incentives in current climate. A comparison of maize yields in northern and southern states suggests minimal historical adaptation to heat, as varieties that are more frequently exposed to temperatures above 30°C exhibit similar sensitivities to varieties grown in the North (Schlenker and Roberts, 2009). A major challenge in developing drought and heat tolerance is that traits that confer these often reduce yields in good years, and growers and seed companies have little economic incentive to accept this trade-off given current markets and insurance programs. Another persistent challenge is the decade or more lag between initial investments and seed release. In short, adaptation could offer large benefits, but only if formidable technical and institutional barriers are overcome. To put the challenge in context, global cereal demand is expected to rise by roughly 1.2% per year (FAO, 2006), so that adapting to 1°C global warming (or avoiding 11% yield loss) is equivalent to keeping pace with roughly 9 years of demand growth. The corresponding expected impact of 2°C global warming is 25%, or roughly 20 years of demand growth.

Future development of new varieties that perform well in hot and dry conditions may also promote adaptation, but again the extent to which this will help remains unclear. Breeders and geneticists must continually weigh trade-offs between producing ample yield under stressful conditions and producing high yields under favorable conditions (Campos et al., 2004). At the higher warming levels considered in this report, it will be increasingly difficult to generate varieties with a physiology that can withstand extreme heat and drought while still being economically productive.

Although most studies have focused on crops, effects of climate change on livestock, aquaculture, and fisheries have also been considered in recent years. Livestock in parts of the world are raised mainly on grain and oilseed crops, in which case impacts will largely follow from the prices of these commodities and the costs of cooling or losing animals during heat waves. In other cases livestock depend on grazing pasture and rangeland grasses, which follow a similar pattern to crops in that temperate regions will see modest gains up to ~2°C local warming, although forage quality may decrease with higher CO_2 (Easterling et al., 2007). Although livestock systems are vulnerable in tropical areas, they may become increasingly relied upon as a strategy to cope with greater risks of crop failures (Thornton et al.,

2009b). As with livestock, impacts on fisheries are still very uncertain, but a recent study suggests that if global average warming were to be 2°C, catch potential could rise by 30-70% in high latitudes and fall by up to 40% in the tropics, as commercial species shift away from the tropics as the ocean warms (Cheung et al., 2010).

Food Prices and Food Security

One of the strengths of a global food system is that shortfalls in one area can be offset by surpluses in another. Models of the global food economy suggest that trade will represent an important but not complete buffer against climate change-induced yield effects (Easterling et al., 2007). Specifically, the comparative advantage will shift toward regions currently below optimum temperatures for cereal production (e.g., Canada) and away from hot tropical nations, with greater flows of food trade from north to south. On average, studies suggest small price changes for cereals up to 2.5°C global temperature increase above pre-industrial levels, with significant increases for further warming, but there is considerable uncertainty around these estimates (see Box 5.2).

Implications of climate change for hunger, or the more technical term—food insecurity—follow in part from price changes, but also depend critically on how sources of income and other aspects of health are affected by climate. A useful rule of thumb provided by early studies suggested that malnourishment would rise by roughly 1% for each 2-2.5% rise in cereal prices (Rosenzweig, 1993). These and subsequent analyses often make untested assumptions about the ability of poor tropical nations to maintain economic growth in the face of declining agricultural productivity. For example, many African countries rely on agriculture for half or more of all economic activity, and losses in productivity could dampen purchasing power. Conversely, where price rises are greater than yield losses, households dependent on agricultural income could see net gains in food security. In general, rural and urban workers with little or no landholdings are the most vulnerable to price shocks. A new generation of models that explicitly account for income sources among poor populations is emerging but yet to provide robust insights. Also important could be climate-induced changes in the incidence of diarrheal and other diseases, which inhibit food security by reducing utilization of nutrients in food.

BOX 5.2 CLIMATE CHANGE IMPACTS ON GLOBAL CEREAL PRICES

Several modeling groups have analyzed future changes in global cereal markets in response to climate change. All operate by making estimates of yield responses in each region, and then inputting these into a model of global trade that computes the optimal mix of crop areas in different regions and the market-clearing price. Five models summarized by the recent IPCC report suggests small price changes for warming up to 2.5°C, and a nonlinear increase in prices thereafter (Easterling et al., 2007). Two important caveats relate to these estimates, however.

First, the yield changes used in these models usually assume considerable levels of farm-level adaptations, which substantially reduce impacts. For example, in one prominent study cereal prices rose by 150% for a 5.2°C global mean temperature rise if farm-level adaptations were not included. When changes in planting dates, cultivar choices, irrigation practices, and fertilizer rates were simulated, these price changes were reduced to roughly 40% (Rosenzweig and Parry, 1994). Other studies often do not estimate impacts without adaptation, making it difficult to gauge assumptions. The costs of adaptation are also not considered in these studies, or reflected in price changes.

Second, most assessments have not adequately quantified sources of uncertainty. Although different climate scenarios are often tested, processes related to crop yield changes and economic adjustments are often implicitly assumed to be perfectly known. An additional source of uncertainty is potential competition with bio-energy crops for suitable land, which could limit the ability of croplands to expand in temperate regions as simulated by most trade models.

5.2 COASTAL EROSION AND FLOODING

Our knowledge of the links between atmospheric concentration limits, trajectories toward equilibrium temperature change, and sea level rise is fraught with uncertainty. As reported in Section 4.8, it is therefore only possible to offer a range of sea level rise between 0.5 and 1.0 m through 2100.

Moving down the causal chain to consider coastal erosion and flooding adds yet another layer of complication because both are driven primarily by storm surges, land-use decisions, and other processes whose intensities and frequencies change from place to place. These changes alter the characters of associated risks even if changes in the intensities and frequencies of the storms, themselves, cannot be projected. The social and economic ramifications of these physical manifestations of climate change depend critically on patterns of future development and population growth. It is, therefore, extremely difficult to offer credible broad-based estimates of vulnerabilities and potential adaptation costs. At best, in fact, we can offer only suggestive ranges of aggregate risk and more quantitative estimates only for specific locations.

Figure 5.2 offers a portrait of the geographic spread of deltas and mega-deltas where mega-cities are at the greatest risk from rising seas—these are the "hot-spots" of "key vulnerabilities" in the coastal zone. Ericson et al. (2006) estimated that nearly 300 million people currently inhabit a sample of 40 such deltas with an average population density of 500 people per km^2.

Translating this observation into projections of future vulnerabilities, Table 5.1 shows the sensitivity of estimates of populations subject to coastal flooding in 2080 to assumptions about socioeconomic development as described in the SRES scenarios—sensitivity generated by differences across the scenarios in population growth and by differences in assumptions about economic development and therefore the capacity to adapt. Figure 5.3 emphasizes the importance of adaptation when it suggests, for example that 1 m of sea level rise could put between 10 and 300 million more people at risk of coastal flooding each year. It is important to note, in interpreting this figure, that the likelihood of inundation from coastal storms may not be proportional with sea level rise. Moreover, the consequences of these storm events calibrated in millions of people in jeopardy from coastal flooding depend on local population densities and geographic features. The result of the

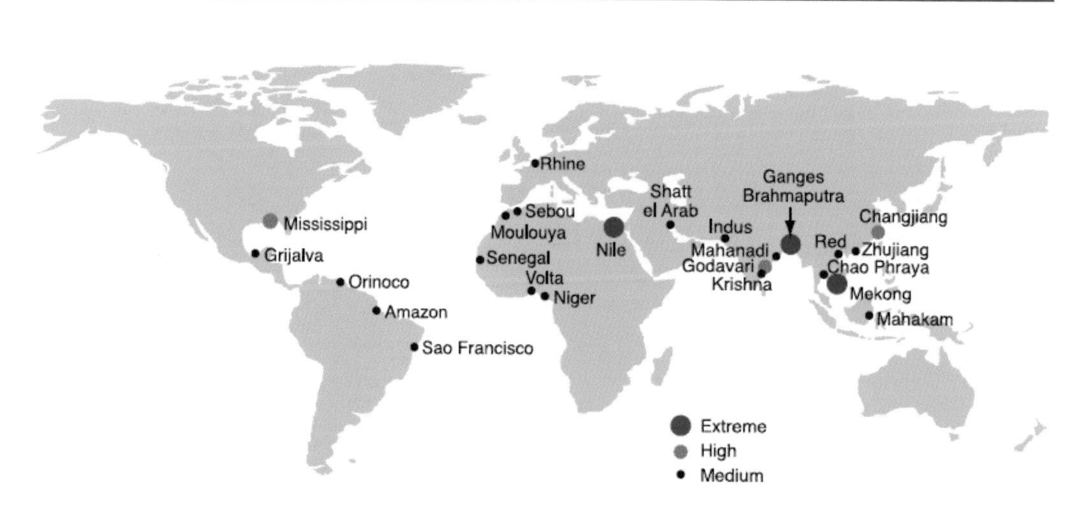

FIGURE 5.2 Relative vulnerability of coastal deltas as shown by the indicative population displaced by current sea-level trends to 2050 (Extreme=>1 million; High=1 million to 50,000; Medium=50,000 to 5,000; following Ericson et al., 2006). Source: Nicholls et al. (2007: Figure 6.6).

TABLE 5.1 Regional Distribution of Population Subject to Coastal Flooding in 2080 along Alternative SRES Scenarios.

Region	1990 (baseline)	SRES scenarios (and sea-level rise scenario in metres)			
		A1FI (0.34)	A2 (0.28)	B1 (0.22)	B2 (0.25)
Australia	1	1	2	1	1
Europe	25	30	35	29	27
Asia	132	185	376	180	247
North America	12	23	28	22	18
Latin America	9	17	35	16	20
Africa	19	58	86	56	86
Global	197	313	561	304	399

* *Area below the 1 in 1,000 year flood level.*

NOTE: Population estimates by region assume proportional population growth within coastal regions. Source: Nicholls (2004) as displayed in Nicholls et al. (2007: Table 6.5).

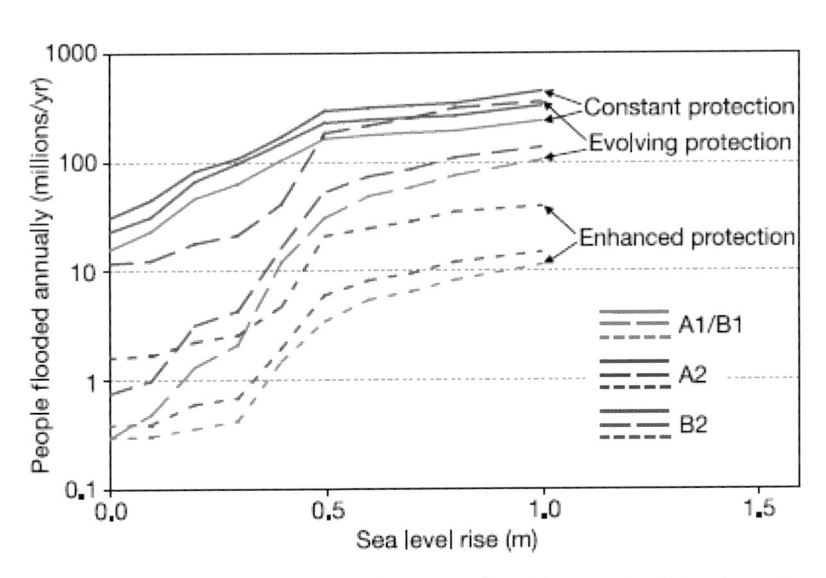

FIGURE 5.3 Estimates of people flooded in coastal areas attributable to sea level rise along alternative SRES scenarios. Estimates of the number of additional people in jeopardy from coastal flooding along alternative SRES development scenarios for three gross categories of adaptation intensities are displayed. Constant protection envisions maintaining current practices, evolving protection envisions increasing protection as local economies grow to preserve the current pattern with respect to national GDP, and enhanced protection envisions accelerating the pace of adaptation so that increasing resources are devoted to protection. Source: Nicholls et al. (2007: Figure 6.8) derived from Nicholls and Tol (2006).

confluence of these complications is a noticeable threshold of accelerating risk around 0.3 m of sea level rise that is captured even by global aggregates regardless of adaptation effort. As a result, even 50 cm of SLR could put between 5 and 200 million more people annually at risk of flooding.

Tol (2007) used a specific integrated assessment model of his own creation to portray aggregate measures of erosion that parallel estimates of populations facing complete displacement and/or significant economic loss derived from economically efficient abandonment and/or growing protection costs across developed and developing countries. Figure 5.4 calibrates his results graphically in relation to sea level rise; they were derived from a socioeconomic portrait that was crafted to be consistent with the IS92a emissions scenario for which seas rise by roughly 60 cm through 2100. This work suggests that 50 cm of sea level rise could permanently displace up to 4 million people and cause more than 250,000 km^2 of wetland and dry-land to be lost to erosion worldwide (with 90% of these losses projected to occur in developing countries). The human faces behind the global displacement results portrayed here can, of course, be seen in examples of erosion from coastal storms and rising seas. In the Arctic, Newtok, Alaska is already preparing for complete displacement, for example, and several neighboring towns face the same fate in the near future. Meanwhile, many small island states like Tuvalo, the Maldives, and the Cook Islands foresee similar futures this century if sea level rise continues.

Geographic detail for physical processes like erosion and inundation

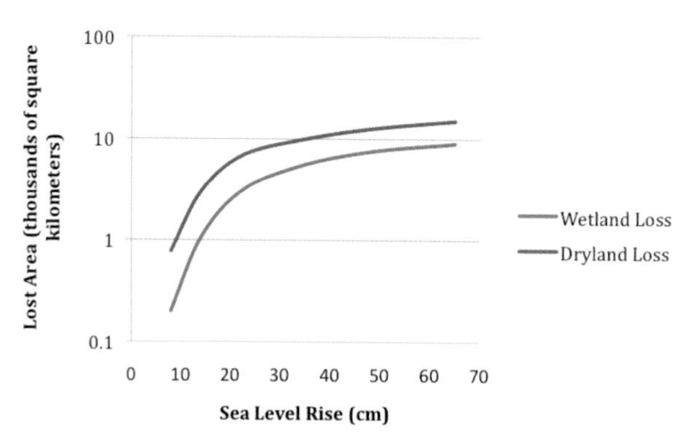

FIGURE 5.4 Losses attributable to sea level rise. Estimates of wetland and dry-land losses for developed (Panel A) and developing countries (Panel B) correlated with sea level rise along a socioeconomic scenario that tracks IS92a. Source: Derived directly from Tol (2007) as depicted in Nicholls et al. (2007: Figure 6.10).

TABLE 5.2 Selected Losses from Sea Level Rise and Associated Erosions across Asia

SLR Rise (from 2000 levels)	Location	Magnitude	Source
0.3 m	China	$81.4 \cdot 10^3$ km^2	Du and Zhang (2000)
	Huanghe-Huaihe Delta	$21.3 \cdot 10^3$ km^2	
	Changjiang Delta	$54.5 \cdot 10^3$ km^2	
	Zhujiang Delta	$5.5 \cdot 10^3$ km^2	
1.0 m	Japan	$2.3 \cdot 10^3$ km^2	Mimura and Yokoki (2004)
1.0 m	Korea	1.2% area	Madsen and Jakobsen (2004)
1.1 m	India and Bangladesh	478 km^2 (11%)	Loucks et al. (2010)
1.2 m	India and Bangladesh	1,396 km^2 (33%)	
0.3 m	India and Bangladesh	4,015 km^2 (96%)	

from sea level rise has been emerging over the past decade. Table 5.2, for example, offers estimates for several locations in Asia. Some are located in important deltas in China where modest sea level rise of 0.3 meters would cause significant loss of land area from inundation and erosion; others are located in eastern and southeastern Asia where 1 m of sea level rise would cause significant loss of land and protective mangroves in addition to putting many people at risk of displacement. The final entry reports recent estimates of associated loss in the habitat of the only tiger population in the world (*panthera tigris*) that is adapted to living in mangroves; Loucks et al. (2010) report that a nonlinear decline to extinction (at 30 cm) would begin around 15 cm of sea level rise.

Turning to specific locations within the United States, where it is possible to focus attention on downstream impacts and the potential adaptation, Figure 5.5 first depicts coastal vulnerability to erosion across the mid-Atlantic region at the end of the century for three sea level rise scenarios. Enormous variability from site to site along the coastline is clearly displayed; and so it is obvious that potential risks and the potential for adaptation can be expected to be equally diverse.

5.3 STREAMFLOW

Runoff is defined as the difference between precipitation and the sum of evapotranspiration and storage change on or below the land surface. On long term balance, it must be balanced by precipitation minus evapotranspiration, which also equals atmospheric moisture convergence. Streamflow is

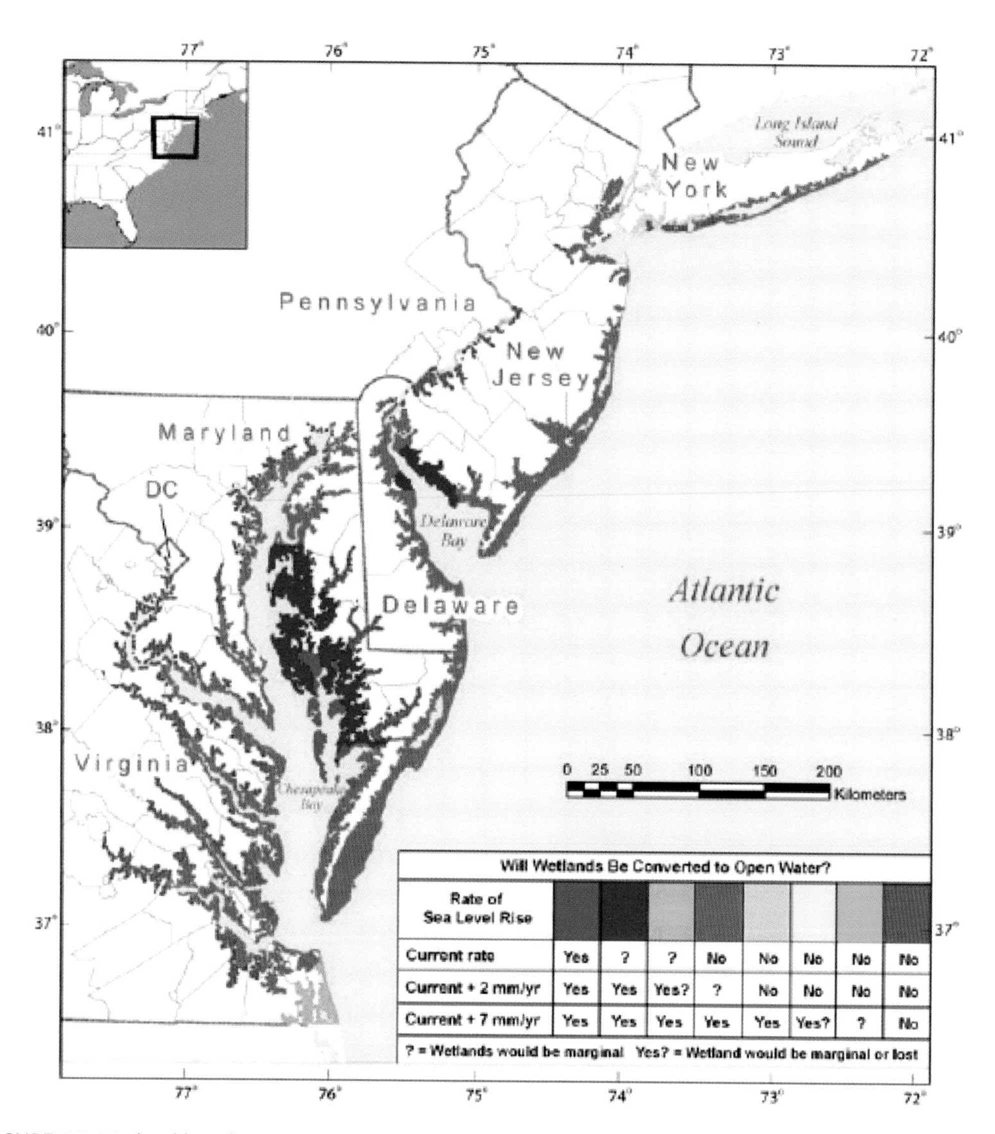

FIGURE 5.5 Wetland loss from erosion attributable to sea level rise along three alternative scenarios. Confirmation of the hypothesis that wetland and marshes would be covered to open water by the end of the century along feasible sea level rise scenarios (current trend up to almost 70cm by 2100. Source: USCCSP (2009: Figure ES.2).

runoff that moves through the channel system to a given point (in practice, often a stream gauge at which streamflow measurements are made).

Runoff is a key index of the availability of freshwater, a quantity that is essential for human life. Although on average only about 5 liters per day of water are necessary for adult survival, water use in industrialized countries is much higher—around 200 liters/day in the United States and 150 liters in Europe. On the other hand, the average for Africa is only about 10 liters, or double the minimum for survival. Total water consumption for agriculture globally is about 10 times that for municipal use (Shiklomanov, 1999). Groundwater, an important source of water in many parts of the world, makes up an additional estimated 25 percent of total global water withdrawals (International Water Management Institute, 2007). Groundwater recharge, although difficult to estimate except on a local basis, is roughly related to the excess of precipitation over evapotranspiration, and so very approximately can be taken as a fraction of annual runoff. Hence, understanding how runoff will change in a future climate is the key to understanding how water availability for human use might change.

In climate models, runoff is represented by land surface models, which have the primary purpose of partitioning net radiation at the land surface into latent, sensible, and ground heat flux. A secondary purpose (which is linked to modeling of the surface energy balance, because evapotranspiration, or equivalently latent heat, is common to the energy and water balances) is to partition precipitation into infiltration and runoff. Runoff is also produced in most models by parameterizations of subsurface hydrologic processes, albeit crudely in most models. Although many models do not consider the transformation of runoff to streamflow, a few do. On long-term (average annual) balance, runoff is roughly equal to streamflow, ignoring channel processes (such as groundwater interactions), which in most cases have modest effect.

Past attempts to understand the sensitivity of runoff and streamflow to a changing climate have followed two general pathways. The most common is to use river basin hydrological models, which typically are forced with precipitation and other surface atmospheric variables, and to prescribe differences in the forcings that reflect the effects of climate change. This approach, sometimes termed downscaling, has been applied in one of three ways. The first is the so-called delta method, which adjusts precipitation and temperature by factors or shifts that reflect climate model-estimated differences between current and future climate. An example of this approach is the study by Hamlet and Lettenmaier (1999) of the effects of climate change on the water resources of the Columbia River Basin. The second approach is statistical downscaling, which "trains" a relationship between current climate (precipitation, surface air temperature, and other variables) produced by a climate model and observations, in such a way as to correct for climate model biases and spatial and temporal scale mismatches. The same relationship is then applied to future climate simulations. For both current and future climate, the downscaled climate model output is used to force a river basin hydrology model. Examples of this approach are the study by Hayhoe et al. (2007) of climate change impacts on the northeastern United States, the Maurer et al. (2007) study of California water resources, and Christensen and Lettenmaier's (2007) study of Colorado River water resources. The third approach is dynamical downscaling, in which a regional climate model (RCM) is nested within a global model to produce more spatially resolved climate model output. Although dynamical downscaling may be preferred on theoretical grounds (because it is based on physical rather than statistical relationships; its application in practice has been problematic because the computational burden is high, hence the number of RCM runs that can be performed for different global models is small. Furthermore, practical constraints, like the fact that global model output sufficient to provide requisite variables at the RCM domain boundary often are not archived, or are not archived at sufficient vertical levels in the atmosphere and time resolution to meet RCM needs. Finally, Wood et al. (2004) showed that even after dynamical downscaling, application of statistical post-processing, similar to that used in statistical downscaling methods, was necessary to remove RCM bias prior to using RCM output to force a hydrological model.

The second pathway is direct analysis of global climate model runoff predictions. This approach usually has not been favored for river basin studies, because of the scale mismatch between the spatial resolution of the global models (typically several degrees latitude by longitude) and the river basin scale. However, for analysis of large river basins, Milly et al. (2005) (and replots of the Milly et al. [2005] results at the scale of the U.S. hydrologic regions in the USCCSP Synthesis and Assessment Product 4.3 [USCCSP, 2008b]) have found this approach to be useful, especially because it avoids the issues associated with downscaling and inconsistencies in fluxes at the river basin scale. Seager et al. (2007), in an analysis of projected changes in runoff over the U.S. Southwest, also analyzed GCM runoff directly.

Despite the fact that direct analysis of GCM runoff has not been widely performed, we use this approach here for two reasons. First, post processing of global model output by statistical downscaling and hydrological modeling results in a mismatch in the surface water balance between the hydrologic model that produces streamflow simulations and the global model. Second, although some (but not all) of these issues are resolvable via dynamical downscaling, the logistical requirements for so doing make dynamical downscaling infeasible for analysis of the broad suite of GCM output in the IPCC AR4 (and upcoming AR5) archives.

We have used as our motivation here the Milly et al. (2005) study, which analyzed multidecadal runoff from 12 GCMs for mid-21st century as compared with early 20th century runoff (24 separate model runs were analyzed; however, some models had multiple ensembles, which were averaged). An important difference, however, is that for reasons articulated in Chapter 2 we performed our analysis in a way that links projected runoff changes with changes in the global mean temperature. Accordingly, we proceeded as follows. We extracted IPCC AR4 archived runoff output along with surface air temperature, precipitation, and evapotranspiration for 23 models for which global monthly output for (in most cases) 1870 through 2100 was archived. For each model and emissions scenario, we computed the global mean temperature for each year. We then computed a 30-year moving average of the global mean temperatures, and from this time series, we determined the year at which the global mean temperature was larger than for 1985 (1971-2000 mean) by 0.5°C, 1°C, 1.5°C, and 2.5°C. We then computed 30-year moving averages of runoff for each model, emissions scenario (A2, A1B, and B1), and grid cell, and overlaid the runoff values for 1985 and the year in which the moving average global temperature rise was the given amount (0.5, ... 2.5) to the U.S. hydrological regions. We then computed runoff changes as percentages per degree C of global warming, and took the median over models and global temperature changes of 1°C, 1.5°C, and 2°C. We also computed equivalent standard errors of the medians (see Table 5.3). Figure O.2 shows the results. In general, runoff decreases over most of the country, with the exception of the northeast and the northwest. Closer examination of similar results (not shown) for precipitation and evapotranspiration indicate that (a) changes in runoff result from a combination of slight increases in precipitation (in the multi-model median) over most of the country with the exception of the Southwest, and increasing evapotranspiration over all but the Southwest (where lower precipitation apparently limits moisture available for evapotranspiration); (b) the general patterns of changes (runoff sensitivities per degree C) are roughly constant over the range 1-2°C global temperature rise; and (c) the patterns of changes and sensitivites per degree C are similar for the different emissions scenarios (hence justifying pooling of the results over emissions scenarios).

We also conducted a similar analysis for global runoff changes, which are also shown in Figure O.2. The figure shows (a) increases (in the multi-

TABLE 5.3 Median Runoff Sensitivities Relative to the Period 1971-2000 per Degree Global Average Temperature Change, the Equivalent Standard Error of the Median, and Fraction of Positive Minus Negative Estimates (FPN) for the U.S. Hydrologic Regions and Alaska

Hydrologic Region	Median (%)	Equivalent Std Error[a] (%)	FPN[b]
1 New_England	1.7	0.6	0.47
2 Mid-Atlantic	1.0	0.7	0.29
3 South_Atlantic-Gulf	-2.0	1.1	-0.18
4 Great_Lakes	1.7	1.3	0.15
5 Ohio	1.5	1.0	0.24
6 Tennessee	-2.8	1.1	-0.24
7 Upper_Mississippi	1.2	1.2	0.09
8 Lower_Mississippi	-6.4	1.7	-0.53
9 Souris-Red-Rainy	1.8	1.6	0.21
10 Missouri	-2.0	1.1	-0.12
11 Arkansas-White-Red	-8.4	1.4	-0.56
12 Texas-Gulf	-7.4	2.1	-0.56
13 Rio_Grande	-12.2	2.3	-0.47
14 Upper_Colorado	-6.3	1.7	-0.62
15 Lower_Colorado	-6.1	2.4	-0.44
16 Great_Basin	-5.1	1.4	-0.44
17 Pacific_Northwest	1.2	0.9	0.24
18 California	-3.3	2.2	-0.29
19 Alaska	9.3	0.9	0.91
Colorado[c]	-6.2	1.6	-0.59

NOTE: The runoff sensitivity is taken as the average sensitivity (as described below) for IPCC AR4 GCM output from A2, A1B, and B1 simulations for each basin as derived from 30-year runoff averages centered on the years for which the global average (for each of 23 models and the 3 emissions scenarios) were 1°C, 1.5°C, and 2°C minus the 30-year model average runoff for 1971-2000, divided by the global temperature change. The total number of model and temperature change pairs was 68, consisting of 23 models for 1°C temperature change, 23 models for 1.5°C temperature change, and 22 models for 2°C temperature change. The equivalent standard error is an estimate of the standard error of the median (Hojo,1931), where "equivalent" pertains to the fact that the estimate is strictly correct only for normally distributed variates.

[a] $1.25\sigma/\sqrt{n}$ where σ is estimated as one-half the inner 68-percentile range.

[b] Fraction of pairs (of 68 total) with inferred positive runoff sensitivities minus fraction with inferred negative sensitivities. FPN therefore ranges from –1 when 68 inferred sensitivites are negative, to 1 when all are positive.

[c] Upper and Lower Colorado combined.

model median runoff sensitivity) across almost all of Eurasia, high-latitude areas, and most of Australia; (b) decreases across much of the United States, southern Europe, and Africa; and (c) (not shown) strong agreement as to the direction of runoff changes at high latitudes and southern Europe, little agreement across most equatorial areas, and only modest agreement over much of the United States (for the United States, see also the fraction positive minus fraction negative in Table 5.3).

Finally, there are some apparent differences in the pattern of runoff changes shown for the United States in Figure O.2 relative to the re-plots of Milly et al. (2005) results in USCCSP (2008b). In general, those results showed more of an east-west divide in runoff changes across the United States, with modest increases in the east, and decreases across most of the west, which were most severe in the Colorado River and Great Basins. We extracted from our multi-model suite the same 12 models analyzed by Milly et al. (from the approximately 20 that were available to us), and analyzed the results in the same way as did Milly et al. (2041-2060 minus 1900-1970 means). When so analyzed, the results were quite similar to those shown in USCCSP (2008a). The differences in patterns in our results as compared with USCCSP (2008b) and Milly et al. (2005) apparently come about because of (a) differences in the 1900-1970 based period (used by Milly et al. [2005] versus 1971-2000 base period that we used, and (b) differences in the number of models considered (nearly double in our analysis relative to those available to Milly et al. (2005). As in Milly et al. (2005) and USCCSP (2008a), we have limited our analysis to annual runoff volumes.

Our overall conclusion therefore is that streamflow in many temperate river basins outside Eurasia will decrease as global temperature increases, with the greatest decreases in areas that are currently arid or semi-arid. Streamflow across most of the United States will decrease, although there is considerable disagreement among models aside from the Southwest, where most models project decreases, and Alaska, where most models project increases. Runoff sensitivities are approximately constant (per degree of global warming) in the range 1-2°C warming. There is strong agreement among models that runoff in the Arctic and other high-latitude areas, including Alaska, will increase.

5.4 FIRE

The impact of climate change on the frequency of large fires and extent of area burned by wildfires depends mainly on the type of vegetation (fuel) and the future weather and climate. In many regions that have been examined, the projected climate change will cause large changes in wildfire frequency and extent. In a broad sense, wildfires will increase in regions that are dominated by forests that are already prone to fire in the current climate, while warming (without a sufficient increase in precipitation) will cause a decrease in wildfires in some shrub and grassland (fuel limited) regions that are prone to fire in the present climate.

The probability of wildfire depends on the availability of fuel, the moisture content of the fuels, and the likelihood of ignition. Ignition of most large wildfires is by lightning, and analyses of the global lightning strike and fire frequency data suggests that ignition is a limiting variable for wildfires in only a few areas on the planet (Krawchuck et al., 2009). Once fire is initiated, the extent of a fire burn depends on fuel distribution and moisture content, wind, and topography.[1] Hence, vegetation type, weather, and the antecedent weather history (climate) are the main environmental parameters that determine the frequency and extent of wildfire.

Comprehensive spatial and temporal records of wildfire and of climate variables are available for the western United States. Analyses of these data have led to a conceptual model that links wildfires to climate variability and vegetation type. For example, Westerling and Bryant (2008) found that in California and surrounding states, wildfires in "wet regions" (defined as when climatological soil moisture exceeded 28% of capacity) tended to be reported as forest fires, while wildfires in dry regions tended to be reported as grassland and shrub fires. They further found that "wet" regions were much more prone to fire in months when the maximum temperature exceeded 23°C. Hence, they called these wet regions "energy limited"—positing the probably of fire increases with increasing temperature because the moisture content of the fuel also decreases with increasing temperature. They noted that wildfires in dry regions were *positively* correlated with the *previous* year's precipitation—a result previously known from the study of Swetnam and Betancourt (1998)—and posited that increased moisture in the previous year allowed for an increase in biomass production in grassland environments that was then available to burn in the following summer. Hence, they called these dry regions "moisture limited"—although in the sense of moisture limiting the mass of fuel available to burn, rather than the incidence of fire. Mid-elevation regions in the Sierra were classified as "energy limited," while southern California was classified as "moisture limited."

Littell et al. (2009) develop empirical models for wildfire area burned throughout the western United States. For predictors, they use seasonal and antecedent climate variables including temperature and precipitation, and the Palmer Drought Severity Index (PDSI) as a proxy for soil moisture. Littell et al. formulate empirical models for zones of similar vegetation and

[1] Fire suppression by management is found in many studies to explain much less variance in total area burned than does the variability in weather and climate. For example, climate variables account for more than 80% of the variance in annual area burned in the boreal forests of western North America between 1960-2002 (Balshi et al., 2009); climate variables and vegetation type account for two-thirds of the variance in area burned in the western United States (Littell et al., 2009).

geography (ecoprovinces) that encompass most of the western United States[2] They develop empirical models using the interannual records for the time period 1977-2003. The empirical models account for, on average, two-thirds of the variance of the wildfire area burned in the 16 ecoprovinces that comprise the western United States (cross-validated). Littell et al. find that in most of the forested western United States, wildfire area burned is best explained by dry and warm conditions in the seasons immediately preceding the fire, presumably because such climate conditions enhance evapotransporation and reduce fuel moisture. In contrast, fire area burned in the southwest United States and in other arid ecoprovinces is strongly dependent on the increased precipitation or positive PDSI (moisture) in seasons preceding the fire season, consistent with the long-standing hypothesis that extra biomass is produced in the seasons prior to the season that experiences an unusually large area burned in these regions.

Together, the studies by Littell et al. (2009), Westerling and Bryant (2008), and others demonstrate that the role of climate variability in regulating area burned is mainly a complex function of the climate (temperature, precipitation, and wind), vegetation type, and orography. Nonetheless, the climate and fire data for the United States are sufficient in spatial and temporal coverage to formulate sensible and remarkably skillful predictive models or incidence of wildfire and wildfire area burned. An example is provided in Figure 5.6 from Littell et al. (2009: Figure 1), which shows the observed wildfire area burned for the entire historical record (1916-2003) and that predicted by empirical models that are trained and cross-validated using the data from 1977-2003.

The results from the aforementioned studies strongly suggest that empirical models of wildfire should provide a skillful projection of how climate change due to increasing greenhouse gases will impact wildfire across wide regions of the globe, including the western United States and Canada. In these regions, much of the area burned is due to large wildfires that result mainly from climate variability, with in-season summer temperature, precipitation, and soil moisture being the predominant controls on fire. The probability of summer averaged temperature and humidity is relatively straightforward to quantify (precipitation perhaps a bit less so), and empirical and hydrologic models needed to project summer soil moisture are quite mature.

To date, only a few studies have examined how the projected climate

[2]An ecoprovince is a large-scale region that is characterized by a common vegetation distribution, orography, and landscape structure (Bailey 1995). Most of the western half of the continental United States is covered by 16 of Bailey's ecoprovinces.

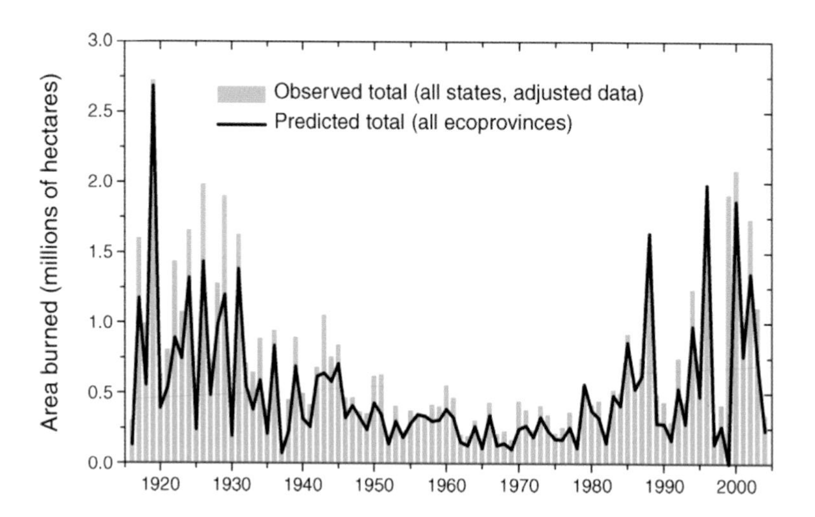

FIGURE 5.6 Observed and reconstructed wildfire area-burned for 11 western U.S. states (bars) and reconstructed (line) for the period 1916-2004. Source: Littell et al., 2009: Figure 1)

changes will impact wildfires. Shown in Figure 5.7 are the projected changes in the probability of large (> 200 ha) wildfires in 2070-2090 compared to the reference period 1961-1990 from one such study (Westerling and Bryant, 2008) that used the output from two of the AR4 climate models (GFDL and PCM) that were forced with two of the emission scenarios (B1 and A2). There are two important lessons to be learned from this plot. First, two different climate models give two very different answers for California: the PCM projects a subtle change in large wildfires everywhere, while the GFDL model projects large increases in the probability of large wildfires throughout northern California. These differences are mainly due to differences in the projected mean climate: the GFDL model features large temperature increases and precipitation decreases throughout California, while the PCM has modest temperature increases and subtle changes in precipitation. Hence, the full suite of AR4 climate models must be employed to project the probability of wildfire changes—throughout the globe. Second, although both models show increasing temperature and decreasing precipitation throughout the state of California, the likelihood of a large fire *increases* in northern California (especially in the Sierra) and *decreases* in southern California in both

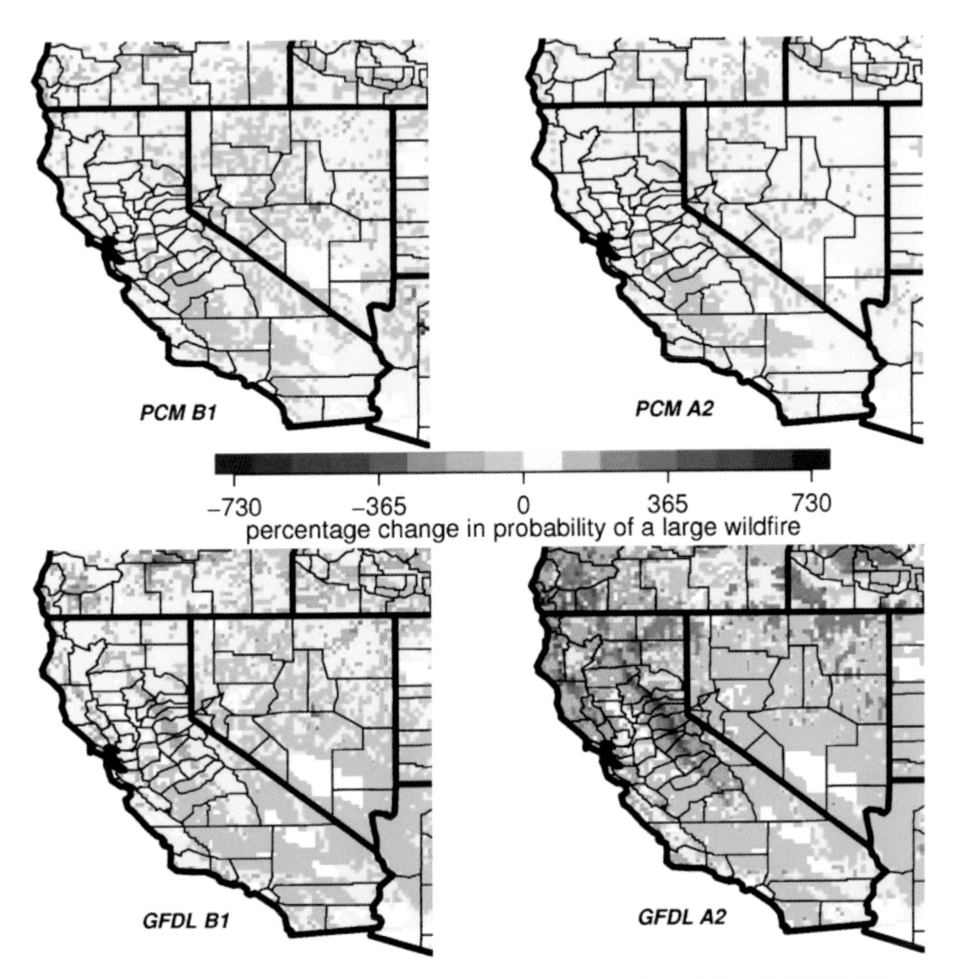

FIGURE 5.7 The projected changes in the probability of large (> 200 ha) wildfires in 2070-2090 compared to the reference period 1961-1990 using the output from two of the AR4 climate models (GFDL and PCM) that were forced with two of the emission scenarios (B1 and A2). Source: Westerling and Bryant (2008: Figure 7).

models. This reflects the fundamental differences in the climate impacts on fire regime in these regions due to differences in vegetation type.

The change in the probability of the wildfire area burned in the western United States due to increased greenhouse gases can be estimated using the methods outlined in Littell et al. (2009), modified to use temperature and precipitation as the predictor variables. Shown in Figure 5.8 is the change

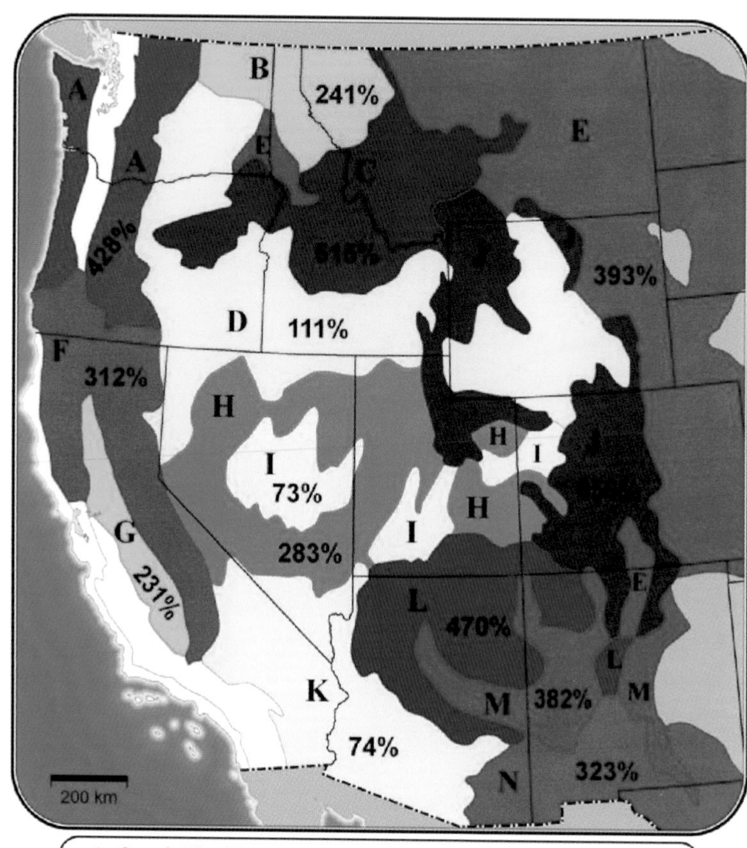

FIGURE 5.8 Map of changes in area burned for a 1°C increase in global average temperature, shown as the percentage change relative to the median annual area burned during 1950-2003. Results are aggregated to ecoprovinces (Bailey, 1995) of the West. Changes in temperature and precipitation were aggregated to the ecoprovince level. Climate-fire models were derived from NCDC climate division records and observed area burned data following methods described in Littell et al. (2009). Sourcce: Figure from Rob Norheim.

in median annual averaged area burned in the western United States due to a 1°C global average temperature increase, using the temperature and precipitation patterns from the corresponding figures in Chapter 4. The increase in median annual area burned ranges from 73% to over 600%, depending on the ecoprovince. Aggregating all 14 ecoprovinces in which fire is most sensitive to temperature variations, the net area burned by the median fire increases from 572,000 ha for the reference period 1970-2003, to 1,800,000 ha with a 1°C global average temperature increase.

Other investigators using different climate models and a variety also find thata climate change will increase the risk of wildfire throughout the western United States, so long as fuel is not limiting. For example, Spracklen et al. (2009) used output from the GISS AGCM coupled to a slab ocean forced by the A1B emission scenario with an empirical model very similar to that used by Littell et al. (2009); they concluded that climate changes projected for the mid-21st century would result in a 54% increase in total annual area burned compare to that in the late 20th century, and burn areas doubling in the Rockies and Pacific Northwest.

Similar studies have been done to examine how climate change will affect forest wildfires in other regions of the world, including Australia, New Zealand, the southern Mediterranean, and in boreal forests of Canada (IPCC, 2007d). In general, these studies report wildfires will increase over the course of the century due to projected climate changes from one or more of the AR4 climate models using one or more of the marker emission scenarios.

As the time horizon increases, a further complication arises in projecting wildfires because of the importance of fuel availability and quality (moisture content) for determining the likelihood and size of wildfires. Systematic climate changes will inevitably alter the distribution of the vegetation, and significant changes in the vegetation would greatly affect the potential for wildfires and the wildfire area burned. For example, if climate change further dries already arid grasslands, then the grasslands will wither to deserts and fire will no longer be supported. On the other hand, if climate change changes the distribution of pest and pathogens such that forests become diseased, then additional fuel will be made available due to forest dieback (see, e.g., discussion of the bugs in the BC/AK spruce) and fires will be enhanced until the extra fuel is exhausted and replaced by woodlands or grasslands. To date, all of the statistical models for wildfire do not admit changes in vegetation type due to the ecosystem response to climate change.

5.5 INFRASTRUCTURE

Infrastructure and Society

Infrastructure provides a broad range of human, environmental, and economic services including buildings, transportation, waste removal and treatment, water lines, communications, and electric power grids designed to improve and sustain our society and our quality of life (Kirshen et al., 2008b). The importance of infrastructure to an industrialized economy is reflected in the magnitude of its investments: in 2007, for example, the U.S. Bureau of Economic Analysis valued the stock of all public non-defense fixed assets in the United States at approximately $8.2 trillion (Heintz et al., 2009).

Changing risk of heat waves and droughts, storms and floods, and rising sea levels are just a few of the hazards climate change poses to infrastructure (Kraas, 2008). These extreme events confront human systems and constructs with weather conditions far outside their accustomed range (Wilbanks et al., 2007). The exposure of society and infrastructure to climate change is exacerbated by the fact that much of the world's population growth over the next few decades is likely to occur in urban areas, as they double in size from 3 to more than 6 billion from 2007 to 2050 (UN, 2008). At the same time, however, this sector has a greater ability to adapt to changing conditions than many others, as humans created the systems being affected by climate change.

In the past, other human stressors on infrastructure—from rapidly expanding urban populations to deteriorating and aging systems, many of them operating on time scales of years rather than decades—meant that climate change was rarely considered as a key influence on infrastructure costs. However, climate changes in recent decades have increased awareness regarding the risk of significant and costly impacts to infrastructure: whether from melting permafrost in the Arctic affecting roads, pipelines, and buildings, or from storm surges in the Gulf potentially flooding and damaging homes, cities, highways, and rail lines (USGCRP, 2009; Figure 5.9).

Infrastructure response to climate change is not always continuous but can be step-wise as system failure may occur above a certain threshold: whether an underpass tends to flood when more than 2.5 inches of rain falls in a 24-hour period, for example, or train rails warp only when temperatures rise above a given threshold. Local conditions can further magnify the susceptibility of cities to climate-related impacts (Wilbanks et al., 2007). High-latitude and coastal areas are uniquely vulnerable to climate change.

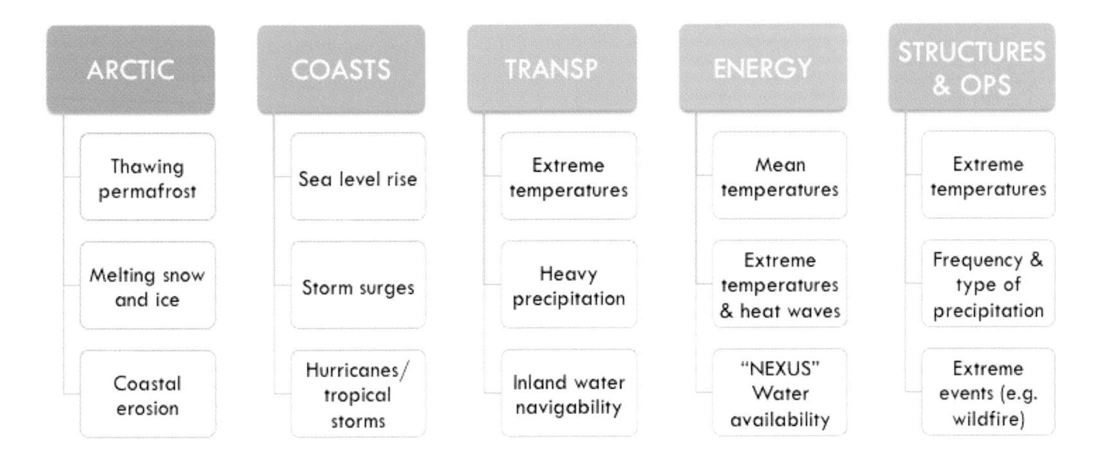

FIGURE 5.9 Climate drivers of impacts on vulnerable regions (Arctic, coasts) and sectors (transportation, energy, and buildings).

In the Arctic, protective shore ice is forming later in the year and breaking up earlier, allowing autumn storms to batter the unprotected coasts, eroding the coastline and endangering its inhabitants (ACIA, 2005). Travel over frozen ground has been cut from 7 to 4 months per year, isolating many communities (USGCRP, 2009). Under an approximate global average temperature change of 2°C, melting of the continuous permafrost area in the Arctic characterizes infrastructure across nearly half of the Arctic land area as "high risk" and significant proportions of Arctic coastline as susceptible to significant erosion (ACIA, 2005).

At the same time, however, decreasing sea ice in the Arctic could open the Northern Sea Route, greatly reducing the distance required to transport goods between Asia, North America, and Europe. Under the same 2°C global temperature increase, the navigation season could last up to 3 months per year before the end of the century (ACIA, 2005).

Coastlines

Low-lying coastal delta areas, many of them home to mega-cities and other densely populated areas, are also at risk. Impacts from coastal erosion and flooding can be driven by sea level rise and storm surge as well as by land-use decisions and other processes characteristic of, and which can only be determined for, a given location. Two examples of coastal impacts are discussed here: New York City and the Sacramento-San Joaquin Delta of California.

New York City (NYC) is both a mega-delta and a mega-city located along the eastern seaboard of the United States. Return times of current 100-year and 50-year coastal storms for NYC are strongly correlated with global and local sea level rise (Figure 5.10, based on Kirshen et al., 2008a). Based on historical records, the projected frequency of coastal storms can be translated into an estimate of the number of buildings in downtown NYC that would be damaged by storms characterized by their current FEMA-based return times (Figure 5.11, based on NYCOEM, 2009).

New York City has responded to these risks by considering a range of adaptation options to reduce the vulnerability of its critical public and private infrastructure over time (NPCC, 2009). One option envisions creating a dynamic process by which FEMA systematically redraws its 100-year flood maps as sea level rises, so insurance markets could appropriately spread the risk. Other plans envision constructing retractable barricades to protect New York from an impending storm—much like the barrage on the Thames that protects London (where plans are underway to build a new barrier based on observed and estimated sea level rise); the barrier that protects St. Petersburg, Russia; and multiple devices now being installed to protect Venice at the border of its lagoon and the Adriatic Sea. The city is considering adjust-

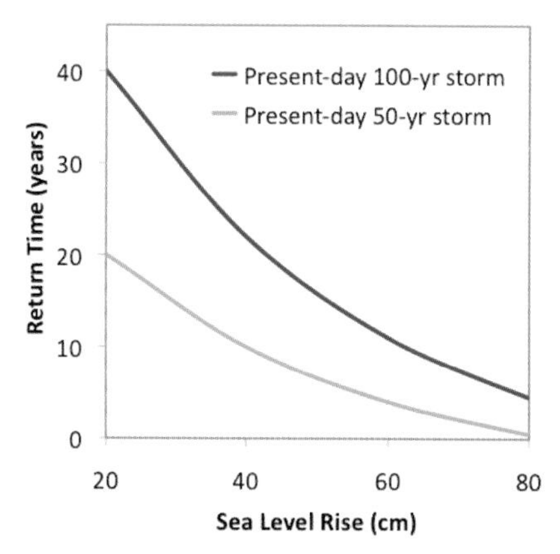

FIGURE 5.10 Projected return time of coastal storms relative to future sea level rise for New York City (Based on Kirshen et al., 2008a).

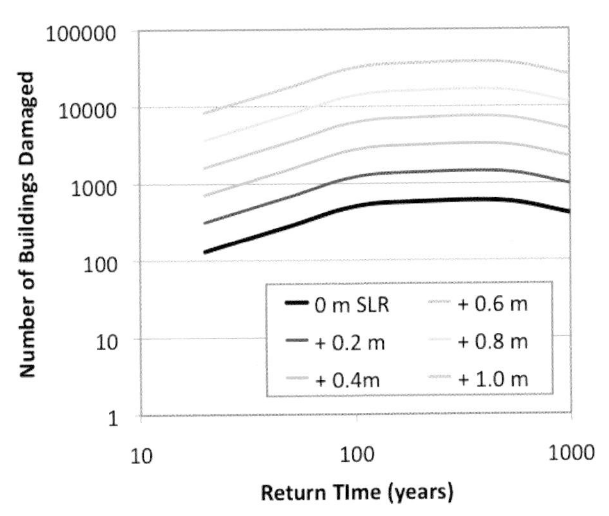

FIGURE 5.11 Expected number of buildings that would be damaged in the present-day downtown New York City, for various types of storms for different sea level trajectories. Based on information presented in Figure 5.10 that includes more detail for intermediate return times and damage estimates from NYCOEM (2009).

ments in building codes and zoning regulations designed to maintain the specific "climate (risk) protection levels" embedded in existing legislation and operating procedures.

On west coast of the United States, the Sacramento-San Joaquin Delta region is home to farms, roads, oil and gas lines, and extensive housing developments (Figure 5.12). The entire Delta is below mean water level, much of it more than 15 feet below. The magnitude of impacts from sea level rise and storm surge breaching of the levees currently protecting the region is estimated at tens of billions of dollars over the next few decades alone (CALFED, 2009).

The California Bay-Delta Authority is exploring options to divert large amounts of water away from the region and/or cut losses in certain areas by deciding ahead of time what to replace and what to leave. Other regions, such as the metropolitan Boston area, have also identified areas likely to flood, and they are exploring the efficacy of prevention and adaptation options such as preserving coastal wetlands and relocating water treatment plants away from the coasts (Kirshen et al., 2008a).

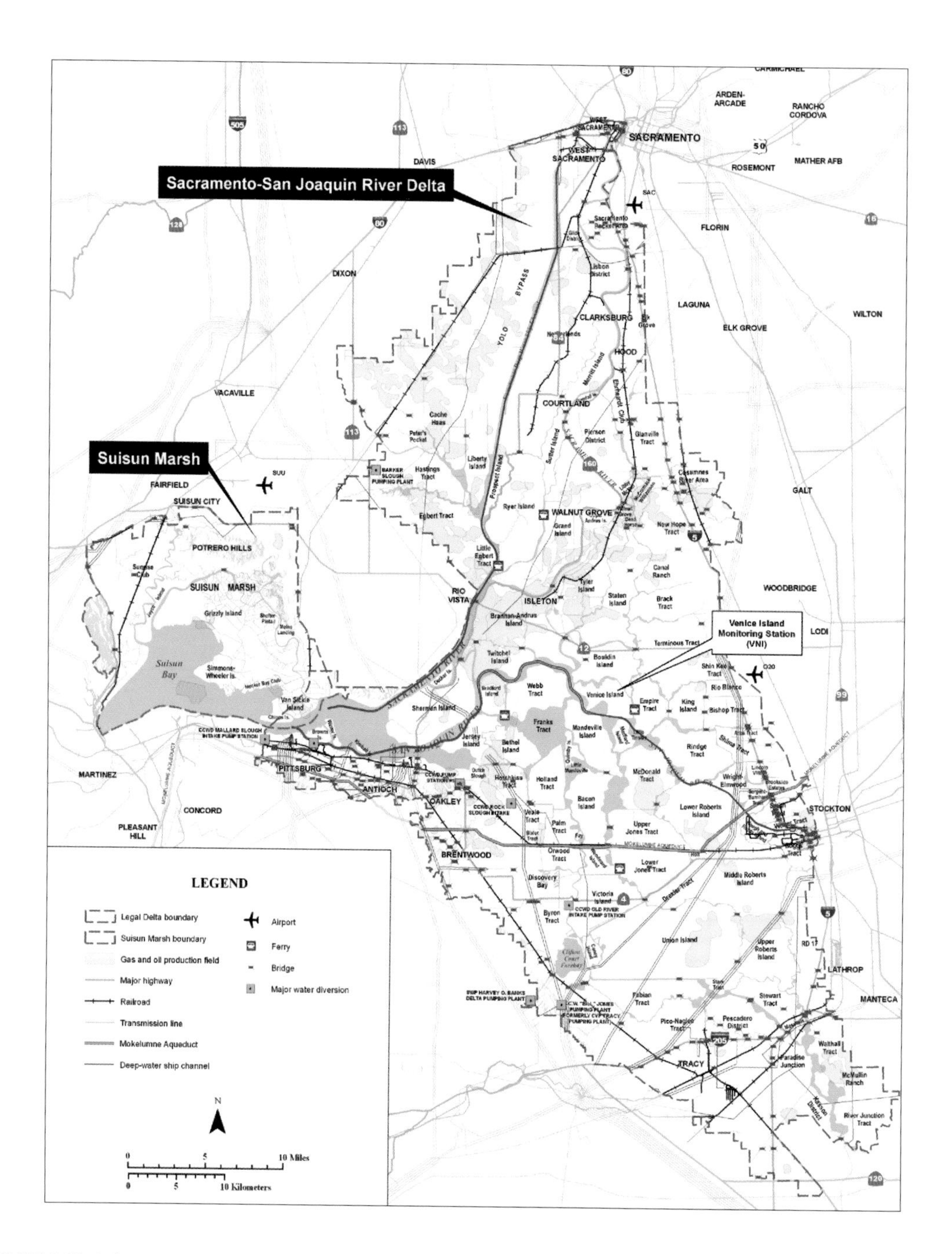

FIGURE 5.12 Infrastructure at risk from sea level rise, storm surge, and levee failure in the Sacramento-San Joaquin Delta includes highways, roads, and rail lines; gas and oil production fields and pipelines; and homes and farms. Source: CALFED (2009).

Transportation, Buildings, and Structures

Impacts on transportation and built structures are primarily the result of changes in temperature and precipitation beyond what the materials used to build the roads, rails, bridges, and buildings were designed to withstand. Transportation can also be impacted by extreme temperatures, heavy rainfall events, and persistent freeze/thaw conditions.

Climate patterns based on the past century, traditionally used by transportation planners to guide their operations and investments, may no longer provide a reliable guide to the future (NRC, 2008). In many northern cities, for example, standard building procedures do not require air conditioning. As temperatures increase, many of these may require expensive retrofits so they can continue to be used in extreme heat.

Across the United States, the average number of days per year with very heavy precipitation has increased significantly over the past 50 years, with the largest increases (58 and 27 percent, respectively) occurring in the Northeast and Midwest regions (USGCRP, 2009). This trend is expected to continue in the future across many regions, as extreme precipitation events increase (Tebaldi et al., 2006). In both the Midwest and Northeast, climate change is also expected to increase the amount of rain that falls in winter and spring, when frozen and/or saturated ground increases flood risk, while decreasing summer and autumn rainfall (Hayhoe et al., 2008, 2010). Increased frequency of winter and spring rainfall, combined with more frequent precipitation extremes, could lead to higher peak streamflows, particularly under higher temperature change and toward the end of the century (Cherkauer et al., 2010). Infrastructure impacts in Chicago are highlighted in Box 5.3.

Energy

Climate change is likely to affect energy demand, production, and reliability (Wilbanks et al., 2007). Although warmer winter temperatures are expected to reduce demand for heating energy, observed correlations between daily mean near-surface air temperature and electricity demand suggest that warmer summer temperatures and more frequent, severe, and prolonged extreme heat events will likely increase demand for cooling energy, particularly as use of air conditioning increases around the world (see Figure 5.14).

The impact of increasing temperatures on air conditioning demand could have a disproportionately large effect in already heavily air-conditioned

BOX 5.3 INFRASTRUCTURE IMPACTS IN CHICAGO

One of the few detailed analysis of potential economic costs of a range of climate change scenarios for transportation infrastructure has been conducted for the city of Chicago (Hayhoe et al., 2010).

Chicago has experienced impacts from weather extremes, when city streets buckled and rail lines warped during the record-breaking 1995 heat wave; or in 1996, when 17 inches of rain fell in a single 24-hour period, with a total estimated cost of flood losses and recovery of $645 million (Changnon et al., 1999).

Impacts of changes in mean and extreme temperature and precipitation on Chicago's public transit, maintenance and construction of roads, highways, bridges, and canals, and airport operations are estimated at $3.3 million under an approximate 2°C change in global mean temperature and $5 million under a 4°C change (Figure 5.13a). Estimates for building-related expenses under higher temperature change are likely to be 10 times greater (conservatively $20 million) than under lower (Figure 5.13b). Values are based on only the sensitivities city officials were aware of based on past conditions and events.

regions, such as the Southwest (Miller et al., 2008). Projected increases in peak electricity demand in particular raise concerns regarding electricity shortages, as risk of shortages may increase with both mean and extreme temperatures.

Electricity generation by fossil and nuclear power plants requires a large supply of water, which may become limited during periods of drought. Although thermoelectric power plants in the United States have not yet had to reduce operations due to insufficient water supply, future water shortages are expected to affect power production in a number of states (Wilbanks et al., 2007). The reliability of energy supplies can be affected by the frequency of droughts and accompanying heat waves, which increase the temperature of the cooling water beyond what the plant may be permitted to release.

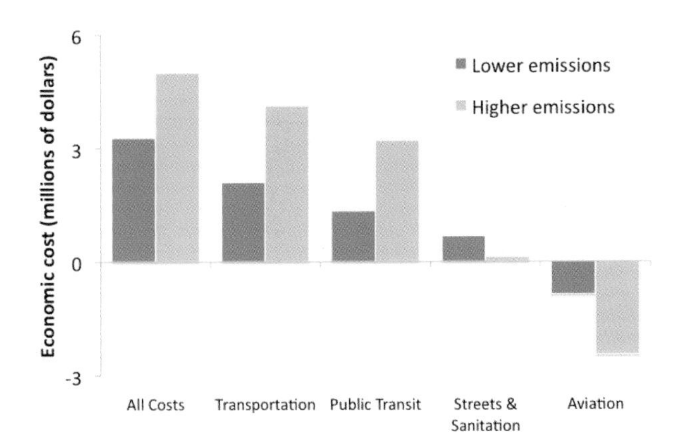

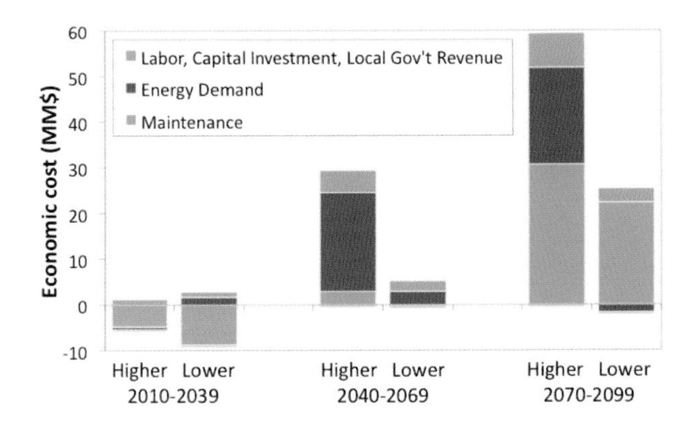

FIGURE 5.13 Economic cost of changes in mean and extreme temperature and precipitation on (a) transportation and (b) all city infrastructure in Chicago, by 2070-2099 as compared to 1961-1990. Source: Hayhoe et al. (2010).

Hydropower, which accounts for 75% of renewable power generation in the United States and the majority of all electricity supply for many nations in South America and Northern Europe, is the most directly sensitive to water availability. The relationship between hydropower generation and precipitation tends to be proportional, with a 1% change in precipitation resulting in approximately 1% change in power generation. However, projecting future climate effects on hydropower generation is limited by uncertainties in precipitation projections. For example, recent projections of hydropower generation in the Sierra Nevada Mountains of California ranged from –10% to +10%, depending on the climate model used (Vicuña et al., 2008). Similarly, projections for the Pacific Northwest ranged from +2% to –30%, depending on the precipitation projections (Markoff et al., 2008).

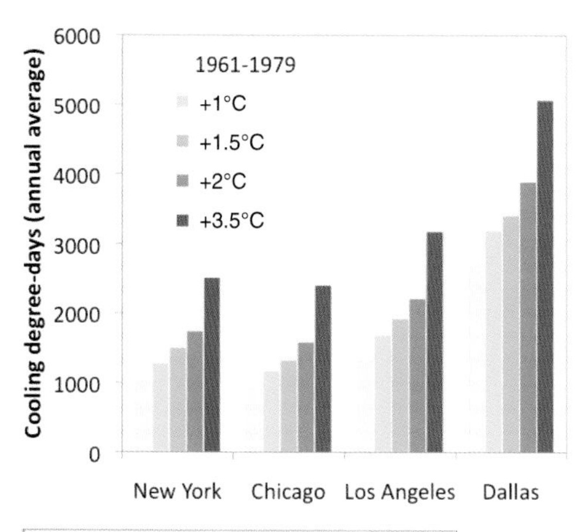

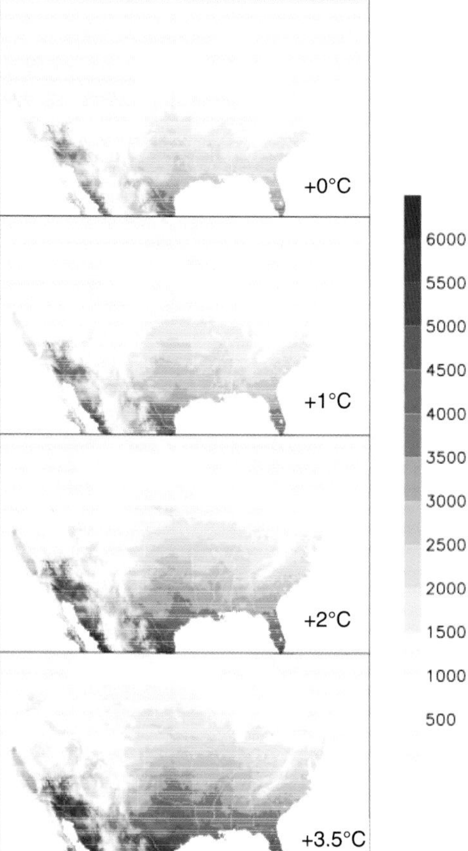

FIGURE 5.14 Heating/cooling degree-days. Projected changes in heating and cooling degree-days for: (a) four U.S. cities by global temperature change, and for (b) the continental United States (Based on USGCRP, 2009). A first approximation for heating and cooling demand is provided by estimates of projected changes in heating and cooling degree-days (Rosenthal et al., 1995; Amato et al., 2005). Projections of the increase in cooling energy use range from 5% to 20% per 1°C increase in temperature for residential buildings and about 9 to 15% per 1°C for commercial buildings. The relationship between cooling energy demand and temperature is nonlinear, with greater increases in demand occurring at higher temperatures (Wilbanks et al., 2007).

5.6 HEALTH

Increasing temperatures alter the risk of direct and indirect weather-related impacts on human health, from cardiovascular and respiratory illnesses to infectious diseases (Patz et al., 2005). Quantifying the impact per degree of global temperature change, however, is complicated by confounding factors such as human behavior and socioeconomic conditions that affect exposure, transmission, and other aspects of risk (Patz et al., 2005). Here, we discuss three main aspects of health-related risks likely to be affected by climate change: heat-related illness and death, vector-borne disease, and health concerns related to poor air and water quality.

Heat-Related Illness and Death

Temperature extremes such as heat waves and periods of extreme cold are known to produce elevated rates of illness and death (McGeehin and Mirabelli, 2001). Together, these accounted for 75% of all deaths due to natural disasters from 1979-2004 (Thacker et al., 2008). From the 1970s through the 1990s, heat-related mortality in the United States declined due to acclimatization and increased use of air conditioning, then flattened out during the past decade (Sheridan et al., 2009).

In the future, extreme heat days and heat wave frequency, intensity, and duration is projected to increase with global mean temperature, while the frequency and intensity of winter cold is projected to decrease (Tebaldi et al., 2006; IPCC, 2007a). Under a 2°C increase in global mean temperature by end-of-century, for example, the average number of days per year with maximum temperatures exceeding 38°C or 100°F across much of the south and central United States is projected to increase by a factor of 3. Under a 3.5°C increase, the number of days is projected to increase by 5 to nearly 10 times historical levels (Figure 5.15).

Although the response of illness and death rates to changing heat extremes can be modified by acclimatization and adaptation strategies (Ebi et al., 2004), it is clear that risks of heat-related mortality increase with temperature, while cold-related mortality risks decrease (Gamble et al., 2008). Prolonged periods of extreme heat with little relief at night can have devastating effects on urban populations (Basu and Samet, 2002), increasing the risk of both illness and death due to heat stress (Martens, 1998; McGeehin and Mirabelli, 2001; Schär et al., 2004).

Temperature-related mortality is strongly linked to a wide variety of social, economic, and behavioral factors, even in industrialized nations such

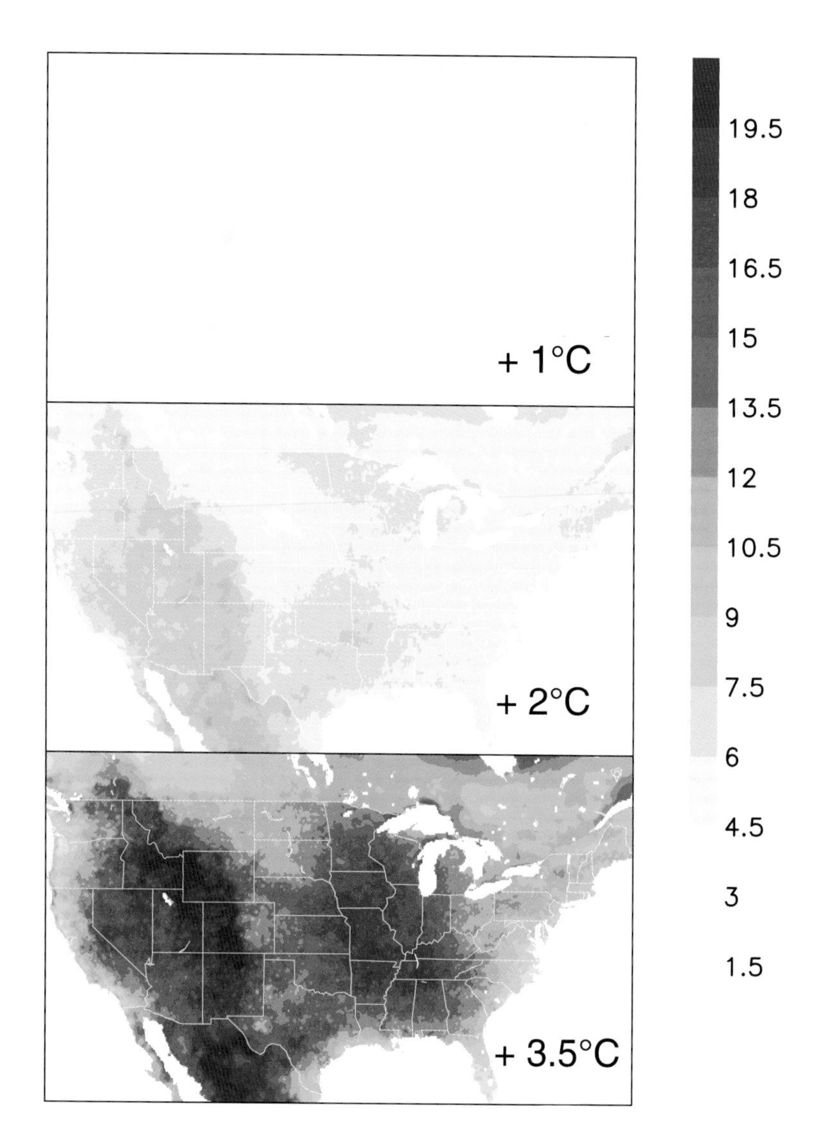

FIGURE 5.15 Projected increase in heat wave duration index, in number of days per event. Defined after Frich et al. (2002) and Tebaldi et al. (2006) as the longest period each year with at least 5 consecutive days during which daily maximum temperature is at least 5°C higher than the climatological (1961-1990) average for that same calendar day. Projected changes are for 20-year periods during which mean global mean temperature increased by 1°C, 2°C, and 3.5°C, respectively relative to the 1961-1979 average.

BOX 5.4 HEAT-RELATED MORTALITY IN CHICAGO

The Chicago heat wave of 1995 is estimated to have been responsible for 692 heat-related deaths within the city of Chicago itself (Kaiser et al., 2007). As global mean temperatures increase, 1995-like conditions are projected to become more frequent in Chicago. Under a 2°C change in global mean temperature, annual average mortality rates are projected to equal those of 1995. Under a 4°C change in global mean temperature, annual average mortality is projected to be twice 1995 levels, with 1995-like heat waves occurring as frequently as three times per year (Hayhoe et al., 2010).

---

as the United States (e.g., Basu, 2009; Ekamper et al., 2009). Since observed mortality rates are more responsive to changes in high temperature than low temperature extremes, at the global scale an overall increase in heat-related deaths is expected to exceed the projected decrease in cold-related deaths (Medina-Roman and Schwartz, 2007).

Several recent severe heat waves have focused public attention on the health risks associated with extreme heat, including a 1995 heat wave in Chicago (see Box 5.4) and the unprecedented European heat wave of 2003, estimated to be responsible for approximately 70,000 excess deaths across 16 European countries (Robine et al., 2008). Past events such as these can be used as case studies for evaluating the potential impacts of future climate conditions. For example, observations of the health effects of a heat wave in California in July 2006 showed that heat-related mortality increased 9% for every 5.5°C increase in apparent temperature, with that specific event causing an estimated 160-505 deaths, a 6-fold increase in the number of heat-related visits to the emergency room, and 10-fold increase in heat-related hospitalizations (Knowlton et al., 2009; Ostro et al., 2009).

Using an analogue approach, Kalkstein et al. (2008) estimated the number of heat-related deaths that might occur in U.S. cities if they experienced conditions similar to the severe Paris heat wave of 2003. Compared to the hottest summers on record, heat-related deaths in St. Louis and New York are projected to increase by 29% and 155%, respectively. Other cities, including Washington, Philadelphia, and Detroit, could experience 2-11% more heat-related deaths, compared to their hottest summers on record. This range of response illustrates the variation among cities in their sensitivity to extreme heat, which might result from factors such as baseline climate conditions (cooler cities may be more susceptible to heat), demographic patterns, and acclimatization measures such as air conditioning.

Heat watch-warning systems presently in operation in major U.S.,

Italian, and Canadian cities have already been shown to save a number of lives when coupled with effective intervention plans (Ebi et al., 2004). These systems, coupled with well-developed intervention activities, represent an important way to lessen the potential for drastic increases in heat-related mortality in coming decades due to climate change.

Long-Term Impacts of Heat Stress

Many of the impact studies summarized in this section and elsewhere throughout this report focus on future projections for the next few decades; nearly all of the impact studies in the literature, regardless of focus area, confine estimates of potential impacts to this century. This limits quantification of impacts based on GCM simulations driven by the SRES scenarios to those associated to global mean temperature changes of 4°C or less, depending on the GCMs and scenarios used to generate future projections.

One recent study, however, attempts to extrapolate beyond this threshold by using long-term simulations reaching a global temperature change of 10°C relative to 1999-2008 (Sherwood and Huber, 2010). Recognizing that most of the impacts studied to date depend on assumptions about future conditions that cannot be reliably simulated with our current state of knowledge, the authors instead focus on an impact that depends on a basic physiological threshold: specifically, that associated with the need of the human body to dissipate the heat it generates. This can happen only if the temperature to which the skin is exposed is lower than the skin temperature itself.

The authors consider future statistics of annual maximum wet bulb temperatures exceeding average human skin temperature of 35°C as a measure of heat stress on humans and other mammals, and illustrate how, assuming carbon emissions and climate change continue unchecked beyond this century, the magnitude of land area that could become uninhabitable due to heat stress could be greater than that lost to sea level rise over the same time frame. Although exact projections of changes in inhabitable land area over multiple centuries are sensitive to key uncertainties in carbon cycle, climate sensitivity, and ocean-atmosphere dynamics discussed in Chapters 3 and 4, this study raises an important point: namely, that hard-wired, physiological limitations on human welfare have not previously been acknowledged in evaluating the long-term risks associated with a given pathway of carbon emissions.

Pests and Disease

Changes in climate can affect the spread of illness and diseases such as malaria and West Nile Virus, carried by animal hosts and mosquito, midge, fly, or tick vectors. Increasing temperatures and changes in precipitation patterns can accelerate current trends of increased risk of exposure to disease by expanding the geographic range and/or populations of the vectors (Ogden et al., 2008). In addition to the potential for more vectors to be present, warmer temperatures increase the virus replication rate, increasing the efficiency of virus transmission (Mellor, 2000).

Despite initial projections suggesting the potential for dramatic future increases in the geographic range of malaria and other infectious diseases under climate change (e.g., Martin and Lefebvre, 1995), subsequent studies have highlighted how the complexity of the systems—involving viral, bacterial, plant, and animal physiology, as well as sensitivity to changes in climate extremes, including precipitation intensity and temperature variability—challenges attempts to resolve the influence of historical climate change on observed trends in disease incidence and develop future projections specific to any particular level of global temperature change. This is even true for what many previously considered the poster child for the influence of climate change on infectious diseases, the spread of malaria throughout the East African highlands (e.g., Pascual et al., 2006).

If historical contributions are difficult to resolve, prediction of future trends is even more so. Most recent projections suggest that the ranges of malaria and other diseases may shift, but increases in some areas will likely be balanced out by decreases in others, resulting in little net increase in area (Lafferty, 2009). The potential for genetic mutation, the influence of changing agricultural techniques, and increased transportation and trade between regions previously not in regular contact are all confounding factors that have been cited as complicating the influence of climate change on a given disease (Gould and Higgs, 2009).

It is also important to remember that good sanitation and insect control programs can limit disease spread even under suitable climate conditions (Lafferty, 2009). In 1882, malaria in the continental United States extended from the Gulf of Mexico to Minnesota. Draining of swamps, improved pesticides, and better management of water resources contributed to the eradication of malaria in the United States shortly after World War II (Oaks et al., 1991). Replicating methods proven successful in the past may be one way to reduce the risk associated with climate change and vector-borne disease.

Air and Water Quality

Increasing temperatures and changing precipitation intensity can also affect air and water quality. Densely populated areas with warm summers tend to have high levels of ozone precursor emissions (nitrogen oxides and volatile organic compounds). These precursor species react in the presence of sunlight, with key reactions proceeding faster at higher temperatures. Ground-level ozone, the primary component of smog, is a respiratory irritant that decreases lung function and may increase the development of asthma in children (Tagaris et al., 2009).

In the future, warmer temperatures and changes in atmospheric circulation patterns may bring oppressive summer weather patterns earlier in the year (see Section 4.4), accelerating the formation rates of tropospheric ozone and increasing the length of time photochemical smog remains over any given location. Barring significant decreases in precursor emissions, tropospheric ozone exceedences could become more common over most densely populated areas in the United States, China, and elsewhere around the world with warm summers (Mickley et al., 2004; Tao et al., 2007; Lin et al., 2008). A study of 50 U.S. cities projected that a 1.6 to 3.2°C increase in local temperature would lead to an average 4.8 ppb increase in ozone levels by 2050, with the greatest increases occurring in cities with already high ozone concentrations (Bell et al., 2007). Ozone levels were projected to exceed the 8-hour regulatory standard an average of 5.5 more days per summer, an increase of 68% over current conditions. Although ozone mortality was projected to increase by 0.11 to 0.27% on average, certain cities exhibited greater sensitivities (e.g., increases of 5% for New York City; Knowlton et al., 2008). Given the sensitivity of ozone levels to precursor emissions, climate effects on ground-level ozone and particulate matter may depend less on direct temperature effects and more on the effectiveness of pollution control measures and climate-driven changes in natural emissions sources (Ebi and McGregor, 2008).

Increasing frequency of heavy downpours is already occuring and is projected to continue to occur in many parts of the world, including the eastern United States (Tebaldi et al., 2006; Karl et al., 2009). These increase the risk of water contamination and spread of water-borne bacterial diseases, with the risk being exacerbated by rising temperatures (Vorosmarty et al., 2000). Current and future deficiencies in watershed protection, infrastructure, and storm drainage systems will likely increase the risk of contamination events as climate variability increases (Rose et al., 2000).

The length of the pollen season has already increased due to both earlier onset of flowering (Parmesan, 2007) and lengthening of the season for late-flowering species (Sherry et al., 2007). In the future, increasing temperatures may also affect the production, toxicity, and/or pollen-producing capacity of allergenic and toxic plants, including ragweed and poison ivy (Mohan et al., 2008; Shea et al., 2008; Ziska et al., 2009). Allergy incidence may also be increased by interactions between airborne allergens and air pollution (D'Amato and Cecchi, 2008). Uncertainties in future emissions and the coupling between climate, air quality, and ecosystem models, however, render speculative any determination of the degree of change (Bernard and Ebi, 2001; Gamble et al., 2008).

5.7 ECOLOGY AND ECOSYSTEMS

Terrestrial Species and Climate Change: What Species Are Up Against

As the climate changed throughout the past millennia, species shifted to track temperature, precipitation, and other weather factors (Graham and Grimm, 1990; Overpeck et al., 1992). The geographic range of any species includes only areas where individuals can endure the extreme temperature and water stress occurring at those locations (Gordon,1982; Chown and Gaston, 1999). Indeed, the ranges of various songbirds in North America are limited by the amount of metabolic energy an individual must exert to stay alive (Root, 1988a,b). Warming in the late-Quaternary resulting in species tracking the changing climate gradient (Graham and Grimm, 1990). Additionally, these species differentially tracked their own unique set of climatic factors, which resulted in many species occurring in unexpected areas (Graham and Grimm, 1990) and new species groups being formed (Overpeck et al., 1992; Hobbs et al., 2009).

Today species continue to shift and change with current climate change. Additionally, if we remain on the "Business as Usual" scenario, then on a sustained global basis the expected rate of change in temperature in the next few decades could be higher than most species have endured over millennia (Hoeg-Guldberg et al., 2007; Loarie et al., 2009; NRC, 2009). In addition the face of the planet is very different from how it has ever been before because people have altered it considerably by constructing various structures on the landscape, such as cities, farms, and roadways. These frequently make movement of species across an area difficult. This is particularly true for species that are slow moving, such as turtles, but it is also true for animals that can move more quickly, including birds, bats, and butterflies. Even with these impediments, many species are changing with the globally changing climate. Indeed, with an increase in the average global temperature of only ~0.6°C, many species around the globe were found to be making many significant changes (Root and Schneider, 2002; Parmesan and Yohe, 2003; Root et al., 2003). Species primarily exhibit two different types of changes: the timing of various temperature-related events, such as blooming or egg laying; and shifting their ranges to formerly cooler regions, such as poleward and up in elevation for terrestrial species and descending deeper in the oceans for marine species (Root and Schneider, 2002; Parmesan and Yohe, 2003; Root et al., 2003; Root et al., 2005; Parmesan, 2006; Fox et al., 2009). Species are exhibiting other types of changes, but they are not reported as often as these two changes. Included in this "other" category are, for example, behavioral changes, changes in size and shape, and genetic changes.

Change in Seasonal Timing

The timing of different seasonal activities (phenology) of many species is shifting in concert with the changing climate. Common changes in the timing of spring activities include the timing of migration in various species such as birds, mammals, fish, and insect species (e.g., MacMynowski and Root, 2007; Ogden et al., 2008) and in budding, blooming, and leafing in plants (e.g., Menzel et al., 2006; Crimmins et al., 2009; Primack and Miller-Rushing, 2009). For example, in the case of plants, several trees in an urban to rural setting in Ohio exhibited an earlier trend in the leafing out in the spring (Shustack et al., 2009). The same is true for the timing of cherry blossoms in Japan, for which data exist from the ninth century (Primack et al., 2009). Satellite data unequivocally confirm the earlier greening of the biosphere (Myneni et al., 1998; White et al., 2009). Of course individual examples are not sufficient to draw generalized conclusions, thus meta-analyses have been performed to look for consistent patterns around the globe over a large number of species that have been observed to have been changing. For all Northern Hemisphere species reported in the literature with more than a 10 year record showing phenological changes in the spring, the average number of days changed in the spring observed over the last 30 years of the 20th century was ~15.5 days or ~5 days per decade earlier timing for those species showing a phenological change in the spring (Root et al., 2003). When both the species changing and those not changing in the same areas are included in the study calculation, then the number of days drops strongly to ~2 days per decade (Parmesan and Yohe, 2003). The increase in global average temperatures over that time period was ~0.4°C, although changes in land temperatures, especially at higher latitudes, were some factor of two larger (IPCC, 2007a).

Contrary to what has been found for spring phenology, trends in autumn phenologies are not as clear. In the spring, animals are getting ready to breed and are driven to breed as earlier as possible. Therefore, these species arrive earlier as the springtime warms, which in turn means they need earlier availability of required resources such as food. If a needed resource is not shifting in concert with a species exhibiting shifts, this could cause some decline in breeding success, for example. In the autumn some leaf fall is delayed, and some migrating animals move their southern migration date earlier than it was before, while others migrate south at roughly the same time that they always have migrated, and some stay longer before migrating (Miholcsa et al., 2009; Schummer et al., 2010).

Mismatch in Timing

Over evolutionary time species have formed predator-prey relationships that are being disrupted because the predator and prey do not necessarily respond to warming by shifting in concert. For example, the common cuckoo (*Cuculus canorus*), a migrant, is a brood parasite laying its eggs in the nests of other birds that migrate either short or long distances. With climate change the short-distance migrants are arriving significantly earlier now than in the past and they are nesting before the cuckoo has arrived. Consequently, the short-distant migrants have reduced brood parasitism (Saino et al., 2009). This is an example where there are both winners and losers with climate change—the short-distance migrants winning and the cuckoo losing. What is "better" or "worse" is of course a value judgment, yet if one species is aided whereas others are disadvantaged to the point of population collapses and perhaps extinction, then the net effect could mean a local loss of species.

Range Shifts

As the climate has changed species have shifted their range because they are attempting to stay in areas where they are exposed to the same regional ambient temperatures. For terrestrial species this means moving toward the pole and up in elevation. From 1914 to 1920, Joseph Grinnell systematically surveyed a 60 to 3,300 m elevation gradient in Yosemite National Park. He kept meticulous field notebooks, and these have been valuable in studying changes in species over a century, because the same area was resurveyed from 2003-2006. Roughly half of the 28 mammals surveyed moved up in elevation by an average of around 500 m. For example, one of the species shifting up in elevation in response to climatic warming is the

American pika (*Ochotona princeps*) (Moritz et al., 2008). In 1917 Grinnell found it up to 2,400 m (Grinnel, 1917), but in 2004 it was recorded up to 2,900 m. The same type of analysis with similar results has also been done on birds (Tingley et al., 2009). Changes in the species are consistent with an increase in minimum January temperature by about 3°C.

In prehistoric times many species communities were composed of species that today are not found together (Overpeck et al., 1992). These so-called no analog communities foreshadowed what we are now observing; many species are moving differentially, which means the various biotic interactions existing within some species communities are changing. For example, the predator-prey cycle will be broken if one species moves into areas where the other does not occur. A situation that could cause concern is if the prey is a pest on our crops or a disease vector, and they no longer are held in check by the predator. Even if mutalistic interactions are disrupted, such as an insect that commonly pollinates a crop shifts, another insect pollinator most likely will move in, but perhaps not immediately. In several locations around China people are having to hand pollinate apple and pear trees to counter the loss of productivity due to a significant decline in the abundance of natural pollinators (Partap et al., 2001).

Changes in Size and Shape, Genetics, and Behavior,

As Earth has warmed many species are showing different types of concurrent changes. Hadly (1997) documented body size changes of a small mammal during historical (natural) climate shifts. Temporal fluctuations of body size in woodrats are so precise that Smith and Betancourt (1998) labeled them "paleo-thermometers." Indeed, two rodent species in southern New Mexico captured from 1991 to 1998 had multiple structures, such as hind leg, ear, and body length, showing strong correlation with a time-appropriate climate variable (Wolf et al., 2009).

Few studies have found a change in genetics with climate change. This could be that relatively few scientists have looked for such a change. Bradshaw and Holzapfel (2008), however, have examined the pitcher-plant mosquito (*Wyeomyia smithii*), and they did find genetic change (Bradshaw et al., 2006).

Species most affected by reductions in Arctic sea-ice extent are those with limited distributions and specialized feeding habits that depend on ice for foraging, reproduction, and predator avoidance, including the ivory gull (*Pagophila eburnean*), Pacific walrus (*Odobenus rosmarus divergens*), ringed seal, hooded seal (*Cystophora cristata*), narwhal (*Monodon monoceros*), and polar bear (*Ursus maritimus*); see Post et al. (2009) and Gilg et al. (2009).

Synergetic Forces

Given the changes humans have made to the planet, many more factors than climate change are affecting species (Laurance and Useche, 2009). These include such things as land-use change, invasive species, introduced chemicals, and hunting. Additionally, changes to a system can feedback on other systems, which in turn may cause species more stress, for example, loss of coastal marshes causing increased erosion. Managers have been working for decades to help species combat stresses, for example, by setting up refuges for ducks where they are provided artificial lakes, food, and hunting protection. Coping with one or even two of these stresses can be difficult for species, but now adding the ubiquitous stress of a changing climate has managers and others concerned about possible population declines, and in extreme cases, extinction.

Certain traits can make a species more prone to extinction. These include the size of the species population (large population size is much more stable than a small size); the size of the range of the species (a species with a larger range size is less likely to become rare and then extinct); and species that specialize on a particular trait, such as a particular prey item or a particular tree for nesting. In the last case, such dependent species are at greater risk of extinction due to their reliance on a particular item that could itself be negatively changed by climate change. Such change can actually cause a cascade of changes through the species communities due to the interconnectedness experienced by species within an ecosystem.

Extinction

The average lifespan of a plant or animal species is ~7 million years, which is the number of years a species persists from when the species arises via speciation to when it goes extinct (Lawton and May, 1995). The best estimate of how many plants and animals are extant compared to all plants and animals that have ever existed is around 2%. Consequently extinction is certainly a normal occurrence. The background level of extinctions (not human induced) differs with different taxa and types of species. In marine species it is estimated to be ~0.1 to 1 Extinction per Million Species Years (E/MSY). The number for mammals is similar, being ~0.2-0.5 E/MSY (Foote, 1997; Alroy, 1998; Regan et al., 2001; MEA, 2005; May, 2010).

Determining historic extinction rates is difficult because of the time it takes for a species committed to extinction—extinction will occur without some major change occurring, such as human management—to reach the point when we can document that the individuals in the population are no

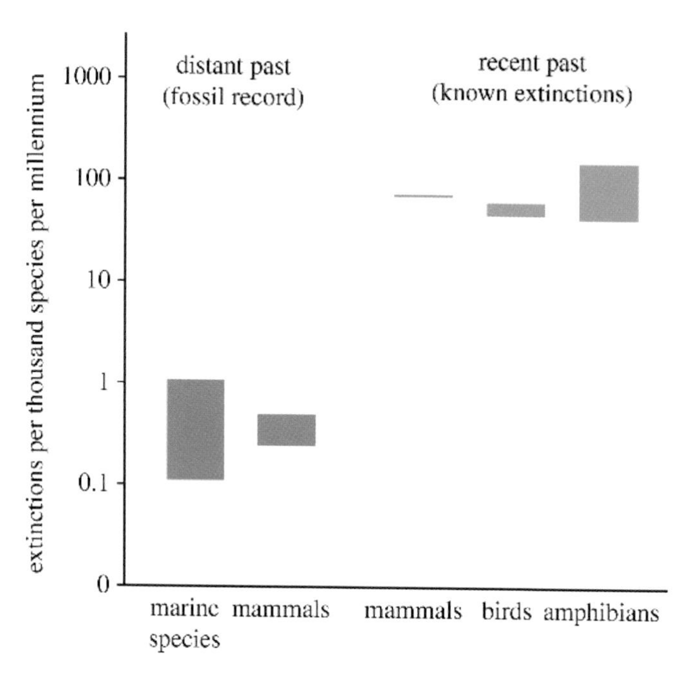

FIGURE 5.16 Species extinction rates, expressed as extinctions per 1,000 species per 1,000 years. The fossil record tells us that in the distant past less than one mammal went extinct every 1,000 years. In the recent past the number of extinctions has increased 3 orders of magnitude higher than that of the fossil record. SOURCE: May (2010).

longer replacing themselves. These species are called functionally extinct. When that assistance is not possible, there is still a time lag due to humans not being comfortable declaring a species extinct. Indeed, there is a "50-year rule" that says a species cannot be said to be extinct until it has not been found for 50 years. Even so, current rates of extinctions can be calculated at least for well-studied species, i.e., mammals, birds. and amphibians. The number of species in these three taxa is around 21,000, and over the past century ~100 species in those taxa are known to have gone extinct, giving an extinction rate of ~50 E/MSY (MEA, 2005). Adding the functionally extinct species, the rate of "extinction" increases to ~200 (May, 2010; see Figure 5.16).

What about future extinctions? For many decades now we have known the importance of three factors in contributing to a species becoming a concern with regard to extinction (Rabinowitz et al., 1986): (1) The overall range size is small enough for a major perturbation to influence greatly all individuals in the species (Harris and Pimm, 2008). For example, the whooping cranes (*Grus americana*) while on their wintering ground in Texas are at high risk if there is an oil spill or powerful hurricane because the entire population winters in a small inlet along the coast; (2) The overall population size is small, which can be detrimental because breeding in a small population can result in inbreeding (Liao and Reed, 2009). This, in turn, can increase the expression of deleterious genes, which decreases the number of fit young being produced, and the loss of genetic variability, which is needed to allow adaptation to novel habitats; and (3) The population depends on some factor or species that could be at risk from disturbances, such as pollution, poaching, or climate change, among others. For example, the red-cocked woodpecker (*Picoides borealis*) must nest in 70- to 120-year-old pines, most of which in the late 1800s and mid 1900s were cut down for various uses. Another example involves plant-eating insects, of which there are estimated to be 213,830 to 547,500 species committed to extinction because the range of their host plants are shrinking in size because of warming (Fonseca, 2009).

When climate change is one of the main forces acting on a species, then distance to closest cool refuge (Figure 5.17) is also as important as

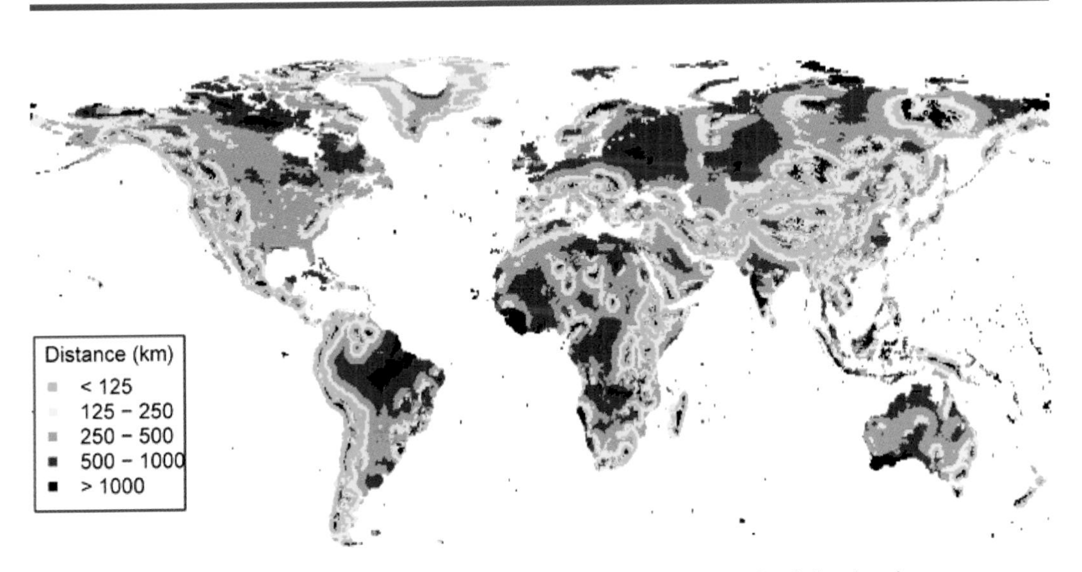

FIGURE 5.17 Map showing the distance to potential cool refuges, where cool is defined as the temperatures in 2100 are equal to or cooler than the temperatures in the 1960s. Used 0.5 x 0.5 latitude-longitude blocks. Source: Wright et al. (2009).

range size, population size, and the vulnerability of a population depending on some limited factor at risk. Reaching the cool refuge is crucial. For example, Sinervo et al. (2010) reported lizards, which have body temperatures determined by habitat temperatures, on four different continents are having to spend more time than before escaping warming temperatures by going into burrows or other locally cooler places. This means they have less time to forage and hence do not have the energy reserves needed to successfully reproduce, which of course can lead to population decline and possibly extinction. These species and others could probably avoid extinction by moving into new other areas in the region that are cooler (e.g., up in elevation). If a species in question has a maximum dispersal distance that is shorter than the distance to the refuge, or if the species is not a good colonizer and fails to become established in the refuge once it gets there, then the species will be in trouble unless it gets human assistance. Species that are in trouble unless aided by humans are called functionally extinct. Examples of functionally extinct species are those that are on oceanic islands or mainland islands, such as oasis within a desert, because unless they fly, dispersing to a cooler refuge is not possible. For example, the Akiapolaau (*Hemignathus munroi*) on the Hawaiian Islands, Aruba island rattlesnake (*Crotalus durissus unicolor*), and the black robin (*Petroica traverse*) on Chatham Islands off the coast of New Zealand are all critically endangered and will go extinct without help from humans.

Hubbell and co-workers (2008) looked at possible extinction rate in the Brazilian Amazon due to land-use change. He estimated that a total of 11,210 tree species occur in the Brazilian Amazon. About 30% of these trees (~3,250) have populations greater that 1 million, and consequently, do not currently face extinction. Fifty percent (~5,300) of all the species have populations less than 10,000. Roughly 35% of these have a high probability of going extinct even without the effects of climate change. This would also result in insects and other species that depend on these tree species to also be facing extinction. Climate change and other drivers could exacerbate the vulnerability of these species of concern.

Deutsch and co-authors (2008) investigated the physiological tolerances of species from pole to pole. They found, as expected, that the extent of physiological tolerance was wider at the higher latitudes and narrower near the equator. From this they reasoned that the tropical species could very well have a more difficult time as the globe warms, because the temperature to be experienced by species in the tropics could well be higher than that ever experienced by the tropical species. Temperate, boreal, and arctic species will also be subjected to warmer temperatures, but those temperatures, for

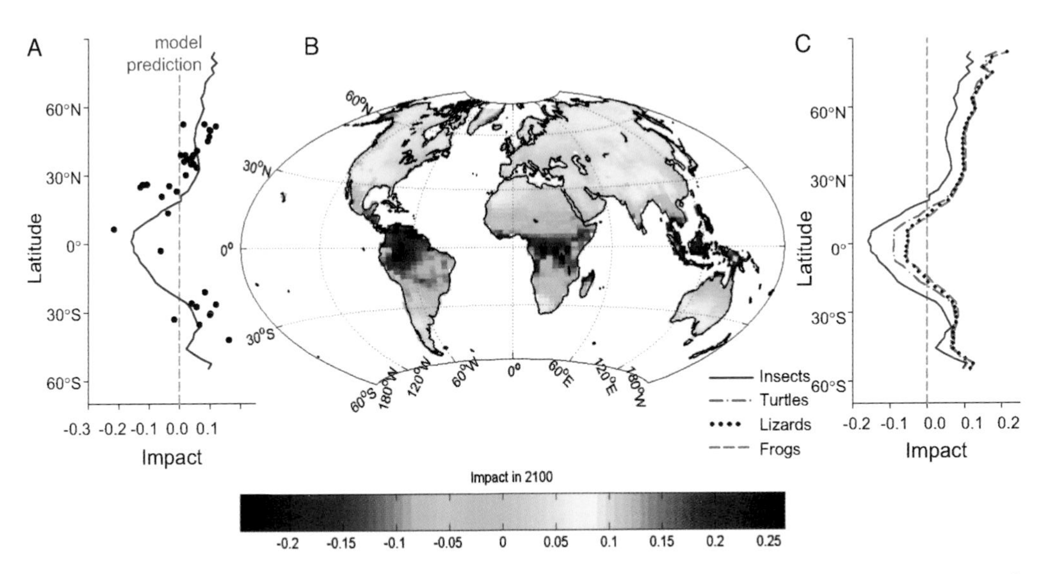

FIGURE 5.18 Predicted impact of warming on the performance of animals with body temperatures equilibrated to ambient (ectothermic or cold-blooded). On both panels the red line is the model output. On the right panel the dots are the intrinsic growth rates for each insect species examined. On the left panel other terrestrial ectotherms, including frogs, toads, lizards, and turtles are plotted. Source: Deutsch et al. (2008).

a while at least, will be within the variability of temperatures experienced. Hence, tropical trees, and other studies will highly likely find the same to be true for tropical animals and are at higher risk of not being resilient to the incipient warming (Figure 5.18).

Using the A1B scenario from IPCC (2007a), Wright and co-authors (2009) found that "75% of the tropical forests present in 2000 will experience mean annual temperatures in 2100 that are greater than the highest mean annual temperature that supports closed-canopy forest today" (p. 1418). As long as there is a cool refuge within dispersal distance and as long as dispersal is possible this increase may not be a big problem. The most common cool refuges for plants are to migrate up in elevation. In the tropics, however, such elevational relief is not available, making cooler refuges for many plants and animals living in the tropics inaccessible. Around 1,200 mammals have ranges restricted to the low-latitude tropics with no easy access to cooler refuges as the globe warms. Another group of mammals at risk of extinction are those about 650 species that have ranges smaller than

3,130 km^2. Without aid from humans these species are probably not going to be able to persist.

Extinction is irreversible. Choices among stabilization targets can be expected to determine the scope of future extinction (e.g., types of species, geographic regions, etc.) that could be caused by climate change, or alternatively the scale of protective adaptation measures such as species management that could be considered to avoid extinctions.

5.8 BIOLOGICAL OCEAN

Impacts of CO_2, pH, and Climate Change in the Ocean's Biology

Marine ecosystems will be affected by climate change via physical changes in ocean properties and circulation (Sections 4.1, 4.4, and 4.7), ocean acidification via altered seawater chemistry from rising atmospheric CO_2 (Section 4.9), and sea-level rise via coastal habitat loss. Some of the key potential impacts will involve changes in the magnitude and geographical patterns of ecological and biogeochemical rates and shifts in the ranges of biological species and community structure (Boyd and Doney, 2002). Impacts are expected to include both direct physiological impacts on organisms through, for example, altered temperature, CO_2, and nutrient supply, and indirect effects through altered food-web interactions such as changing seasonal timing (phenology) of phytoplankton blooms or disruptions in predatory-prey interactions.

Primary production by upper-ocean phytoplankton forms the base of the marine food-web and drives ocean biogeochemistry through the export flux of organic matter and calcareous and siliceous biominerals from planktonic shells. Plankton growth rates for individual species are temperature dependent and tend to increase under warming up to some threshold. When viewed in aggregate, plankton community production rates approximately follow an exponential curve in nutrient replete conditions, which would suggest increasing global primary productivity over this century as sea surface temperatures increase (Sarmiento et al., 2004). In most regions of the ocean, however, primary production rates are limited by nutrients such as nitrogen, phosphorus, and iron. Diatoms, a key shell-forming group of phytoplankton, are also limited by silicon. The rates of many other biological processes, such as bacterial respiration and zooplankton growth and respiration, also speed-up as temperature rises, the integrated effect at the ecosystem level is difficult to predict from first principles. Warming also occurs in conjunction with other factors (rising CO_2, altered ocean circulation), and the potential synergistic or antagonistic effects of multiple stressors must be considered.

Satellite observations indicate a strong negative relationship, at interannual time scales, between marine primary productivity and surface warming in the tropics and subtropics, most likely due to reduced nutrient supply from increased vertical stratification (Behrenfeld et al., 2006). Satellite data also indicate that the very lowest productivity regions in subtropical gyres expanded in area over the past decade (Polovina et al., 2008), although these trends may be due to interannual variability (Henson et al., 2010). Numerical models project declining low-latitude marine primary production in response to 21st century climate warming (Sarmiento et al., 2004; Steinacher et al., 2010) (Figure 5.19). Warmer, more nutrient-poor conditions in the subtropics could enhance biological nitrogen fixation (Boyd and Doney, 2002), an effect that may be amplified by higher surface water CO_2 levels (Hutchins et al., 2009). The situation is less clear in temperate and polar waters, although there is a tendency in most models for increased production due to warming, reduced vertical mixing, and reduced sea-ice cover. For example, the rapid warming and sea-ice retreat along the West Antarctic Peninsula has lead to a poleward shift in the region of strong seasonal primary production that has impacts for higher trophic levels including seabirds (Montes-Hugo et al., 2009). In most open-ocean regions, however, the climate signal in primary production and other ecosystem properties may be difficult to distinguish from natural variability for many decades (Boyd et al., 2008; Henson et al., 2010). Changes in atmospheric nutrient deposition (nitrogen and iron) linked to fossil-fuel combustion and agriculture also can alter marine productivity but mostly on regional scales near industrial and agricultural sources (Duce et al., 2008; Krishnamurthy et al., 2009).

Subsurface oxygen levels likely will decline due to warmer waters (lower oxygen solubility) and altered ocean circulation, leading to an enlargement of open-ocean oxygen minimum zones and stronger coastal oxygen depletion in some regions (Keeling et al., 2010; Rabalais et al., 2010). Low subsurface O_2, termed hypoxia, occurs naturally in open-ocean and coastal environments from a combination of weak ventilation and/or strong organic matter degradation. Dissolved O_2 gas is essential for aerobic respiration, and low O_2 levels negatively affect the physiology of higher animals leading to so-called "dead-zones" where many macro-fauna are absent. Coastal hypoxia can lead to marine habitat degradation and, in extreme cases, extensive fish and invertebrate mortality (Levin et al., 2009; Rabalais et al., 2010). Expanded open-ocean oxygen minimum zones would increase denitrification and may contribute to increased oceanic production of the greenhouse gas

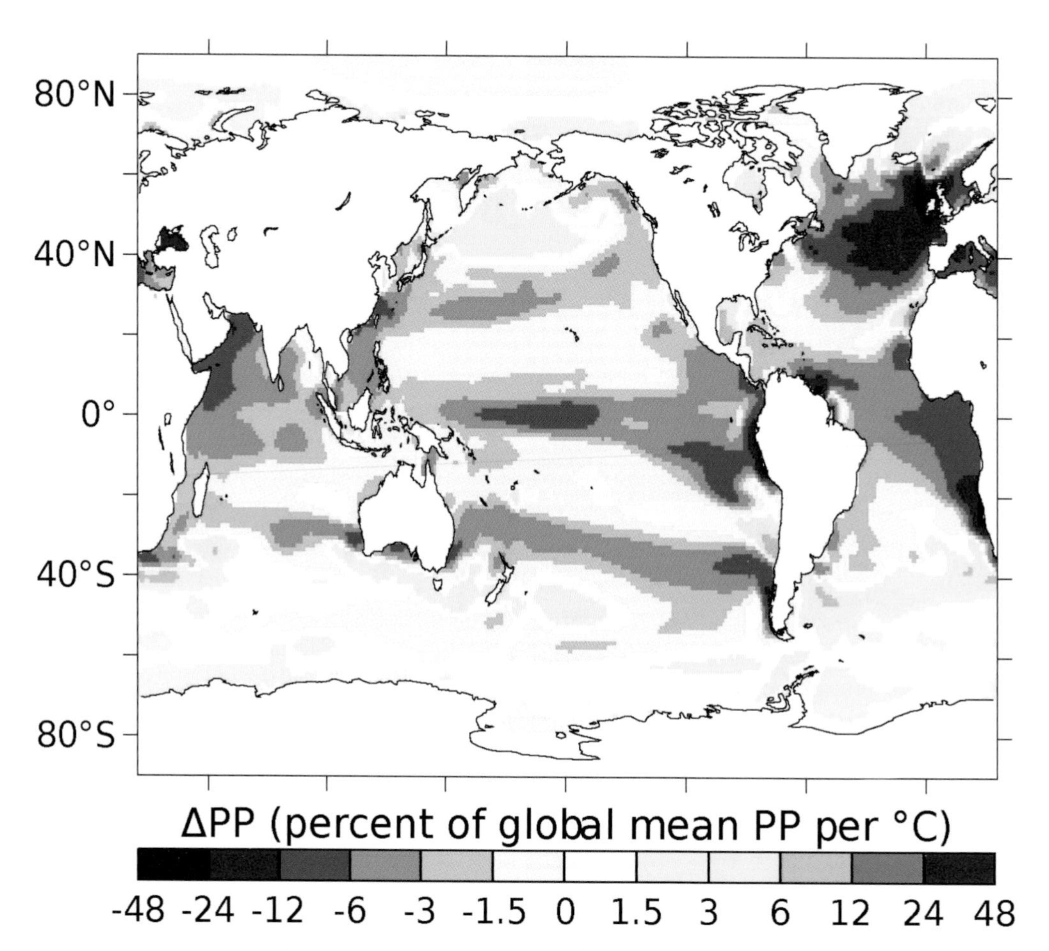

FIGURE 5.19 Model projected change in vertically integrated annual mean primary production (PP) relative to pre-industrial conditions (decadal mean 1860-1869) for the end of the 21st century under SRES A2. The changes represent the difference between 2090-2099 and 1860-1869 (decadal means). Multi-model means have been computed for four coupled ocean-atmosphere models using regional skill scores as weights. Where no observation-based data is available to calculate skill scores (e.g., in the Arctic) the arithmetic mean of the model results is shown. The magnitude of the primary production changes are shown in percent normalized to global mean areal primary production rate and are presented for a nominal increase in global mean surface air temperature of 1°C. Source: Steinacher et al. (2010).

nitrous oxide (N_2O). The organic matter respiration that generates hypoxia also elevates CO_2, and multiple stressors of warming, deoxygenation, and ocean acidification magnify physiological and microbial responses (Pörtner and Farrell, 2008; Brewer and Peltzer, 2009).

Open-ocean deoxygenation has been observed in the thermocline of the North Pacific and tropical oceans over decadal periods, perhaps due to natural climate variability (Mecking et al., 2008). Models project long-term reductions of 1-7% in the global oxygen inventory and expansions of open-ocean oxygen minimum zone over the 21st century (Frölicher et al., 2009; Keeling et al., 2010). The duration, intensity, and extent of coastal hypoxia has also been increasing substantially over the last half-century, but primarily due to elevated fertilizer run-off and atmospheric nitrogen deposition that contribute to coastal eutrophication, enhanced organic matter production, and export and subsurface decomposition that consumes O_2. Climate change could accelerate coastal hypoxia via surface warming and regional increases in precipitation and river runoff that increase water-column vertical stratification; on the other hand, more intense tropical storms could disrupt stratification and increase O_2 ventilation (Rabalais et al., 2010). Expanding coastal hypoxia is also induced in some regions by reorganization in ocean-atmosphere physics. Off the Oregon-Washington coast, increased wind-driven upwelling is linked to the first appearance of hypoxia, and even anoxia, on the inner-shelf after five decades of hypoxia-free observations (Chan et al., 2008). Further south in the California Current System, the depth of hypoxic surface has shoaled along the coast by up to 90 m (Bograd et al., 2008). The same physical phenomenon, along with the penetration of fossil-fuel CO_2 into off-shore source waters, are introducing waters corrosive to aragonite ($\Omega < 1$) onto the continental shelf (Feely et al., 2008). There is conflicting evidence on how coastal upwelling may respond to climate change, and impacts may vary regionally (Bakun et al., 2010).

Laboratory and mesocosm experiments indicate that many marine organisms are sensitive to elevated CO_2 and ocean acidification, with both positive and negative physiological responses (Fabry et al., 2008; Doney et al., 2009a,b; NRC, 2010). The projected rates of change in global ocean pH and Ω over the next century are a factor of 30-100 times faster than temporal changes in the recent geological past, and the perturbations will last many centuries to millennia. Although there are spatial and temporal variations in surface seawater pH and saturation state, projected future surface water pH values for the open-ocean are below the range experienced by contemporary populations, and the ability of marine organisms to acclimate or adapt to the magnitude and rate of change is unknown.

The largest identified negative impacts are on shell and skeleton growth by calcifying species including corals, coralline algae, and mollusks. Corals utilize the aragonite mineral form of calcium carbonate, and the rate of coral calcification declines with falling aragonite saturation state even when waters remain supersaturated, and corals appear to need saturation states ($\Omega > 3$) for healthy growth (Langdon and Atkinson, 2005; Kleypas and Yates, 2009). Decreased calcification is observed for corals with symbiotic zooxanthella (photosynthetic algae living within coral animals), and CO_2 fertilization of zooxanthella does not alleviate acidification effects. Studies of net community calcification rates for coral reef ecosystems indicate that overall net calcification also decreases with rising CO_2 (Silverman et al., 2007), and model studies suggest a threshold of about 500-550 ppm CO_2 where coral reefs would begin to erode rather than grow, negatively impacting the diverse reef-dependent taxa (Silverman et al., 2009). Observed physiological responses for mollusks, such as pteropods, oysters, clams, and mussels, include reduced calcification, increased juvenile mortality and reduced larval settlement, and smaller, thinner, and malformed shells (Orr et al., 2005; Green et al., 2009; Miller et al., 2009). Crustaceans also utilize calcium carbonate in their shells, but the response to elevated CO_2 is less well-understood with studies reporting both increased and decreased calcification rates (Fabry et al., 2008). Decreased calcification rates with rising CO_2 are observed as well for key planktononic calcifiers including foramaniferia and most strains or coccolithophores.

Some organisms may benefit in a high-CO_2 world, in particular photosynthetic organisms that are currently limited by the amount of dissolved CO_2. In laboratory experiments with elevated CO_2, higher photosynthesis rates are found for certain phytoplankton species, seagrasses, and macroalgae, and enhanced nitrogen-fixation rates are found for some cyanobacteria (Hutchins et al., 2009). Indirect impacts of ocean acidification on non-calcifying organisms and marine ecosystems as a whole are possible but more difficult to characterize from present understanding. A limited number of field studies that have been carried out in mostly benthic systems with naturally elevated CO_2 are broadly consistent with the laboratory studies in terms of predicted changes in community structure (e.g., decrease in calcifiers; increase in non-calcifying algae) (Hall-Spenser et al., 2008; Wootton et al., 2008). Polar ecosystems also may be particularly susceptible when surface waters become undersaturated for aragonite, the mineral form used by many mollusks including pteropods, which are an important prey species for some fish. Socioeconomic impacts from degraded fisheries and other marine resources are possible but poorly known at this point (Cooley and Doney, 2009).

Based on historical survey data, the geographic range of many marine species has shifted poleward and into deeper waters due to ocean warming (Perry et al., 2005; Nye et al., 2009). Model projections indicate that poleward expansion and equatorial contraction of geographical ranges for particular species will continue and that at any particular location the frequency of replacement of "cool-water" species by "warm-water" species will likely increase. Individual marine species will be impacted differentially; for example pelagic fish ranges may be impacted more than demersal ranges, which will lead to changes at the community and ecosystem model. Few studies have looked comprehensively across many marine taxa and geographic regions, but a recent pilot model projection suggests the potential for significant changes in community structure in the Arctic and Southern Ocean biodiversity due to invasion of warm water species and high local extinction rates in the tropics and subpolar domains. Fish stock size may either grow or decline due to altered primary production, prey abundance, and temperature-dependent growth rates, the trend for each species depending on its particular biology and habitat (Brown et al., 2010; Hare et al., in press). Complex predation and competition interactions may reverse the expected responses for some species (Brown et al., 2010). Climate change may also disrupt larval dispersal and development patterns as well as existing predator-prey interactions through altered currents and seasonal phenologies for spawning and plankton blooms (Parmesan, 2006).

Specific marine habitats may be particularly sensitive to changing climate. Rising sea-level would impact, and in many cases degrade, coastal wetlands and estuaries, coral reefs, mangroves, and salt-marshes through inundation and enhanced coastal erosion rates; these coastal environments serve as important nursery habitats for larval and juvenile life-stages. Regional impacts depend on local vertical land movements and would be exacerbated where the inland migration of ecosystems is limited by coastal development and infrastructure. The thermal tolerance of many coral species is limited, and over the past several decades, warmer sea surface temperatures have led to widespread tropical coral bleaching events (loss of algal zooxanthella) and increased coral mortality. Warming and more local human impacts have been associated with declines in the health of coral reef ecosystems worldwide. Bleaching can occur for sea surface temperature changes as small as +1-2°C above climatological maximal summer sea surface temperatures, and more frequent and intense bleaching events are anticipated with further climate warming (e.g., Veron et al., 2009). Sea-ice dependent species are also at risk, and rapid warming in the Arctic and parts of Antarctica has resulted in substantial shifts in whole food-webs (Ducklow et al., 2007; Montes-Hugo et al., 2009).

5.9 ILLUSTRATIVE ADDITIONAL FACTORS

There are many more climate impacts that could be very important but are not as well understood as those described above. Some illustrative examples are briefly provided here.

National Security

Many processes could plausibly connect climate change to national security concerns. For instance, military experts have pointed to the potential for climate-induced food and water shortages to contribute to political instability, which can then be exploited by extremists (CNA Corporation, 2007). The potential for mass migrations associated with resource shortages or flooding are also potential "threat multipliers." Climate changes will also likely affect military operations, such as via inundation of low-lying military bases, and introduce new geopolitical dilemnas, such as the opening of sea routes in the Arctic.

Yet perhaps because of the complex nature of national security threats and the paucity of relevant data, there are relatively few quantitative examples that document the climate sensitivity of phenomena related to national security. Some empirical evidence suggests an important role for climate in domestic and international conflict. Long-term fluctuations of global wars and death rates since 1400 are correlated with shifts in temperature (Zhang et al., 2007a). In Africa, civil wars since 1980 have been roughly 50% more likely in years 1°C warmer than average (Burke et al., 2009). Precipitation decreases are also associated with conflict in Africa, although projected rainfall changes are not large relative to historical variability (Miguel et al., 2004; Hendrix and Glaser, 2007).

Obviously more work is needed to advance understanding of national security threats from climate change. Specifically, although the implications of climate change for resource scarcity are uncertain, the complex relationship between resource scarcity and conflict is even more tenuously understood (Barnett, 2003; Nordås and Gleditsch, 2007). At the same time, military experts routinely caution that waiting for quantitative precision can be very risky, and intuition alone is often used to make major strategic decisions for national security (CNA Corporation, 2007).

Dynamic Vegetation

Changes in climate and CO_2 beyond 2100 will likely be sufficient to cause large-scale shifts in natural ecosystems. Although relatively few modeling studies extend beyond 2100, many show substantial changes occurring by 2100 that indicate the potential for even greater changes in the following centuries. Indeed, some major shifts such as the expansion of shrublands in Arctic regions are already evident in recent decades (Sturm et al., 2001; Tape et al., 2006), and this shift is consistent with results from warming experiments in the region (Walker et al., 2006). One important consequence of this expansion is that the resulting decrease in surface albedo can amplify local summer warming in the future by a factor of two or more (Chapin et al., 2005).

Major biome shifts also appear likely in some temperate and tropical regions by 2100 (Scholze et al., 2006). The Eastern part of the Amazon rainforest, for example, may shift to a seasonally dry forest or even a savanna due to likely rainfall decreases in the dry season by 2100 (Cox et al., 2004; Malhi et al., 2009). Beyond 2100, these shifts become more likely. High CO_2 levels will likely promote expansion of vegetation into currently barren tundra and desert ecosystems, because of higher water-use efficiencies, which again would amplify local warming because of albedo effects (Bala et al., 2006).

Most models used to simulate future vegetation changes rest on strong empirical relationships between current climate and the distribution of major biomes. Less is known about how transitions between equilibrium states occur, and for instance whether deep roots of established trees limit their sensitivity to climate shifts. Another source of uncertainty in projections of vegetation change is potential interactions with local land use, which for instance could accelerate regional climate change in tropical forests (Malhi et al., 2008). Despite these uncertainties, higher emissions scenarios will almost certainly result in climate shifts that are large enough to cause major vegetation shifts by 2100 and beyond.

Some Climate Changes Beyond 2100

More is known about the very long term (millennia) and the present century, but there is a gap in understanding and more limited knowledge of climate system behavior over the next few centuries. Here we present two examples of areas where information on the next few centuries is available.

Circulation

About half of the AR4 climate models were used to project the future climate beyond 2100 to 2200. These simulations were performed using the

A1B emission scenario until 2100, and then holding the forcing fixed to 2200. In some of the models, the MOC was seen to maintain an equilibrium strength that was similar to that projected for 2100; in other models the MOC strengthened somewhat from their nadir early in the 22nd century. Too few of the higher end AR4 climate models have been integrated far enough into the future to assess whether persistently high greenhouse gas concentrations will cause a permanent change to the strength of the MOC.

Sea Ice Beyond 2100

Climate model simulations suggest that in the decades following 2100 the Arctic may be perennially ice-free (Winton, 2006a,b; Eisenman and Wettlaufer, 2009). However, on the millennium scale the system may oscillate between being totally ice-covered or having ice only along the land margins (Ridley et al., 2008). Only two IPCC models predict a year-round ice-free Arctic in the decades after 2100. However, these are the models that have the most sophisticated sea ice components. The scenario within which these models lose their Arctic ice is the 1% per year CO_2 increased to quadrupling, a concentration of 1,120 ppm, after which although atmospheric CO_2 is kept constant temperatures continue to rise. These models, one initiated from pre-industrial conditions and the other from present-day conditions are run for nearly 300 years; quadrupling occurs at 140 years. Both models exhibit a gradual linear decline in September sea-ice loss becoming ice free when the average polar temperature is –9°C. In March, the transition to an ice-free state is also linear until polar temperatures reach –5°C, at which point one model experiences an abrupt transition, associated with that model's ice-albedo feedback mechanism, while in the other it remains linear and the ocean heat flux plays a larger role. The temperature at which the Arctic becomes ice free in these models is 13°C above present-day values (Winton, 2006a,b). The ice-albedo, convective cloud, and ocean heat transport feedbacks all play necessary roles in the loss of the winter sea ice (Abbot et al., 2009b). However, the ice-albedo feedback plays a key role.

While an ice-free Arctic may present new economic opportunities it will also likely have profound impacts on climatological and ecological systems locally and globally. A few of these are mentioned here. From the physical standpoint, the loss of Arctic sea ice means that the mediating influence of sea ice on energy flux exchanges between the atmosphere and ocean will no longer prevail and the Arctic atmosphere will warm. Model studies suggest that these two impacts will affect the effectiveness of the overturning in the thermohaline circulation (e.g., Broecker, 1997; Lemke et al., 2007; Levermann et al., 2007), an impact with global consequences for climate variability. The reduced latitudinal temperature gradients that result from the Arctic warming will modify the atmospheric circulation dynamics in the Northern Hemisphere. Mid-latitude storm tracks may shift (e.g., Deser and Teng., 2008), the westerlies may weaken, and storm intensities may decrease poleward of 45 N (e.g., Royer et al., 1990; Honda et al., 1999). Large-scale pressure systems such as the Azores High (Raymo et al., 1990) as well as the Asian monsoon and the Hadley Cell circulation systems may be affected (Liu et al., 2007).

Along with these impacts on the atmospheric and oceanic circulation, loss of Arctic sea ice has the potential to enhance the rates of surface melt of Greenland's glaciers. Present-day enhanced melting of Greenland's ice sheet is associated with increased advection of ocean heat onto the ice sheet from a warmer ocean, resulting in enhanced melt (e.g., Rennermalm et al., 2009). The warmer ocean surface temperatures that will occur in the absence of sea ice can be expected to enhance the rates of warming. The increased melt will contribute to sea level rise.

Sea ice in the Arctic is of major ecological importance; it is a habitat for a variety of species. An ice-free Arctic will promote large scale changes in Arctic marine ecosystems. Already in the Arctic, loss of sea ice has been associated with polar bear population decrease (e.g., DeWeaver, 2007); seasonal or perennial loss of sea ice will only exacerbate this situation. Sea ice protects the shorelines from erosion and helps maintain continuous permafrost. Lawrence et al. (2008b) show that loss of Arctic sea ice speeds the degradation of permafrost. Warming of the permafrost has already led to the destabilization of infrastructure in the Arctic, and removal of the protective cover of ice has already led to increased shoreline erosion (IPCC, 2007a,b); this can only worsen as sea ice cover is lost. Additionally, warming of the permafrost may lead to the emission of methane to the atmosphere, which has the potential to enhance greenhouse gas-related warming (Macdonald, 1990).

6

Beyond the Next Few Centuries

6.1 LONG-TERM FEEDBACKS AND EARTH SYSTEM SENSITIVITY

Earth System Sensitivity: On the Brink of the Anthropocene

Settled agricultural civilization arose during the interglacial period known as the Holocene, which has extended over the past 10,000 years. To geologists of the future, the evolution of the present interglacial will look different from that of the previous major interglacial—the Eemian, which set in 130,000 years ago. During other interglacials (which have occurred about every 100,000 years for more than the past 1 million years), CO_2 reached a peak value of about 300 ppm and thereafter began to fall; in this interglacial, CO_2 will instead rise by an amount that will be determined by human activities. We are now entering a new geological epoch, called the Anthropocene, during which the evolution of Earth's environment will be largely controlled by human activities, notably emissions of carbon dioxide from deforestation and fossil fuel burning. The Anthropocene will leave an imprint in the geological record as distinctive as other events that today's geologists find significant enough to merit a name. Actions taken within this century will determine whether the Anthropocene climate anomaly represents a small deviation from the Holocene climate, or a major shift with a duration of many thousands—perhaps even hundreds of thousands—of years. Over such long time scales, determining the climate response requires consideration of Earth System Sensitivity.

Earth System Sensitivity involves a number of processes that are less well understood than those involved in fast-feedback climate sensitivity or transient climate response. The challenges are compounded by the slow nature of these feedbacks, which makes them difficult to study through current observations of the changing climate. Study of paleoclimate provides a window into the operation of these slow processes, but the endeavor there is hampered by imprecision in our knowledge of past climate and past greenhouse gas concentrations. The principal known processes involved in Earth System Sensitivity are:

- The carbon cycle, including ocean carbon uptake and release, terrestrial carbon uptake or release, and release of methane by destabilization of clathrates stored in permafrost or in sea-floor sediments
- Major land ice sheets (such as those of Greenland and Antarctica)
- Vegetation changes affecting albedo and the hydrological cycle
- Changes in atmospheric chemistry that may affect aerosol formation and methane concentration
- Changes in atmospheric dust loading

Because the climate system is being pushed into uncharted territory without any precise past analogue, it is possible that the Earth system is subject to additional as-yet unidentified feedbacks. Although all of the above feedbacks have been implicated in past climate changes (as reviewed, e.g., in Lunt et al., 2010), the following discussion will focus on the first two.

As net cumulative CO_2 emissions increase, the amount by which the global temperature exceeds the peaks of the past 2 million years increases. Moreover, the length of time over which the climate is substantially warmer than previous interglacials becomes longer, allowing more time for slow components of the climate system to respond. The very long-term human imprint on climate can be assessed by computing the warming remaining after many centuries, taking into account only the climate sensitivity applied to the CO_2 remaining after allowing for uptake of carbon emissions by land and ocean. The resulting warming would be affected further by the additional feedbacks involved in Earth System Sensitivity, but examining the basic long-term warming gives an indication of the magnitude of climate change upon which these feedbacks act.

The uncertainty in the future course of climate is affected both by uncertainties in climate sensitivity and uncertainties in the carbon cycle. The joint effects of these uncertainties are presented in Figure 6.1. Some of the carbon cycle models included in the calculation sequester a moderate amount of carbon in land ecosystems during the early centuries, but none produces a significant long-term carbon release from land or marine sedimentary carbon pools. The effect of such a release would need to be taken into account by explicitly adding it in to the cumulative emissions directly produced by fossil fuel burning and land-use changes.

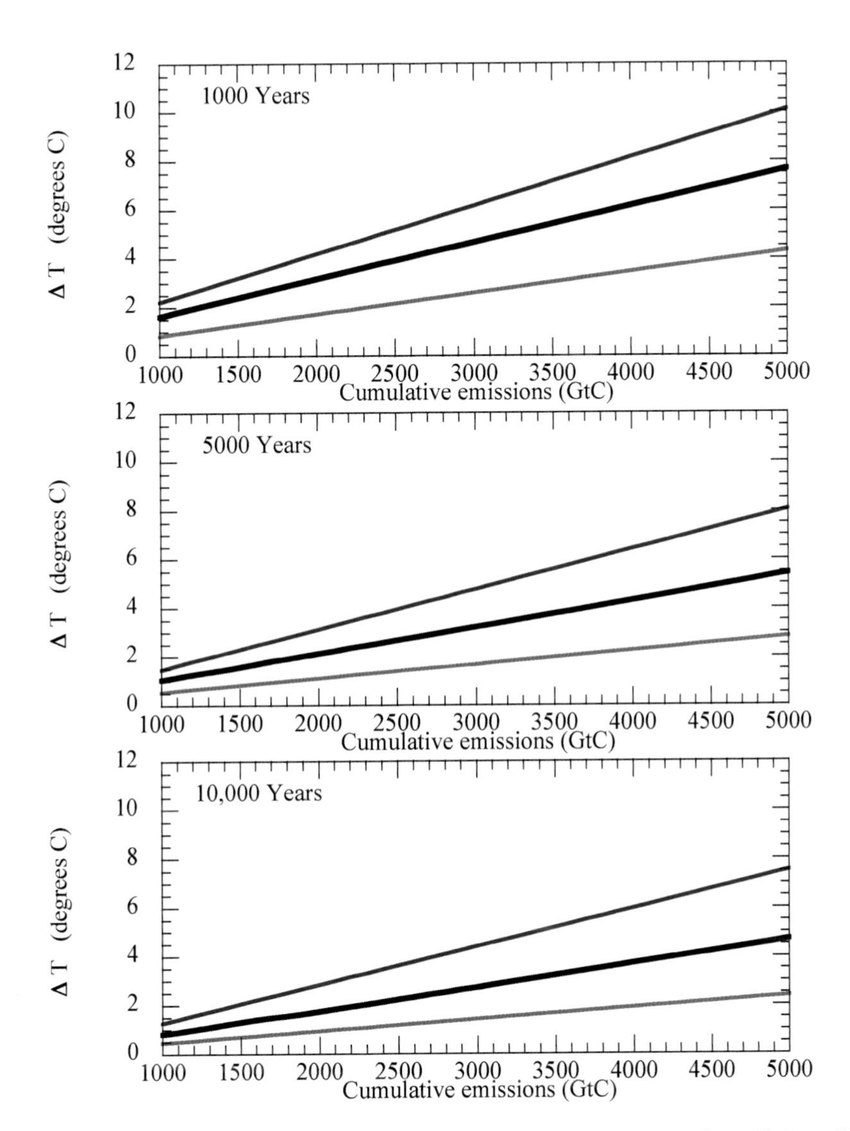

FIGURE 6.1 Range of very long-term warming obtained by applying the range of equilibrium climate sensitivity in Table 3.1 to the LTMIP ensemble of carbon-cycle models discussed in Archer et al. (2009). The upper red curve gives the maximum, the heavy black curve the median, and the lower green curve the minimum warming over all combinations of climate sensitivity and carbon-cycle models. These results incorporate the uptake of CO_2 by land and ocean, but do not include other Earth System Sensitivity feedbacks such as vegetation change or ice sheet response. See Methods appendix for details of the calculation.

With 1,000 GtC cumulative emissions the median estimate of the warming falls to 1.6°C at 1,000 years. The high end warming is still above 2°C at this time, though it falls below 2°C by 5,000 years. With 2,500 GtC cumulative emissions the median warming is still 2.7°C at 5,000 years, while the high-end warming exceeds 4°C. If cumulative emissions reach 5,000 GtC, even the lowest estimated warming remains above 2°C after 10,000 years, while the high end is above 7.5°C. In all cases, the warming decays very little between 5,000 years and 10,000 years, and in fact the warming remaining after 10,000 years would take 100,000 years or more to recover under the slow action of silicate weathering processes, even in the absence of destabilizing long-term Earth System feedbacks.

The large range of possible very long-term warming exhibited in the preceding discussion is due in large measure to the uncertainty in climate sensitivity. To what extent do past climate variations help to constrain this spread? The study of the instrumental record of climate provides, at best, a window into transient climate sensitivity. To obtain observational constraints on equilibrium climate sensitivity or Earth System Sensitivity, one must look into the more distant past. There are many ways to make use of the past climate record as a guide to the future, and in evaluating the published results, one must take care to distinguish the kind of climate sensitivity being estimated, which categories of climate forcings are regarded as feedbacks, and which categories of climate forcings are regarded as known or diagnosed forcings to be used in determining the sensitivity of the rest of the climate system.

Part of the cooling during the Last Glacial Maximum was due to a reduction in atmospheric CO_2, and this can be used to estimate climate sensitivity. To accomplish this, one must estimate and subtract out the portion of climate change forced by changes in Earth's orbit, by growth of the Northern Hemisphere ice sheets, and by dust radiative effects. In the end, this provides an estimate of climate sensitivity rather than Earth System Sensitivity, because the non-CO_2 forcings are treated diagnostically instead of as feedbacks. On this basis, Hargreaves and Annan estimate a most likely climate sensitivity corresponding to $\Delta T_{2x} = 2.5C$, with low likelihood that ΔT_{2x} exceeds 6°C. When additional observational constraints are incorporated, the maximum likely value is reduced to 4°C, in line with the range seen in the IPCC ensemble of models. Crucifix (2006) cautions, however, that the sensitivity of Earth's climate to reductions in CO_2 may not be a good indication of the sensitivity to increases.

The warm climates of Earth's more distant past provide our most important guide as to Earth System Sensitivity. There are three times of particular interest: the Pliocene period extending from 5.3 to 2.6 million years ago and preceding the initiation of the Pleistocene glacial/interglacial cycles; the Miocene period from 23 to 5.3 million years ago, during which Antarctic glaciation was initiated; and the Paleocene/Eocene boundary about 56 million years ago, at which point a massive release of carbon dioxide (presumably from land ecosystems) caused a massive warming of an already warm and largely ice-free climate.

The importance of the Pliocene is that the carbon dioxide concentrations at the time were only moderately greater than at present (380-425 ppm), but the climate was substantially different from today's climate (Lunt et al., 2010). Antarctica was already glaciated, but the Northern Hemisphere (particularly Greenland) was free of large ice sheets. Northern Hemisphere mid-latitude and high-latitude temperatures were considerably greater than today's. Tropical temperatures were not much warmer than modern values, although the characteristic east-west temperature gradient of the Pacific may have been much weaker, consisting of a permanent El Niño pattern (Wara et al., 2005). It is difficult to estimate global mean temperatures directly from proxy data, but using a combination of simulations and marine proxies, Lunt et al. (2010) estimate that the Pliocene global mean temperature was 3°C warmer than at present. The Pliocene provides at least a hint that the Earth System Sensitivity is such that a doubling of atmospheric CO_2, or perhaps even less, could cause a transition to a largely ice-free Northern Hemisphere, provided the CO_2 remains high long enough. In a model-based diagnosis of Pliocene climate feedbacks, Lunt et al., 2010 estimate that Earth System Sensitivity (not counting carbon cycle feedbacks) is 1.45 times the basic climate sensitivity.

The Paleocene-Eocene Thermal Maximum (PETM) provides one of the most worrying indicators of Earth System Sensitivity. This event begins at the end of the Paleocene, during which Earth was already in a warm, globally ice-free state. Carbon isotope data indicate that at this time a rapid release of isotopically light carbon occurred, and that the climate warmed globally by about 4°C. The most consistent picture at present is that the isotopically light carbon comes from a release of about 3,000 GtC of presumably land-based organic carbon, rather than from a destabilization of methane clathrates. For a review of the PETM and estimates of the amount of organic carbon release, the reader is referred to Zeebe et al., 2009. The essential challenge posed by the PETM is that one must explain an already warm Paleocene climate (presumably caused by elevated CO_2), at the same time as accounting for the additional warming caused by release of additional carbon, which must not exceed the limits allowed by data. The problem is that the radiative effect of CO_2 is logarithmic in CO_2 concentration, so that if one starts with a high level of CO_2 in order to explain the warm Paleocene, one needs an unrealistically large amount of additional CO_2 to double or quadruple CO_2 and give the requisite warming. On the other hand, if the sensitivity of climate to CO_2 is very high, then one can explain the Paleocene temperature with a smaller CO_2 concentration, and also get the required PETM warming from the release of a smaller amount of carbon to the atmosphere. This appears to demand a very high climate sensitivity, at least at the top of the IPCC range, and perhaps beyond (Pagani et al., 2006). Methane or other, currently unknown, radiative forcing agents may have affected the Pliocene climate and the PETM warming, but for now the simplest explanation of the PETM would appear to be that climate sensitivity is very high.

Impacts of anthropogenic global warming are quite sensitive to tropical warming, and the warm climates of the past shed some light on this issue as well. It has occasionally been proposed that the tropics are subject to a thermostat of one sort or another that limits tropical warming, but such proposals have been found to have no basis in physics (Williams et al., 2009). Moreover, the paleoclimate record of the Eocene and Paleocene provides direct support for the possibility of tropical temperatures considerably in excess of those prevailing today (Huber, 2008). Uncertainties in past CO_2 concentrations, however, make it impossible to say whether current general circulation models overestimate or underestimate tropical climate sensitivity. It is generally recognized, however, that general circulation models have difficulty reproducing the low meridional temperature gradient prevailing in past warm climates (Pierrehumbert, 2002; Huber and Sloan, 2001). This suggests that the Earth system is subject to feedbacks amplifying polar warming, which are not adequately represented in current models (Abbot et al., 2009a).

The Potential for Large Biogeochemical Emissions: Evidence and Time Scales

Emissions of greenhouse gases could be augmented in a warmer world due to releases of gases from biogeochemical processes, such as methane from methane hydrates both in permafrost at high latitudes and under the deep ocean, enhanced nitrous oxide emissions from soils, and increased release of carbon dioxide from warming peat, soils, and the biosphere (Denman et al., 2007). Some of these sources could be very large, raising the issue of the risk of substantial contributions to climate change. For example, some estimates suggest that the carbon reservoir in the form of methane stored in permafrost is of the order of 7.5-400 GtC (Brook et al., 2008), while that under the sea floor could amount to 500-2,500 GtC (Buffett and Archer, 2004; Milkov, 2004; Brook et al., 2008). Field studies have demonstrated remarkably large local emissions of methane in association with warming and melting of permafrost at particular Arctic sites, and there is evidence for substantial trace gas emissions at some times in the distant past in the paleoclimatic record (e.g., Walter et al., 2006).

Thus there is a potential for great risk, and this attracts the interest of scientists, the public, media, and policy makers. At present, our assessment is that it is not possible to quantify these risks. A challenge for climate science is not only to evaluate the physics and chemistry of the underlying processes, but also to explain why local observations or evidence from past climates do not necessarily imply that these factors are important for current and future anthropogenic climate changes.

Methane concentrations are currently about twice their pre-industrial levels; they were nearly stable for about a decade in 1997-2006, but began to increase again in 2007 (Rigby et al., 2008; Dlugokencky et al., 2009). Many studies establish ongoing permafrost retreat (Lemke et al., 2007) as well as considerable warming in the Arctic in both 2007 and 2008. But while global methane observations suggest a contribution from an increased source in the Arctic in 2007, there was no significant Arctic contribution to the methane increase in 2008 (Dlugokencky et al., 2009). Thus the currently warm Arctic does not seem to be a consistent source of methane that is significant on the global scale compared to other sources (which include wetlands, agriculture, animal husbandry, and waste processing, see Denman et al., 2007). One factor influencing methane release from permafrost is the amount of liquid water present, which controls whether decomposition is aerobic or anaerobic. This implies that not only thermal but also hydrological conditions are involved in whether or not conditions favor methane releases reaching the atmosphere on a large enough scale to be significant; similarly methane released from the sea floor can be degraded by bacteria before reaching the surface (Brook et al., 2008). Therefore, methane observations from particular sites, while sometimes dramatic and suggestive, may be insufficient for characterization of the much larger scales needed to understand global methane increases.

The large increase in methane observed at the time of the Younger-Dryas transition about 11,600 years ago is an example of a methane-climate feedback that has attracted significant interest. One recent study using isotopes suggests that the primary methane source at that time was from wetlands rather than permafrost (Petrenko et al., 2009). Several studies suggest a potential to increase current methane concentrations in a warmer world through wetland emissions (by perhaps as much as a doubling for 3°C warming, see Denman et al., 2007, and references therein); such a change, although significant, is far smaller than the potential reservoir in permafrost and illustrates that understanding the suite of contributing sources of trace gases is key to future projections.

A recent study by Fyke and Weaver (2006) studied the potential for significant methane emissions from the sea floor in a warmer world, and showed that the magnitude of the methane source depends upon thermal diffusivity (i.e., how rapidly warmth can be transmitted through the deep ocean and sediments), how large the warming is, and how long it lasts. That study suggests the potential for significant methane-climate feedbacks (up to 10% enhancement in the climate feedback parameter for warming pulses lasting 1,000 years). However, the total calculated methane release was limited, due to the very slow time scales involved compared to the likely duration of human-induced climate changes.

The effect of the release of clathrate methane on climate depends on the form and the time scale of the release of the stored carbon. When methane concentrations are much lower than CO_2 concentrations, as they are at present, the release of a GtC in the form of methane results in much greater radiative forcing than the release of a GtC in the form of CO_2; the precise ratio depends on the atmospheric methane and CO_2 concentrations at the time of release. Although release as methane would lead to a strong transient warming spike, the methane oxidizes to CO_2 on a decadal time scale, reducing the radiative forcing. Nonetheless, the clathrate reservoir is large enough that if a substantial portion of it is released, it would have a substantial radiative effect even after being converted to CO_2.

As an example, let's suppose that 100 GtC is released suddenly into the atmosphere as methane. This would increase the atmospheric methane concentration by 46 ppm over its present value of 1.8 ppm, leading to a radiative forcing of 6 W/m^2, which would cause a transient climate warming of 2.1 to 3.6°C, based on the likely range of transient climate sensitivity. Once oxidized to CO_2, however, the radiative forcing subsides to only 0.6 W/m^2 when applied on top of an ambient CO_2 concentration of 390 ppm. On the other hand, release of, say, 1,000 GtC of clathrate methane would add significantly to the long-term radiative forcing even if the release were slow enough that the carbon accumulated in the atmosphere in the form of CO_2. A slow release of methane would further add to the long-term warming since the release would sustain higher steady-state methane concentrations during the time when the release was occurring.

In addition to carbon stored in the form of methane clathrates there is also a substantial reservoir of organic carbon in near-surface land deposits. The standard estimate of the amount of carbon in the top meter of the land surface, plus living biomass, is about 2,100 gigatonnes of carbon, equivalent to nearly 1,000 ppm of atmospheric CO_2 concentration before allowing for ocean uptake (Trumper et al., 2009). This figure is probably an underestimate of terrestrial organic carbon storage, as it doesn't fully account for carbon storage in peatlands, and there is considerable potential that permafrost systems may store an additional 1,000 GtC or more of organic carbon (Schuur et al., 2008). At all times, microbial respiration is oxidizing some of this carbon and turning it to CO_2, while photosynthesis is storing new carbon in the terrestrial pool. Currently, it is the photosynthetic storage that wins, so that the terrestrial ecosystem provides a moderate net sink of anthropogenic carbon. Are there circumstances when the balance can change and the terrestrial ecosystem instead becomes a net source of atmospheric CO_2?

Terrestrial carbon-cycle modeling was discussed in detail in Section 2.4. Many models predict that terrestrial ecosystems will continue to be a modest sink of anthropogenic carbon, but this conclusion is dependent on highly contested aspects of the CO_2 fertilization effect, and in particular the possible role of nutrient limitation in inhibiting fertilization. At least one carbon-cycle model predicts that terrestrial ecosystems can become a net CO_2 source of carbon by 2050, with eventual releases of up to 5 GtC per year (Cox et al., 2000). The best evidence that the Earth system can indeed release several thousand GtC of organic carbon in the context of a warming environment is provided by the Paleocene-Eocene Thermal Maximum.

The PETM event demonstrates that the Earth system can succumb to destabilizing carbon-cycle feedbacks, in which an organic carbon pool (presumably on land) begins to oxidize rapidly and becomes a source rather than a sink of atmospheric carbon dioxide. In the case of the PETM, the release amounted to 3,000 GtC (Zeebe et al., 2009), which is eight times as much carbon as has been released by all fossil burning to date, and comparable to the higher range of estimates of what might be released by future fossil-fuel burning. The processes that led to the PETM carbon release are not at all understood, and so the risk that anthropogenic global warming could trigger a similar catastrophic release cannot at present be quantified, save to say that the risk is a real one. Estimates of carbon-cycle feedback based on Pleistocene or Holocene carbon dioxide fluctuations suggest much smaller carbon-cycle feedbacks than the PETM (Frank et al., 2010), but the PETM provides the closest analogue to what might happen to the carbon cycle in a climate substantially warmer than the present.

In summary, there is potential for biogeochemical feedbacks to climate change through emissions of trace gases but it cannot be quantified at present. Such releases could add 2,000 GtC or more to the carbon directly put into the atmosphere by fossil-fuel burning and land-use change. If some of this stored carbon is released rapidly in the form of methane, it could lead to a pronounced transient warming spike above and beyond the more persistent warming caused by increases in atmospheric CO_2. Local measurements of releases are important in understanding processes but are currently insufficient to characterize the significance of sources for the global atmosphere. Thresholds for large effects are difficult to establish from paleoclimatic information, since key processes may become important over very slow time scales of many thousands of years in a warmer world, but may take place too slowly to be important on the time scales of relevance for human perturbations (i.e., the very long time scales required for the process to act may exceed the Anthropocene period over which human carbon emissions are expected to significantly warm the atmosphere).

Ice Sheets Beyond 2100

The extent to which the great ice sheets of Greenland and Antarctica will survive the Anthropocene is a question of paramount importance. Paleoclimate reconstructions provide some information regarding the conditions for initiation of these ice sheets, but there are no known instances of complete deglaciation of these ice sheets that could provide ground-truth for estimates of the temperature thresholds for deglaciation and the time required for deglaciation to take place. Therefore, estimates of temperature and duration thresholds must be drawn from ice sheet models, past partial deglaciations of Greenland and Antarctica, and past total deglaciations of the Laurentide or Fenno-Scandian ice sheets. It should be kept in mind that there are many important physical processes that are not well represented in current ice-sheet models, so that any thresholds based on models should be taken as only a general indication of what ice sheets can do, rather than precise, definitive values.

The waxing and waning of large land ice sheets has a profound effect on sea level. Moreover, the major categories of past climate states are often distinguished by the presence of ice near both poles (the Pleistocene and Holocene), near the South Pole only (the Pliocene), or by ice-free conditions in both the Arctic and Antarctic (the Eocene). The glacial/interglacial transitions of the Pleistocene involve the growth and decay of the Laurentide and

Fenno-Scandian ice sheets, but going forward into warmer climates, it is the fate of the Greenland and Antarctic ice sheets that is of prime interest.

The loss of Greenland would make the Anthropocene look something like the Pliocene world, before the Pleistocene glacial/interglacial cycles set in, and there is strong evidence that this could happen in response to anthropogenic CO_2 emissions. Using coupled GCM/ice-sheet models, Ridley et al. (2005) find that Greenland deglaciates almost completely when the global mean warming exceeds 5°C for 1,500 years, and (Vizcaino et al., 2008) find near-total deglaciation when the global mean warming remains above 3.5°C for 4,000 years. Gregory et al. (2004) argue that even lesser warming could lead to deglaciation and cite indications that, once deglaciated, the Greenland ice cap might not recover even if CO_2 were restored to pre-industrial levels. The deglaciation of Greenland is primarily sensitive to Greenland regional summer warming, and models differ greatly as to the relationship of global mean temperature to the local summer warming. Estimates of the global mean warming needed for deglaciation range from a low of 1.9°C to a high of 4.6°C (Meehl et al., 2007). The vulnerability of Greenland found in models is consistent with the absence of Northern Hemisphere glaciation in the Pliocene. Calculated Greenland melt is shown in Figure 6.2. During the Eemian about 130,000 years ago, Arctic temperatures were about 3-5°C warmer than present, and it has been estimated that the loss of ice from Greenland and other Arctic ice fields contributed up to 4 m to sea level rise (Jansen et al., 2007).

In protracted warm conditions, the deglaciation of Greenland proceeds from robust melt ablation, with little need to involve the less well understood aspects of ice flow. Various aspects of ice dynamics that are not currently well represented in glacier models have the potential to allow the deglaciation to proceed much more rapidly, but nothing definitive can be said about the minimum time scale at present.

It is sometimes asserted that anthropogenic CO_2 emission would be beneficial because it could avoid an impending ice age, but this is incorrect. In fact, Berger et al. (2003) and Berger and Loutre (2002) project that Earth's current orbital configuration would make the Holocene interglacial unusually long even without bringing anthropogenic CO_2 into the mix. Earth's orbit would not be expected to produce another ice age for at least 30,000 years (Jansen et al., 2007). Anthropogenic warming is piling warming on top of an interglacial that is already projected to be unusually long, and moreover doing it at a time when the precessional cycle will be swinging into a hot Northern Hemisphere phase over the next 5,000 years. This adds to the prospect that anthropogenic CO_2 emissions could lead to a major

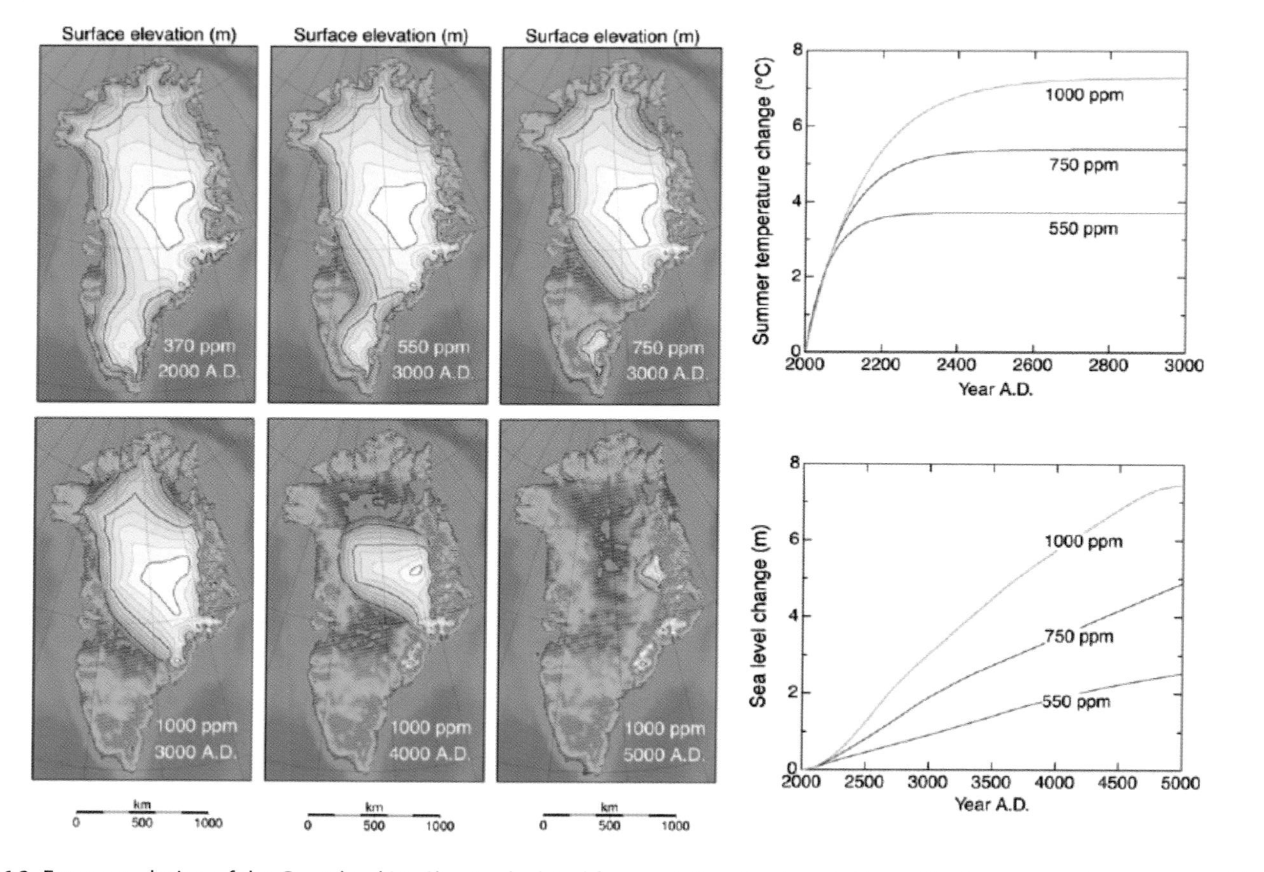

FIGURE 6.2 Future evolution of the Greenland Ice Sheet calculated from a 3D ice-sheet model forced by three greenhouse gas stabilization scenarios. The warming scenarios correspond to the average of seven IPCC models in which the atmospheric CO_2 concentration stabilizes at levels between 550 and 1,000 ppm after a few centuries (Gregory et al., 2004) and is kept constant after that. For a sustained average summer warming of 7.3°C (1,000 ppm), the Greenland Ice Sheet is shown to disappear within 3,000 years, raising sea level by about 7.5 m. For lower CO_2 concentrations, melting proceeds at a slower rate, but even in a world with twice as much CO_2 (550 ppm or a 3.7°C summer warming) the ice sheet will eventually melt away apart from some residual glaciation over the eastern mountains. The figure is based on the models discussed in Huybrechts and de Wolde (1999). Source: Alley et al. (2005).

climate transition. Berger et al. (2003) and Berger and Loutre (2002) note that the Anthropocene may initiate a 50,000-year interruption of Northern Hemisphere glaciation, during much of which time the Northern Hemisphere may be almost ice-free.

During the last interglacial period (the Eemian), global sea level was at least 3 m, and probably more than 5 m, higher than at present. But some studies suggest that the high sea levels during the last interglacial period have been proposed to result mainly from disintegration of the West Antarctic ice sheet, with model studies attributing only 1-2 m of sea level rise to meltwater from Greenland. Cuffey and Marshall (2000) suggest that the Greenland ice sheet was considerably smaller and steeper during the Eemian, and plausibly contributed 4-5.5 m to the sea level highstand during that period. New results from ice cores indicate that a significant ice sheet was covering Greenland during the warm Eemian period and that the reduction of the Greenland ice sheet at most contributed a sea level rise of 1-2 m of the observed 5 m (Dahl-Jensen et al., 2005).

Total loss of the Antarctic ice sheet would make the Anthropocene look something like the largely ice-free pre-Miocene climates that prevailed more than 30 million years ago. Antarctica is less subject to direct ablation by melting than is Greenland, and so the conditions for deglaciation of Antarctica are much more dependent on poorly understood aspects of ice dynamics. Many important processes, including ice streams and ice shelves, are not represented at all in the models used to study the problem. There have been few modeling studies, and little confidence can be placed in the few that have been done. Conventional thinking has it that the current East Antarctic ice sheet would be very difficult to get rid of by anthropogenic warming (Huybrechts, 1993). Indeed, Vizcaino et al. (2008) find that the Antarctic ice sheet *grows* in volume even when the global mean warming exceeds 4°C for 1,000 years. In a study of Antarctic glacial fluctuations, Pollard and Deconto (2005) find that substantial portions of Antarctica deglaciate in under 10,000 years in response to CO_2 concentrations on the order of 840 ppm, but express doubt that their results are applicable to the more extensive Antarctic glacier of today. Modeling studies, however, support the possibility of deglaciation of the West Antarctic ice sheet if subjected to local oceanic warming of as little as 5°C over a few thousand years, and there is moreover good support for episodic deglaciation of the West Antarctic Ice Sheet in Pliocene conditions (Pollard and DeConto, 2009). Based on the pattern scaling given in Section 4.1, this would correspond roughly to a 5°C global mean warming, but much caution should be exercised in applying these transient climate response patterns to long-term climate behavior, especially in view of the great difficulties models have with reproducing Antarctic regional climate.

A great deal of public attention has focused on 350 ppm as a target for CO_2 concentration. The identification of 350 ppm as a danger threshold for CO_2 (Hansen et al., 2008) rests largely on the study of the Miocene initiation of Antarctic glaciation. The assumption here is that Antarctica would deglaciate at CO_2 concentrations similar to those which permitted the initiation of Antarctic glaciation. The CO_2 concentration at the time of Antarctic glacier initiation is not very well known, and 350 ppm is drawn from the lower end of estimates of the concentrations prevailing at the time of initiation of Antarctic glaciation. Beyond that it is likely that the East Antarctic ice sheet, once formed, can survive considerably higher temperatures than those prevailing at the time of its initiation. The paleoclimate argument for 350 ppm as a danger threshold must be considered speculative, but the essential difficulty is that there are no past analogues in which a massive East Antarctic ice sheet has been subjected to temperatures as warm as those that may prevail in the Anthropocene.

It is important to recognize that the hypothetical dangers from a CO_2 concentration of 350 ppm reside in the very long-term climate feedbacks. Therefore, the CO_2 would have to remain above this level for thousands of years in order for these feedbacks to be a major concern. The concentration can exceed 350 ppm at its peak without incurring a risk of triggering the long-term feedbacks, so long as it subsides to 350 ppm or less over a few thousand years. In the carbon cycle model of Eby et al. (2009), cumulative carbon emissions of 850 GtC or less are required to allow the CO_2 concentration to subside to 350 ppm within 5,000 years, but over the range of carbon cycle models with long-term climate and sediment dissolution feedbacks discussed in Archer et al. (2009), two-thirds of the models recover to under 350 ppm in 5,000 years when the cumulative emissions are 1,000 Gt.

If uncompensated by increase in water storage in East Antarctica, the total loss of the Greenland ice sheet would lead to a sea level rise of 7.5 m (Bamber et al., 2001), while the total loss of the West Antarctic Ice Sheet would lead to an additional rise of 5 m (Bamber et al., 2009). The latter is comparable to the contributions of the West Antarctic Ice Sheet to Pliocene and Pleistocene sea level fluctuations as modeled by Pollard and DeConto (2009). Melting of all mountain glaciers and ice caps would add at least 0.7 m to sea level (Bahr et al., 2009), although it should be noted there is considerable uncertainty regarding the thickness of many ice caps in mountainous regions. This would be added to the long-term sea level rise due to thermal expansion (estimated to be 0.2 to 0.6 m per degree C of global warming in IPCC, 2007a), which would amount to 0.6 to 1.8 m for a long-term warming of 3°C or 1.2 to 3.6 m for a long-term warming of 6°C.

6.2 LONG TERM SOCIETAL AND ENVIRONMENTAL ISSUES

Over the coming millennia, some impacts of climate change may settle into new patterns of climate variability with the successful implementation of stabilization policies that cap cumulative emissions and therefore limit increases in global mean temperature. Climate variability could then be distributed around different means (with perhaps different higher moments), but it is possible that societies could become accustomed to these new environments. That world would be different than today, but new conditions could become routine to people living on Earth one or two thousand years from now. Other impacts, however, could continue for many centuries past the date of temperature stabilization.

Rising seas and melting glaciers and/or ice sheets easily fit into this second category of persistent and growing very long-term significance. Figure 6.3 displays, for example, contours of 1 m of sea level rise for Florida, which could occur by 2100 based on Section 4.8. In the longer term, much larger sea level rise is possible over millennia (see Section 6.1). Clearly, increases in risks from inundation, repeated flooding, and coastal erosion that have already been documented in some places for modest sea level rise could therefore continue as the future unfolds and could well be amplified over the long term depending upon the rate at which they occur as the climate system changes.

To get a better understanding of what associated vulnerabilities might look like as the long-term future unfolds, one might contemplate tracking widespread migration over recent time away from areas of exacerbated climate risk, but attribution would be extremely difficult. As noted in Wilbanks et al. (2007: Box 7.2), observed environmental migration is often a temporary reaction to the calamitous ramifications of one extreme event or another. WDR (2010) reports that displaced people (the estimated 26 million people who have moved permanently during recent years) constitute less than 10% of the world's international migrants and that most of these people relocate still live within the same country or, at worst, somewhere in the same region of the world.

IPCC (2007e) reported, in words that were unanimously approved in the plenary as part of the Summary for Policymakers, that

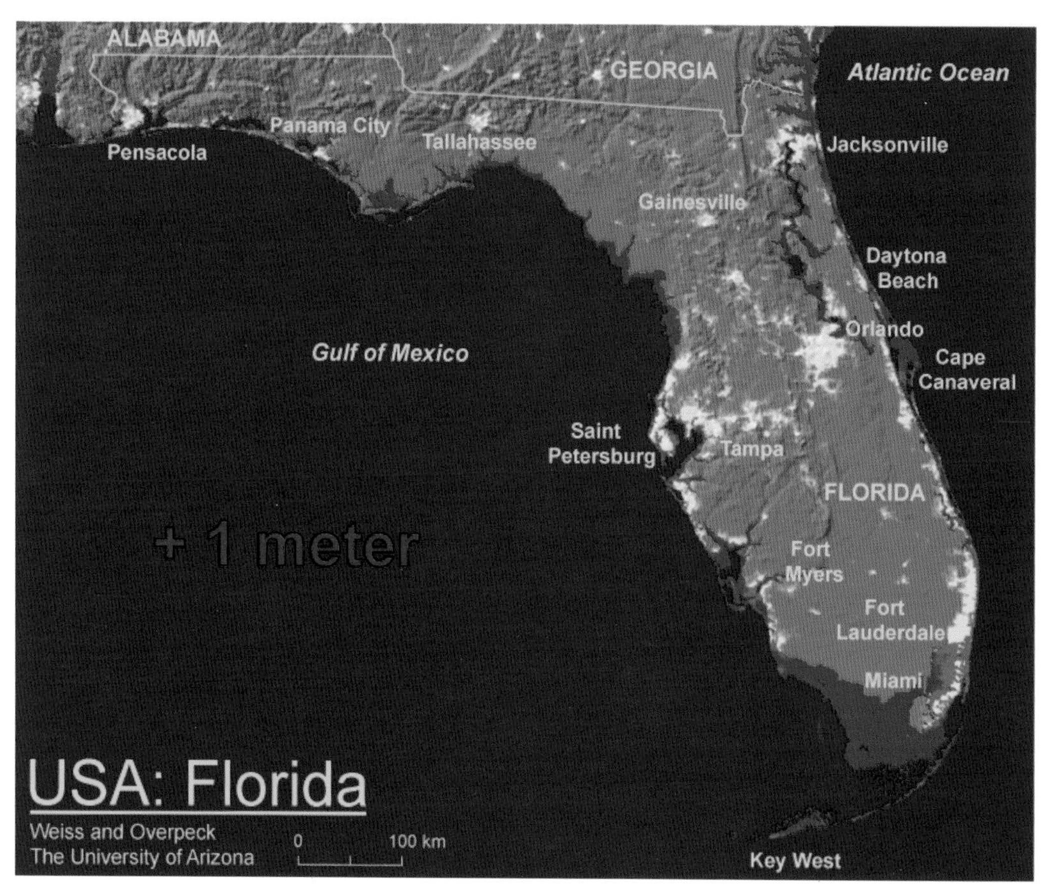

FIGURE 6.3 Effect of sea level rise of 1 m in the area of Florida. Red regions would be subject to inundation at this sea level rise. Source: University of Arizona, see *http://www.geo.arizona.edu/dgesl/research/other/climate_change_and_sea_level/sea_level_rise/sea_level_rise_technical.htm*.

There is high confidence that neither adaptation nor mitigation alone can avoid all climate change impacts; however, they can complement each other and together can significantly reduce the risks of climate change. Adaptation is necessary in the short and longer term to address impacts resulting from the warming that would occur even for the lowest stabilization scenarios assessed. Unmitigated climate change would, in the long term, be *likely* to exceed the capacity of natural, managed and human systems to adapt. The time at which such limits could be reached will vary between sectors and regions. (p. 19).

This finding, drawn in large measure from Yohe et al. (2006), refers to aggregate measures of vulnerability projected over a one-century time scale; it cannot be interpreted as meaning that every sector and every person would be incapable of adapting to preserve their standards of living. It does, however, suggest that responding to climate change and associated climate variability over the very long term could become increasingly more difficult and expensive across developed and developing countries, alike.

To summarize, more nuanced analyses of some sources of vulnerability to climate change that would persist and, indeed, continue to grow over the coming millenia are required to provide useful insight into the consequences of stabilization over the very long term. Since rising seas are the source of one such persistent and growing threat across the world, though, it is entirely plausible that displaced people may be forced to migrate even if temperature increases are capped. They may move within a country or region (like, as reported in Kates et al. [2006], the tens of thousands of people who moved to communities across the United States after Hurricane Katrina and have decided not to return), but they may not.

References

Abbot, D. S., M. Huber, G. Bousquet, and C. C. Walker. 2009a. High-CO_2 cloud radiative forcing feedback over both land and ocean in a global climate model. *Geophysical Research Letters* 36:L05702, **DOI:** 10.1029/2008GL036703.

Abbot, D. S., C. C. Walker, and E. Tziperman. 2009b. Can a convective cloud feedback help to eliminate winter sea ice at high CO_2 concentrations? *Journal of Climate 22* (21):5719-5731, **DOI:** 10.1175/2009JCLI2854.1.

Agrawal, M., M. Auffhammer, U.K. Chopra, L. Emberson, M. Iyngararasan, N. Kalra, M. V. Ramana, V., and A. K. S. a. J. V. Ramanathan. 2008. *Impacts of Atmospheric Brown Clouds on Agriculture, Part II of Atmospheric Brown Clouds: Regional Assessment Report with Focus on Asia.* Nairobi, Kenya.

Ainsworth, E. A., A. D. B. Leakey, D. R. Ort, and S. P. Long. 2008. FACE-ing the facts: inconsistencies and interdependence among field, chamber and modeling studies of elevated CO_2 impacts on crop yield and food supply. *New Phytologist 179* (1):5-9.

Allan, R. P. and B. J. Soden. 2008. Atmospheric warming and the amplification of precipitation extremes. *Science 321* (5895):1481-1484, **DOI:** 10.1126/science.1160787.

Allen, M. R., D. J. Frame, C. Huntingford, C. D. Jones, J. A. Lowe, M. Meinshausen, and N. Meinshausen. 2009. Warming caused by cumulative carbon emissions towards the trillionth tonne. *Nature 458* (7242):1163-1166, **DOI:** 10.1038/nature08019.

Alley, R. B., P. U. Clark, P. Huybrechts, and I. Joughin. 2005. Ice-sheet and sea-level changes. *Science 310* (5747):456-460, **DOI:** 10.1126/science.1114613.

Allison, I., R. B. Alley, H. A. Fricker, R. H. Thomas, and R. C. Warner. 2009a. Ice sheet mass balance and sea level. *Antarctic Science 21* (5):413-426, **DOI:** 10.1017/S0954102009990137.

Allison, I., N. L. B., R. A. Bindoff, R. A. Bindschadler, P. M. Cox, N. de Noblet, M. H. England, J. E. Francis, N. Gruber, A. M. Haywood, D. J. Karoly, G. Kaser, C. Le Quéré, T. M. Lenton, M. E. Mann, B. I. McNeil, A. J. Pitman, S. Rahmstorf, E. Rignot, H. J. Schellnhuber, S. H. Schneider, S. C. Sherwood, R. C. J. Somerville, K.Steffen, E. J. Steig, M. Visbeck, and A. J. Weaver. 2009b. *The Copenhagen Diagnosis, 2009b: updating the world on the latest climate science.* Sydney, Australia: The University of New South Wales Climate Change Research Centre (CCRC).

Alroy, J. 1998. Equilibrium diversity dynamics in North American mammals. In *Biodiversity dynamics: turnover of populations, taxa, and communities,* edited by M. McKinney and J. Drake. 232-287. New York: Columbia University Press.

Amato, A. D., M. Ruth, P. Kirshen, and J. Horwitz. 2005. Regional energy demand responses to climate change: methodology and application to the Commonwealth of Massachusetts. *Climatic Change 71* (1-2):75-201, **DOI:** 10.1007/s10584-005-5931-2.

Angert, A., S. Biraud, C. Bonfils, C. C. Henning, W. Buermann, J. Pinzon, C. J. Tucker, and I. Fung. 2005. Drier summers cancel out the CO_2 uptake enhancement induced by warmer springs. *Proceedings of the National Academy of Sciences of the United States of America* 102:10823-10827, **DOI:** 10.1073/pnas.0501647102.

Arblaster, J. M. and G. A. Meehl. 2006. Contributions of external forcings to southern annular mode trends. *Journal of Climate 19* (12):2896-2905.

Archer, D. 2005. Fate of fossil fuel CO_2 in geologic time. *Journal of Geophysical Research-Oceans* 110(C9):C09S05, **DOI:** 10.1029/2004JC002625.

Archer, D. and Victor Brovkin. 2008. The millennial atmospheric lifetime of anthropogenic CO_2. *Climatic Change 90* (3):283-297.

Archer, D., H. Kheshgi, and E. Maier-Reimer. 1997. Multiple timescales for neutralization of fossil fuel CO_2. *Geophysical Research Letters 24* (4):405-408.

Archer, D., M. Eby, V. Brovkin, A. Ridgwell, L. Cao, U. Mikolajewicz, K. Caldeira, K. Matsumoto, G. Munhoven, A. Montenegro, and K. Tokos. 2009. Atmospheric lifetime of fossil fuel carbon dioxide. *Annual Review of Earth and Planetary Sciences 37*:117-134, **DOI:** 10.1146/annurev.earth.031208.100206.

ACIA (Arctic Climate Impact Assessment). Berner, J., T. V. Callaghan, S. Fox, C. Furgal, A. H. Hoel, H. Huntington, A. Instanes, G. P. Juday, E. Källén, V. M. Kattsov, D. R. Klein, H. Loeng, M. Long Martello, G. McBean, J. J. McCarthy, M. Nuttall, T. D. Prowse, J. D. Reist, A. Stevermer, A. Tanskanen, M. B. Usher, H. Vilhjálmsson, J. E. Walsh, B. Weatherhead, G. Weller, F. J. Wrona, et al. 2005. *Arctic Climate Impact Assessment*. New York City: Cambridge University Press.

Arzel, O., T. Fichefet, and H. Goosse. 2006. Sea ice evolution over the 20th and 21st centuries as simulated by current AOGCMs. *Ocean Modelling 12* (3-4):401-415, **DOI:** 10.1016/j.ocemod.2005.08.002.

Bahr, D. B., M. Dyurgerov, and M. F. Meier. 2009. Sea-level rise from glaciers and ice caps: a lower bound. *Geophysical Research Letters 36:*L03501, **DOI:** 10.1029/2008GL036309.

Bailey, R. G. 1995. *Description of the Ecoregions of the United States* (2nd, rev. and enl. ed.). Washington, DC: U.S. Department of Agriculture, Forest Service.

Bakun, A., D. B. Field, A. Redondo-Rodriguez, and S. J. Weeks. 2010. Greenhouse gas, upwelling-favorable winds, and the future of coastal ocean upwelling ecosystems. *Global Change Biology 16* (4):1213-1228, **DOI:** 10.1111/j.1365-2486.2009.02094.x.

Bala, G., K. Caldeira, A. Mirin, M. Wickett, C. Delire, and T. J. Phillips. 2006. Biogeophysical effects of CO_2 fertilization on global climate. *Tellus Series B-Chemical and Physical Meteorology 58* (5):620-627, **DOI:** 10.1111/j.1600-0889.2006.00210.x.

Ballester, J., F. Giorgi, and X. Rodo. 2010. Changes in European temperature extremes can be predicted from changes in PDF central statistics. *Climatic Change 98* (1-2):277-284, **DOI:** 10.1007/s10584-009-9758-0.

Balshi, M. S., A. D. McGuire, P. Duffy, M. Flannigan, J. Walsh, and J. Melillo. 2009. Assessing the response of area burned to changing climate in western boreal North America using a Multivariate Adaptive Regression Splines (MARS) approach. *Global Change Biology 15* (3):578-600, **DOI:** 10.1111/j.1365-2486.2008.01679.x.

Bamber, J. L., S. Ekholm, and W. B. Krabill. 2001. A new, high-resolution digital elevation model of Greenland fully validated with airborne laser altimeter data. Journal of Geophysical *Research B 106*:6733-6745.

Bamber J. L., R. E. M. Riva, B. L. A. Vermeersen, and A. M. LeBroq. 2009. Reassessment of the potential sea-level rise from a collapse of the West Antarctic Ice Sheet. *Science* 324 (5929):901-903, **DOI:** 10.1126/science.1169335.

Barnett, J. 2003. Security and climate change. *Global Environmental Change-Human and Policy Dimensions 13* (1):7-17.

Barnett, T. P., D. W. Pierce, H. G. Hidalgo, C. Bonfils, B. D. Santer, T. Das, G. Bala, A. W. Wood, T. Nozawa, A. A. Mirin, D. R. Cayan, and M. D. Dettinger. 2008. Human-induced changes in the hydrology of the western United States. *Science 319* (5866):1080-1083, **DOI:** 10.1126/science.1152538.

Basu, R. 2009. High ambient temperature and mortality: a review of epidemiologic studies from 2001 to 2008. *Environmental Health 8* (40): **DOI:** 10.1186/1476-069X-8-40.

Basu, R. and J. M. Samet. 2002. Relation between elevated ambient temperature and mortality: a review of the epidemiologic evidence. *Epidemiologic Reviews 24* (2):190-202, **DOI:** 10.1093/epirev/mxf007.

Battisti, D. S. and R. L. Naylor. 2009. Historical warnings of future food insecurity with unprecedented seasonal heat. *Science 323* (5911):240-244, **DOI:** 10.1126/science.1164363.

Behrenfeld, M. J., R. T. O'Malley, D. A. Siegel, C. R. McClain, J. L. Sarmiento, G. C. Feldman, A. J. Milligan, P. G. Falkowski, R. M. Letelier, and E. S. Boss. 2006. Climate-driven trends in contemporary ocean productivity. *Nature 444* (7120):752-755, **DOI:** 10.1038/nature05317.

Bell, M. L., R. Goldberg, C. Hogrefe, P. L. Kinney, K. Knowlton, B. Lynn, J. Rosenthal, C. Rosenzweig, and J. A. Patz. 2007. Climate change, ambient ozone, and health in 50 US cities. *Climatic Change 82* (1-2):61-76, **DOI:** 10.1007/s10584-006-9166-7.

Bengtsson, L., K. I. Hodges, M. Esch, N. Keenlyside, L. Kornblueh, J. J. Luo, and T. Yamagata. 2007. How may tropical cyclones change in a warmer climate? *Tellus Series a-Dynamic Meteorology and Oceanography 59* (4):539-561, **DOI:** 10.1111/j.1600-0870.2007.00251.x.

Berger, A. and M. F. Loutre. 2002. An exceptionally long interglacial ahead? *Science 297* (5585):1287-1288.

Berger, A., M. F. Loutre, and M. Crucifix. 2003. The Earth's climate in the next hundred thousand years (100 kyr). *Surveys in Geophysics 24* (2):117-138.

Bernard, S. M. and K. L. Ebi. 2001. Comments on the process and product of the health impacts assessment component of the national assessment of the potential consequences of climate variability and change for the United States. *Environmental Health Perspectives 109:*177-184.

Berner, R. 2004. *The Phanerozoic Carbon Cycle*. USA: Oxford University Press.

Bernstein, L., P. Bosch, O. Canziani, Z. Chen, R. Christ, O. Davidson, W. Hare, S. Huq, D. Karoly, V. Kattsov, Z. Kundzewicz, J. Liu, U. Lohmann, M. Manning, T. Matsuno, B. Menne, B. Metz, M. Mirza, N. Nicholls, L. Nurse, R. Pachauri, J. Palutikof, M. Parry, D. Qin, N. Ravindranath, A. Reisinger, J. Ren, K. Riahi, C. Rosenzweig, M. Rusticucci, S. Schneider, Y. Sokona, S. Solomon, P. Stott, R. Stouffer, T. Sugiyama, R. Swart, D. Tirpak, C. Vogel, G. Yohe. 2007. *IPCC, 2007: Climate Change 2007: Synthesis Report. Contribution of Working Groups I, II and III to the Fourth Assessment Report of the Intergovernmental Panel on Climate Change*. Cambridge.

Bindoff, N. L., J. Willebrand, V. Artale, A, Cazenave, J. Gregory, S. Gulev, K. Hanawa, C. Le Quéré, S. Levitus, Y. Nojiri, C.K. Shum, L.D. Talley, and A. Unnikrishnan. 2007. Observations: oceanic climate change and sea level. In *Climate Change 2007: The Physical Science Basis. Contribution of Working Group I to the Fourth Assessment Report of the Intergovernmental Panel on Climate Change*, edited by S. Solomon, D. Qin, M. Manning, Z. Chen, M. Marquis, K. B. Averyt, M. Tignor and H. L. Miller. Cambridge: Cambridge University Press.

Bitz, C. M., P. R. Gent, R. A. Woodgate, M. M. Holland, and R. Lindsay. 2007. The influence of sea ice on ocean heat uptake in response to increasing CO_2 (vol 19, pg 2437, 2007). *Journal of Climate 20* (12):2864-2864, **DOI:** 10.1175/JCLI4294.1.

Blanford, G. J., R. G. Richels, and T. F. Rutherford. 2009. Feasible climate targets: The roles of economic growth, coalition development and expectations. *Energy Economics 31* (SUPPL. 2):S82-S93, **DOI:** 10.1016/j.eneco.2009.06.003.

Boden, T. A., G. Marland, and R. J. Andres. 2009. *Global, Regional, and National Fossil-Fuel CO_2 Emissions*. Oak Ridge, TN: Oak Ridge National Laboratory.

Boe, J., A. Hall, and X. Qu. 2009. September sea-ice cover in the Arctic Ocean projected to vanish by 2100. *Nature Geoscience 2* (5):341-343, **DOI:** 10.1038/NGEO467.

Bograd, S. J., C. G. Castro, E. Di Lorenzo, D. M. Palacios, H. Bailey, W. Gilly, and F. P. Chavez. 2008. Oxygen declines and the shoaling of the hypoxic boundary in the California Current. *Geophysical Research Letters 35* (12): L12607, **DOI:** 10.1029/2008GL034185.

Bonfils, C., B. D. Santer, D. W. Pierce, H. G. Hidalgo, G. Bala, T. Das, T. P. Barnett, D. R. Cayan, C. Doutriax, A. W. Wood, A. Mirin, and T. Nozawa. 2008. Detection and attribution of temperature changes in the mountainous Western United States. *Journal of Climate 21* (23):6404-6424, **DOI:** 10.1175/2008JCLI2397.1.

Bony, S., R. Colman, V. M. Kattsov, R. P. Allan, C. S. Bretherton, J. L. Dufresne, A. Hall, S. Hallegatte, M. M. Holland, W. Ingram, D. A. Randall, B. J. Soden, G. Tselioudis, and M. J. Webb. 2006. How well do we understand and evaluate climate change feedback processes? *Journal of Climate 19* (15):3445-3482.

Boyd, P. W., and S. C. Doney. 2002. Modelling regional responses by marine pelagic ecosystems to global climate change. *Geophysical Research Letters 29* (16):1806, **DOI:** 10.1029/2001GL014130.

Boyd, P. W., S. C. Doney, R. Strzepek, J. Dusenberry, K. Lindsay, and I. Fung. 2008. Climate-mediated changes to mixed-layer properties in the Southern Ocean: assessing the phytoplankton response. *Biogeosciences 5* (3):847-864.

Bradshaw, W. E. and C. M. Holzapfel. 2008. Genetic response to rapid climate change: it's seasonal timing that matters. *Molecular Ecology 17* (1):157-166, **DOI:** 10.1111/j.1365-294X.2007.03509.x.

Bradshaw, W. E., C. M. Holzapfel, and D. Mathias. 2006. Circadian rhythmicity and photoperiodism in the pitcher-plant mosquito: Can the seasonal timer evolve independently of the circadian clock? *American Naturalist 167* (4):601-605.

Brewer, P. G., and E. T. Peltzer. 2009. Limits to marine life. *Science 324* (5925):347-348, **DOI:** 10.1126/science.1170756.

Broecker, W. S. 1997. Thermohaline circulation, the Achilles heel of our climate system: will man-made CO_2 upset the current balance? *Science* 278 (5343):1582-1588.

Brook, E. 2008. Potential for abrupt changes in atmospheric methane. In *Abrupt Climate Change Final Report*, edited by P. Clark, A. Weaver, K. Steffen, E. Cook, and E. Brook. 360-452. Washington, DC: U.S. Climate Change Science Program and the Subcommittee on Global Change Research, Lead agency: U.S. Geological Survey; contributing agencies; National Oceanic and Atmospheric Administration, National Science Foundation.

Brown, C., E. Fulton, A. Hobday, R. Matear, H. Possingham, C. Bulman, V. Christensen, R. E. Forrest, P. C. Gehrke, N. A. Gribble, S. P. Griffiths, H. Lozano-Montes, J. M. Martin, S. Metcalf, T. A. Okey, R. Watson, and A. J. Richardson. 2010. Effects of climate-driven primary production change on marine food webs: implications for fisheries and conservation. *Global Change Biology 16* (4):1194-1212, **DOI:** 10.1111/j.1365-2486.2009.02046.x.

Brown, J., K. Hinkel, and F. Nelson. 2000. The circumpolar active layer monitoring (calm) program: research designs and initial results. *Polar Geography 24* (3):166-258.

Brown, R. D. and P. Mote. 2009. The response of northern hemisphere snow cover to a changing climate. *Journal of Climate 22* (8):2124-2145, **DOI:** 10.1175/2008JCLI2665.1.

Buffett, B. and D. Archer. 2004. Global inventory of methane clathrate: sensitivity to changes in the deep ocean. *Earth and Planetary Science Letters 227* (3-4):185-199, **DOI:** 10.1016/j.epsl.2004.09.005.

Burke, M., E. Miguel, S. Satyanath, J. A. Dykema, and D. B. Lobell. 2009. Warming increases the risk of civil war in Africa. *Proceedings of the National Academy of Sciences of the United States of America 106* (49):20670-20674, **DOI:** 10.1073/pnas.0907998106.

Byrne, R. H., S. Mecking, R. A. Feely, and X. Liu. 2010. Direct observations of basin-wide acidification of the North Pacific Ocean. *Geophysical Research Letters 37* (2):L02601, **DOI:** 10.1029/2009GL040999.

Caldeira K. and J. Kasting. 1993. Insensitivity of global warming potentials to carbon dioxide emission scenarios. *Nature 366* (6452):251-253.

CALFED (California Department of Water Resources). 2009. Delta Risk Management Strategy, Phase I. Available online at: *http://science.calwater.ca.gov/drms/drms_irp.html* (Accessed 12 Jan 2010).

Campos, H., A. Cooper, J. E. Habben, G. O. Edmeades, and J. R. Schussler. 2004. Improving drought tolerance in maize: a view from industry. *Field Crops Research 90* (1):19-34, **DOI:** 10.1016/j.fcr.2004.07.003.

Canadell, J. G., C. Le Quere, M. R. Raupach, C. B. Field, E. T. Buitenhuis, P. Ciais, T. J. Conway, N. P. Gillett, R. A. Houghton, and G. Marland. 2007. Contributions to accelerating atmospheric CO_2 growth from economic activity, carbon intensity, and efficiency of natural sinks. *Proceedings of the National Academy of Sciences of the United States of America 104* (47):18866-18870, **DOI:** 10.1073/pnas.0702737104.

Carleton, A. M., G. John, and R. Welsch. 1998. Interannual variations and regionality of Antarctic sea-ice-temperature associations. In *Annals of Glaciology Vol 27, 1998*, edited by W. F. Budd. 403-408. Cambridge: International Glaciological Society.

Casola, J. H., L. Cuo, B. Livneh, D. P. Lettenmaier, M. T. Stoelinga, P. W. Mote, J. M. Wallace. 2009. Assessing the impacts of global warming on snowpack in the Washington Cascades. *Journal of Climate 22* (10):2758-2772, **DOI:** 10.1175/2008JCLI2612.1.

Cavalieri, D. J. and C. L. Parkinson. 2008. Antarctic sea ice variability and trends, 1979-2006. *Journal of Geophysical Research-Oceans 113* (C7):C07004, **DOI:** 10.1029/2007JC004564.

Cazenave, A., A. Lombard, and W. Llovel. 2008. Present-day sea level rise: A synthesis. *Comptes Rendus Geoscience 340* (11):761-770, **DOI:** 10.1016/j.crte.2008.07.008.

Cazenave, A., D. Dominh, S. Guinehut, E. Berthier, W. Llovel, G. Ramillien, M. Ablain, and G. Larnicol. 2009. Sea level budget over 2003-2008: a reevaluation from GRACE space gravimetry, satellite altimetry and Argo. *Global and Planetary Change 65* (1-2):83-88, **DOI:** 10.1016/j.gloplacha.10.004.

Chan, F., J. A. Barth, J. Lubchenco, A. Kirincich, H. Weeks, W. T. Peterson, and B. A. Menge. 2008. Emergence of anoxia in the California current large marine ecosystem. *Science 319* (5865):920, **DOI:** 10.1126/science.1149016.

Chang, E. K. M. and Y. J. Guo. 2007. Is the number of North Atlantic tropical cyclones significantly underestimated prior to the availability of satellite observations? *Geophysical Research Letters 34* (14): L14801, **DOI:** 10.1029/2007GL030169.

Changnon, S. A. 1999. Record flood-producing rainstorms of 17-18 July 1996 in the Chicago metropolitan area. Part III: Impacts and responses to the flash flooding. *Journal of Applied Meteorology 38* (3):273-280.

Chapin, F. S., M. Sturm, M. C. Serreze, J. P. McFadden, J. R. Key, A. H. Lloyd, A. D. McGuire, T. S. Rupp, A. H. Lynch, J. P. Schimel, J. Beringer, W. L. Chapman, H. E. Epstein, E. S. Euskirchen, L. D. Hinzman, G. Jia, C. L. Ping, K. D. Tape, C. D. C. Thompson, D. A. Walker, and J. M. Welker. 2005. Role of land-surface changes in Arctic summer warming. *Science 310* (5748):657-660, **DOI:** 10.1126/science.1117368.

Cherkauer, K. 2010 forthcoming. Climate impacts on river hydrology.

Cheung, W. W. L., V. W. Y. Lam, J. L. Sarmiento, K. Kearney, R. Watson, D. Zeller, and D. Pauly. 2010. Large-scale redistribution of maximum fisheries catch potential in the global ocean under climate change. *Global Change Biology 16* (1):24-35, **DOI:** 10.1111/j.1365-2486.2009.01995.x.

Chown, S. L. and K. J. Gaston. 1999. Exploring links between physiology and ecology at macro-scales: the role of respiratory metabolism in insects. *Biological Reviews 74* (1):87-120.

Christensen, N. S. and D. P. Lettenmaier. 2007. A multimodel ensemble approach to assessment of climate change impacts on the hydrology and water resources of the Colorado River Basin. *Hydrology and Earth System Sciences 11* (4):1417-1434.

Christensen, N. S., A. W. Wood, N. Voisin, D. P. Lettenmaier, and R. N. Palmer. 2004. The effects of climate change on the hydrology and water resources of the Colorado River basin. *Climatic Change 62* (1-3):337-363.

Christidis, N., P. A. Stott, S. Brown, D. J. Karoly, and J. Caesar. 2007. Human contribution to the lengthening of the growing season during 1950-99. *Journal of Climate 20* (21):5441-5454, **DOI:** 10.1175/2007JCLI1568.1.

Church, J. A. and N. J. White. 2006. A 20th century acceleration in global sea-level rise. *Geophysical Research Letters 33* (1): L01602, **DOI:** 10.1029/2005GL024826.

CNA Corporation. 2007. National Security and the Threat of Climate Change. Available online at: *http://securityandclimate.cna.org/report/National%20Security%20and%20the%20Threat%20of%20Climate%20Change.pdf.*

Comiso, J. C. 2003. Warming trends in the Arctic from clear sky satellite observations. *Journal of Climate 16* (21):3498-3510.

Cooley, S. R. and S. C. Doney. 2009. Anticipating ocean acidification's economic consequences for commercial fisheries. *Environmental Research Letters 4* (2):024007, **DOI:** 10.1088/1748-9326/4/2/024007.

Cox, P. M., R. A. Betts, C. D. Jones, S. A. Spall, and I. J. Totterdell. 2000. Acceleration of global warming due to carbon-cycle feedbacks in a coupled climate model (vol 408, pg 184, 2000). *Nature 408* (6813):750.

Cox, P. M., R. A. Betts, M. Collins, P. P. Harris, C. Huntingford, and C. D. Jones. 2004. Amazonian forest dieback under climate-carbon cycle projections for the 21st century. *Theoretical and Applied Climatology 78* (1-3):137-156, **DOI:** 10.1007/s00704-004-0049-4.

Crimmins, T. M., M. A. Crimmins and C. D. Bertelsen. 2009. Flowering range changes across an elevation gradient in response to warming summer temperatures. *Global Change Biology 15* (5):1141-1152, **DOI:** 10.1111/j.1365-2486.2008.01831.x.

Crucifix, M. 2006. Does the Last Glacial Maximum constrain climate sensitivity? *Geophysical Research Letters 33* (18):L18701, **DOI:** 10.1029/2006GL027137.

Crutzen, P. and E. F. Stoermer. 2000. The Anthropocene. *IGBP Newsletter 41*:17-18.

Cubasch, U., G. A. Meehl, G .J. Boer, R .J. Stouffer, M. Dix, A. Noda, C. A. Senior, S. C. B. Raper, and K. S. Yap. 2001. Projections of future climate change. In *Climate Change 2001: The Scientific Basis: Contribution of Working Group I to the Third Assessment Report of the Intergovernmental Panel on Climate Change*, edited by Y. D. J. T. Houghton, D. J. Griggs, M. Noguer, P. J. Van der Linden, X. Dai, K. Maskell and C. A. Johnson. 526-582. Cambridge: Cambridge University Press.

Cuffey, K. M. and S. J. Marshall. 2000. Substantial contribution to sea-level rise during the last interglacial from the Greenland ice sheet. *Nature 404* (6778):591-594, **DOI:** 10.1038/35007053.

D'Amato, G. and L. Cecchi. 2008. Effects of climate change on environmental factors in respiratory allergic diseases. *Clinical and Experimental Allergy 38* (8):1264-1274, **DOI:** 10.1111/j.1365-2222.2008.03033.x.

Dahl-Jensen, D., O. Bennike and E. Willerslev. 2005. Greenland ice cores tell tales on the Eemian period and beyond. American Geophysical Union, Fall Meeting 2005, Abstract PP43C-03.

Danabasoglu, G. and P. Gent. 2009. Equilibrium climate sensitivity: is it accurate to use a slab ocean model? *Journal of Climate 22* (9):2494-2499.

Denman, K. L., G. Brasseur, A. Chidthaisong, P. Ciais, P.M. Cox, R. E. Dickinson, D. Hauglustaine, C. Heinze, E. Holland, D. Jacob, U. Lohmann, S Ramachandran, P. L. da Silva Dias, S.C. Wofsy and X. Zhang. 2007. Couplings Between Changes in the Climate System and Biogeochemistry. In *Climate Change 2007: The Physical Science Basis. Contribution of Working Group I to the Fourth Assessment Report of the Intergovernmental Panel on Climate Change*, edited by S. Solomon, D. Qin, M. Manning, Z. Chen, M. Marquis, K. B. Averyt, M.Tignor and H. L. Miller. Cambridge: Cambridge University Press.

Deser, C. and H. Teng. 2008. Evolution of Arctic sea ice concentration trends and the role of atmospheric circulation forcing, 1979-2007. *Geophysical Research Letters 35* (2):L02504, **DOI:** 10.1029/2007GL032023.

Dessai, S., X. F. Lu, and M. Hulme. 2005. Limited sensitivity analysis of regional climate change probabilities for the 21st century. *Journal of Geophysical Research-Atmospheres 110* (D19):D19108, **DOI:** 10.1029/2005JD005919.

Dessler, A. E. and S. C. Sherwood. 2009. Atmospheric science: A matter of humidity. *Science 323* (5917):1020-1021, **DOI:** 10.1126/science.1171264.

Deutsch, C. A., J. J. Tewksbury, R. B. Huey, K. S. Sheldon, C. K. Ghalambor, D. C. Haak, and P. R. Martin. 2008. Impacts of climate warming on terrestrial ectotherms across latitude. *Proceedings of the National Academy of Sciences of the United States of America 105* (18):6668-6672, **DOI:** 10.1073/pnas.0709472105.

DeWeaver, E. 2007. Uncertainty in climate model projections of arctic sea ice decline: an evaluation relevant to polar bears. United States Department of the Interior.

Dlugokencky, E. J., L. Bruhwiler, J. W. C. White, L. K. Emmons, P. C. Novelli, S. A. Montzka, K. A. Masarie, P. M. Lang, A. M. Crotwell, J. B. Miller, and L. V. Gatti. 2009. Observational constraints on recent increases in the atmospheric CH4 burden. *Geophysical Research Letters 36*:L18803, **DOI:** 10.1029/2009GL039780.

Domingues, C. M., J. A. Church, N. J. White, P. J. Gleckler, S. E. Wijffels, P. M. Barker, J. R. Dunn. 2008. Improved estimates of upper-ocean warming and multi-decadal sea-level rise. *Nature 453* (7198):1090-U1096, **DOI:** 10.1038/nature07080.

Doney, S. C., V. J. Fabry, R. A. Feely, and J. A. Kleypas. 2009a. Ocean acidification: The other CO_2 problem. *Annual Review of Marine Science 1*:169-192, **DOI:** 10.1146/annurev.marine.010908.163834.

Doney, S. C., W. M. Balch, V. J. Fabry, and R. A. Feely. 2009b. Ocean acidification: A critical emerging problem for the ocean sciences. *Oceanography 22* (4):16-25.

Dore, J. E., R. Lukas, D. W. Sadler, M. J. Church, and D. M. Karl. 2009. Physical and biogeochemical modulation of ocean acidification in the central North Pacific. *Proceedings of the National Academy of Sciences of the United States of America 106* (30):12235-12240, **DOI:** 10.1073/pnas.0906044106.

Du, B. and J. Zhang. 2000. Adaptation strategy for sea level rise in vulnerable areas along China's coast. *Acta Oceanologica Sinica 19* (4):1-16.

Duce, R. A., J. LaRoche, K. Altieri, K. R. Arrigo, A. R. Baker, D. G. Capone, S. Cornell, F. Dentener, J. Galloway, R. S. Ganeshram, R. J. Geider, T. Jickells, M. M. Kuypers, R. Langlois, P. S. Liss, S. M. Liu, J. J. Middleburg, C. M. Moore, S. Nickovic, A. Oschlies, T. Pedersen, J. Prospero, R. Schlitzer, S. Seitzinger, L. L. Sorensen, M. Uematsu, O. Ulloa, M. Voss, B. Ward, and L. Zamora. 2008. Impacts of atmospheric anthropogenic nitrogen on the open ocean. *Science 320* (5878):893-897, **DOI:** 10.1126/science.1150369.

Ducklow, H. W., K. Baker, D. G. Martinson, L. B. Quetin, R. M. Ross, R. C. Smith, S. E. Stammerjohn, M. Vernet, and W. Fraser. 2007. Marine pelagic ecosystems: The West Antarctic Peninsula. *Philosophical Transactions of the Royal Society B-Biological Sciences 362* (1477):67-94.

Dufresne, J. L. and S. Bony. 2008. An assessment of the primary sources of spread of global warming estimates from coupled atmosphere-ocean models. *Journal of Climate 21* (19):5135-5144, **DOI:** 10.1175/2008JCLI2239.1.

Easterling, W. E., P. K. Aggarwai, P. Batima, K. M. Brander, L. Erda, S. M. Howden, A. Kinlenko, J. Morton, J.-F. Soussana, J. Schmidhuber, and F. N. Tubiello. 2007. Food, fibre and forest products. In *Climate Change 2007: Impacts, Adaptation and Vulnerablity. Contribution of Working Group II to the Fourth Assessment Report of the Intergovermental Panel on Climate Change,* edited by O. F. C. M. L. Parry. 273-313. Cambridge: Cambridge University Press.

Ebi, K. L. and G. McGregor. 2008. Climate change, tropospheric ozone and particulate matter, and health impacts. *Environmental Health Perspectives 116* (11):1449-1455, **DOI:** 10.1289/ehp.11463.

Ebi, K. L., T. J. Teisberg, L. S. Kalkstein, L. Robinson and R. F. Weiher. 2004. Heat watch/warning systems save lives: Estimated costs and benefits for Philadelphia 1995-98. *Bulletin of the American Meteorological Society 85* (8):1067-1073, **DOI:** 10.1175/BAMS-85-8-1067.

Eby, M., K. Zickfeld, A. Montenegro, D. Archer, K. J. Meissner, and A. J. Weaver. 2009. Lifetime of anthropogenic climate change: millennial time scales of potential CO_2 and surface temperature perturbations. *Journal of Climate 22* (10):2501-2511, **DOI:** 10.1175/2008JCLI2554.1.

Eisenman, I. and J. S. Wettlaufer. 2009. Nonlinear threshold behavior during the loss of Arctic sea ice. *Proceedings of the National Academy of Sciences of the United States of America 106* (1):28-32, **DOI:** 10.1073/pnas.0806887106.

Ekamper, P., F. van Poppel, C. van Duin, and J. Garssen. 2009. 150 Years of temperature-related excess mortality in the Netherlands. *Demographic Research 21* (14):385-425, **DOI:** 10.1073/pnas.0806887106.

Emanuel, K., R. Sundararajan, and J. Williams. 2008. Hurricanes and global warming—Results from downscaling IPCC AR4 simulations. *Bulletin of the American Meteorological Society 89* (3):347. **DOI:** 10.1175/BAMS-89-3-347.

Emanuel, K. A. 1987. The dependence of hurricane intensity on climate. *Nature 326* (6112): 483-485.

Emori, S. and S. J. Brown. 2005. Dynamic and thermodynamic changes in mean and extreme precipitation under changed climate. *Geophysical Research Letters 32* (17):L17706, **DOI:** 10.1029/2005GL023272.

Ericson, J. P., C. J. Vorosmarty, S. L. Dingman, L. G. Ward, and M. Meybeck. 2006. Effective sea-level rise and deltas: Causes of change and human dimension implications. *Global and Planetary Change 50* (1-2):63-82, **DOI:** 10.1016/j.gloplacha.2005.07.004.

Fabry, V. J., B. A. Seibel, R. A. Feely, and J. C. Orr. 2008. Impacts of ocean acidification on marine fauna and ecosystem processes. *Ices Journal of Marine Science 65* (3):414-432, **DOI:** 10.1093/icesjms/fsn048.

FAO (Food and Agricultural Organization). 2006. World agriculture towards 2030/2050, Interim Report. Food and Agricultural Organization of the United Nations. Rome, Italy. Available online at: *http://www.fao.org/docrep/005/Y4252E/y4252e00.htm.*

Feely, R. A., C. L. Sabine, K. Lee, W. Berelson, J. Kleypas, V. J. Fabry, and F. J. Millero. 2004. Impact of anthropogenic CO_2 on the CaCO3 system in the oceans. *Science 305* (5682):362-366.

Feely, R. A., C. L. Sabine, J. M. Hernandez-Ayon, D. Ianson and B. Hales. 2008. Evidence for upwelling of corrosive "acidified" water onto the continental shelf. *Science 320* (5882):1490-1492, **DOI:** 10.1126/science.1155676.

Fischer, E. M. and C. Schar. 2009. Future changes in daily summer temperature variability: driving processes and role for temperature extremes. *Climate Dynamics 33* (7-8):917-935, **DOI:** 10.1007/s00382-008-0473-8.

Flanner, M. G., C. S. Zender, J. T. Randerson, and P. J. Rasch. 2007. Present-day climate forcing and response from black carbon in snow. *Journal of Geophysical Research-Atmospheres 112* (D11): D11202, **DOI:** 10.1029/2006JD008003.

Flato, G. M. and CMIP contributors. 2004. Sea-ice and its response to CO_2 forcing as simulated by global climate models. *Climate Dynamics 23* (3-4):229-241, **DOI:** 10.1007/s00382-004-0436-7.
Fonseca, C. R. 2009. The silent mass extinction of insect herbivores in biodiversity hotspots. *Conservation Biology 23* (6):1507-1515, **DOI:** 10.1111/j.1523-1739.2009.01327.x.
Foote, M. 1997. Estimating taxonomic durations and preservation probability. *Paleobiology 23* (3):278-300.
Forster, P., V. Ramaswamy, P. Artaxo, T. Berntsen, R. Betts, D. W. Fahey, J. Haywood, J. Lean, D. C. Lowe, G. Myhre, J. Nganga, R. Prinn, G. Raga, M. Schulz and R. Van Dorland. 2007. Changes in atmospheric constituents and in radiative forcing. In *Climate Change 2007: The Physical Science Basis. Contribution of Working Group I to the Fourth Assessment Report of the Intergovernmental Panel on Climate Change,* edited by S. Solomon, D. Qin, M. Manning, Z. Chen, M. Marquis, K. B. Averyt, M.Tignor and H. L. Miller. New York: Cambridge University Press.
Fourier, J. 1827. Mémoire sur les températures du globe terrestre et des espaces planétaires. *Mémoires d l'Académie Royale des Sciences de l'Institute de France VII*:570-604.
Fox, C., R. Harris, S. Sundby, E. Achterberg, J. I. Allen, J. Allen, A. Baker, C. P. D. Brussaard, P. Buckley, E. J. Cook, S. R. Dye, M. Edwards, L. Fernand, P. Kershaw, J. Metcalfe, S. Osterhus, T. Potter, E. Sakshaug, D. Speirs, E. Stenevik, M. S. John, F. Thingstad, B. Wilson. 2009. Transregional linkage in the north-eastern Atlantic: an "end-to-end" analysis of pelagic ecosystems. *Oceanography and Marine Biology: An Annual Review 47*:1-75.
Frank, D. C., J. Esper, C. C. Raible, U. Buntgen, V. Trouet, B. Stocker, and F. Joos. 2010. Ensemble reconstruction constraints on the global carbon cycle sensitivity to climate. *Nature 463* (7280):527-U143, **DOI:** 10.1038/nature08769.
Frich, P., L. V. Alexander, P. Della-Marta, B. Gleason, M. Haylock, A. Tank, and T. Peterson. 2002. Observed coherent changes in climatic extremes during the second half of the twentieth century. *Climate Research 19* (3):193-212.
Friedlingstein, P., P. Cox, R. Betts, L. Bopp, W. Von Bloh, V. Brovkin, P. Cadule, S. Doney, M. Eby, I. Fung, G. Bala, J. John, C. Jones, F. Joos, T. Kato, M. Kawamiya, W. Knorr, K. Lindsay, H. D. Matthews, T. Raddatz, P. Rayner, C. Reick, E. Roeckner, K. G. Schnitzler, R. Schnur, K. Strassman, A. J. Weaver, C. Yoshikawa, and N. Zeng. 2006. Climate-carbon cycle feedback analysis: results from the (CMIP)-M-4 model intercomparison. *Journal of Climate 19* (14):3337-3353.
Frolicher, T. L., F. Joos, G. K. Plattner, M. Steinacher, and S. C. Doney. 2009. Natural variability and anthropogenic trends in oceanic oxygen in a coupled carbon cycle-climate model ensemble. *Global Biogeochemical Cycles 23*:GB1003, **DOI:** 10.1029/2008GB003316.
Fung, I. Y., S. C. Doney, K. Lindsay, and J. John. 2005. Evolution of carbon sinks in a changing climate. *Proceedings of the National Academy of Sciences of the United States of America 102* (32):11201-11206, **DOI:** 10.1073/pnas.0504949102.
Fyke, J. G. and A. J. Weaver. 2006. The effect of potential future climate change on the marine methane hydrate stability zone. *Journal of Climate 19* (22):5903-5917.
Gamble, J., K. Ebi, F. Sussman and T. Wilbanks. 2008. Analyses of the effects of global change on human health and welfare and human systems. United States Climate Change Science Program Synthesis and Assessment Product 4. 6.
Gilg, O., D. Boertmann, F. Merkel, A. Aebischer, and B. Sabard. 2009. Status of the endangered ivory gull, Pagophila eburnea, in Greenland. *Polar Bioliology 32* (9):1275-1286, **DOI:** 10.1007/s00300-009-0623-4.
Gillett, N. P., D. A. Stone, P. A. Stott, T. Nozawa, A. Y. Karpechko, G. C. Hegerl, M. F. Wehner, and P. D. Jones. 2008a. Attribution of polar warming to human influence. *Nature Geoscience 1* (11):750-754, **DOI:** 10.1038/ngeo338.

Gillett, N. P., P. A. Stott, and B. D. Santer. 2008b. Attribution of cyclogenesis region sea surface temperature change to anthropogenic influence. *Geophysical Research Letters 35* (9): L09707, **DOI:** 10.1029/2008GL033670.

Gillett, N. P., A. J. Weaver, F. W. Zwiers, and M. D. Flannigan. 2004. Detecting the effect of climate change on Canadian forest fires. *Geophysical Research Letters 31* (18): L18211, **DOI:** 10.1029/2004GL020876.

Giorgi, F. and R. Francisco. 2000. Evaluating uncertainties in the prediction of regional climate change. *Geophys. Res. Lett. 27*:1295-1298.

Giorgi, F. and X. Bi. 2005. Regional changes in surface climate interannual variability for the 21st century from ensembles of global model simulations. *Geophysical Research Letters 32* (13): L13701, **DOI:** 10.1029/2005GL023002.

Gordon, M. S. 1982. *Animal Physiology: Principles and Adaptations.* New York: Macmillan.

Gould, E. A. and S. Higgs. 2009. Impact of climate change and other factors on emerging arbovirus diseases. *Transactions of the Royal Society of Tropical Medicine and Hygiene 103* (2):109-121, **DOI:** 10.1016/j.trstmh.2008.07.025.

Graham, R. W. and E. C. Grimm. 1990. Effects of global climate change on the patterns of terrestrial biological communities. *Trends in Ecology and Evolution 5* (9):289-292.

Green, M. A., G. G. Waldbusser, S. L. Reilly, K. Emerson, and S. O'Donnell. 2009. Death by dissolution: Sediment saturation state as a mortality factor for juvenile bivalves. *Limnology and Oceanography 54* (4):1037-1047.

Gregory, J. M. 2000. Vertical heat transports in the ocean and their effect an time-dependent climate change. *Climate Dynamics 16* (7):501-515.

Gregory, J. M. and J. F. B. Mitchell. 1997. The climate response to CO_2 of the Hadley Centre coupled AOGCM with and without flux adjustment. *Geophysical Research Letters 24* (15):1943-1946.

Gregory, J. M. and P. M. Forster. 2008. Transient climate response estimated from radiative forcing and observed temperature change. *Journal of Geophysical Research-Atmospheres 113* (D23):D23105, **DOI:** 10.1029/2008JD010405.

Gregory, J. M., P. A. Stott, D. J. Cresswell, N. A. Rayner, C. Gordon, and D. M. H. Sexton. 2002. Recent and future changes in Arctic sea ice simulated by the HadCM3 AOGCM. *Geophysical Research Letters 29* (24):2175, **DOI:** 10.1029/2001GL014575.

Gregory, J. M., P. Huybrechts, and S. C. B. Raper. 2004. Climatology—threatened loss of the Greenland ice sheet. *Nature 428* (6983):616, **DOI:** 10.1038/428616a.

Gregory, J. M., C. D. Jones, P. Cadule, and P. Friedlingstein. 2009. Quantifying carbon cycle feedbacks. *Journal of Climate 22* (19):5232-5250, **DOI:** 10.1175/2009JCLI2949.1.

Grinnell, J. 1917. Field tests of theories concerning distributional control. *American Naturalist 51*:115-128.

Gruber, N., M. Gloor, S. E. M. Fletcher, S. C. Doney, S. Dutkiewicz, M. J. Follows, M. Gerber, A. R. Jacobson, F. Joos, K. Lindsay, D. Menemenlis, A. Mouchet, S. A. Muller, J. L. Sarmiento, and T. Takahashi. 2009. Oceanic sources, sinks, and transport of atmospheric CO_2. *Global Biogeochemical Cycles 23*:GB1005, **DOI:** 10.1029/2008GB003349.

Hadly, E. A. 1997. Evolutionary and ecological response of pocket gophers (Thomomys talpoides) to late-Holocene climatic change. *Biological Journal of the Linnean Society 60* (2):277-296.

Hall-Spencer, J. M., R. Rodolfo-Metalpa, S. Martin, E. Ransome, M. Fine, S. M. Turner, S. J. Rowley, D. Tedesco, and M. C. Buia. 2008. Volcanic carbon dioxide vents show ecosystem effects of ocean acidification. *Nature 454* (7200):96-99, **DOI:** 10.1038/nature07051.

Hamlet, A. F. and D. P. Lettenmaier. 1999. Effects of climate change on hydrology and water resources in the Columbia River basin. *Journal of the American Water Resources Association 35* (6):1597-1623.

Hamlet, A. F., P. W. Mote, M. P. Clark, and D. P. Lettenmaier. 2005. Effects of temperature and precipitation variability on snowpack trends in the western United States. *Journal of Climate 18* (21):4545-4561.

Hanna, H., P. Huybrechts, I. Janssens, J. Cappelen, K. Steffen and A. Stephens. 2005. Runoff and mass balance of the Greenland ice sheet: 1958-2003. *Journal of Geophysical Research 110* (D13):D13108, **DOI:** 10.1029/2004JD005641.

Hanna, E., P. Huybrechts, K. Steffen, J. Cappelen, R. Huff, C. Shuman, T. Irvine-Fynn, S. Wise, and M. Griffiths. 2008. Increased runoff from melt from the Greenland ice sheet: A response to global warming. *Journal of Climate 21* (2):331-341, **DOI:** 10.1175/2007JCLI1964.1.

Hannart, A., J. L. Dufresne, and P. Naveau. 2009. Why climate sensitivity may not be so unpredictable. *Geophysical Research Letters 36*:L16707, **DOI:** 10.1029/2009GL039640.

Hansen, J. and L. Nazarenko. 2004. Soot climate forcing via snow and ice albedos. *Proceedings of the National Academy of Sciences of the United States of America 101* (2):423-428, **DOI:** 10.1073/pnas.2237157100.

Hansen, J., S. Makiko, K. Pushker, D. Beerling, R. Berner, V. Masson-Delmotte, M. Pagani, M. Raymo, D. L. Royer, and J. C. Zachos. 2008. Target atmospheric CO_2: where should humanity aim? *The Open Atmospheric Science Journal* 2:217-231.

Hare, J. A., M. A. Alexander, M. J. Fogarty, E. H. Williams, and J. D. Scott. Forecasting the dynamics of a coastal fishery species using a coupled climate-population model. *Ecological Applications,* in press.

Harris, G. and S. L. Pimm. 2008. Range size and extinction risk in forest birds. *Conservation Biology 22* (1):163-171, **DOI:** 10.1111/j.1523-1739.2007.00798.x.

Hartmann, D. L., J. M. Wallace, V. Limpasuvan, D. W. J. Thompson, and J. R. Holton. 2000. Can ozone depletion and global warming interact to produce rapid climate change? *Proceedings of the National Academy of Sciences of the United States of America 97* (4):1412-1417.

Hayhoe, K. 2010 forthcoming. An integrated framework for quantifying and valuing climate change impacts on urban energy and infrastructure: a Chicago case study.

Hayhoe, K., C. P. Wake, T. G. Huntington, L. F. Luo, M. D. Schwartz, J. Sheffield, E. Wood, B. Anderson, J. Bradbury, A. DeGaetano, T. J. Troy, and D. Wolfe. 2007. Past and future changes in climate and hydrological indicators in the US Northeast. *Climate Dynamics 28* (4):381-407, **DOI:** 10.1007/s00382-006-0187-8.

Hayhoe, K., C. Wake, B. Anderson, X.-Z. Liang, E. Maurer, J. Zhu, et al. 2008. Regional climate change projections for the Northeast USA. *Mitigation and Adaptation Strategies for Global Change 13* (5):425-436.

Hayhoe, K. S. Sheridan, L. Kalkstein, and S. Greene. 2010. Climate change, heat waves, and mortality projections for Chicago. *Journal of Great Lakes Research,* In Press, Corrected Proof, ISSN 0380-1330, **DOI:** 10.1016/j.jglr.2009.12.009.

Hegerl, G. C., F. W. Zwiers, P. Braconnot, N. P. Gillett, Y. Luo, J. A. M. Orsini, N. Nicholls, J. E. Penner, and P. A. Stott. 2007. Understanding and attributing climate change. In *Climate Change 2007: The Physical Science Basis. Contribution of Working Group I to the Fourth Assessment Report of the Intergovernmental Panel on Climate Change,* edited by S. Solomon, D. Qin, M. Manning, Z. Chen, M. Marquis, K. B. Averyt, M. Tignor and H. L. Miller. Cambridge, United Kingdom and New York, NY, USA: Cambridge University Press.

Heintz, J., R. Pollin, H. Garrett-Peltier. 2009. *How Infrastructure Investments Support the U.A. Economy: Employment, Productivity And Growth.* Alliance for American Manufacturing. Washington, DC: University of Massachusetts Political Economy Research Institute.

Held, I. M. and B. J. Soden. 2000. Water vapor feedback and global warming. *Annual Review of Energy and the Environment 25*:441-475.

Held, I. M. and B. J. Soden. 2006. Robust responses of the hydrological cycle to global warming. *Journal of Climate 19* (21):5686-5699.

Held, I. M., M. Winton, K. Takahashi, T. Delworth, F. Zeng, and G. K. Vallis. 2010. Probing the fast and slow components of global warming by returning abruptly to pre-industrial forcing. *Journal of Climate 23* (9):2418-2427, **DOI:** 10.1175/2009JCLI3466.1.

Hendrix, C. S. and S. M. Glaser. 2007. Trends and triggers: climate, climate change and civil conflict in Sub-Saharan Africa. *Political Geography 26* (6):695-715, **DOI:** 10.1016/j.polgeo.2007.06.006.

Henson, S. A., J. L. Sarmiento, J. P. Dunne, L. Bopp, I. Lima, S. C. Doney, J. John, and C. Beaulieu. 2010. Detection of anthropogenic climate change in satellite records of ocean chlorophyll and productivity. *Biogeosciences 7* (2):621-640.

Hidalgo, H. G., T. Das, M. D. Dettinger, D. R. Cayan, D. W. Pierce, T. P. Barnett, G. Bala, A. Mirin, A. W. Wood, C. Bonfils, B. D. Santer, and T. Nozawa. 2009. Detection and attribution of streamflow timing changes to climate change in the western United States. *Journal of Climate 22* (13):3838-3855, **DOI:** 10.1175/2009JCLI2470.1.

Hobbs, R. J., E. Higgs, and J. A. Harris. 2009. Novel ecosystems: implications for conservation and restoration. *Trends in Ecology & Evolution 24* (11):599-605, **DOI:** 10.1016/j.tree.2009.05.012.

Hoegh-Guldberg, O., P. J. Mumby, A. J. Hooten, R. S. Steneck, P. Greenfield, E. Gomez, C. D. Harvell, P. F. Sale, A. J. Edwards, K. Caldeira, N. Knowlton, C. M. Eakin, R. Iglesias-Prieto, N. Muthiga, R. H. Bradbury, A. Dubi, and M. E. Hatziolos. 2007. Coral reefs under rapid climate change and ocean acidification. *Science 318* (5857):1737-1742, **DOI:** 10.1126/science.1152509.

Hojo, T. 1931. Distribution of the median, quartiles and interquartile distance in samples from a normal population. *Biometrika* 23:361-363.

Holland, M. M. and C. M. Bitz. 2003. Polar amplification of climate change in coupled models. *Climate Dynamics 21* (3-4):221-232, **DOI:** 10.1007/s00382-003-0332-6.

Holland, M. M., C. M. Bitz, and B. Tremblay. 2006. Future abrupt reductions in the summer Arctic sea ice. *Geophysical Research Letters 33* (23):L23503, **DOI:** 10.1029/2006GL028024.

Holland, M. M., D. A. Bailey, and S. Vavrus. 2010. Inherent sea ice predictability in the rapidly changing Arctic environment of the Community Climate System Model, version 3, *Climate Dyn*, in press, **DOI:** 10.1007/s00382-010-0792-4.

Honda, M., K. Yamazaki, H. Nakamura, and K. Takeuchi. 1999. Dynamic and thermodynamic characteristics of atmospheric response to anomalous sea ice extent in the Sea of Okhotsk. *Journal of Climate 12* (12):3347-3358.

Hosaka, M., D. Nohara, and A. Kitoh. 2005. Changes in snow cover and snow water equivalent due to global warming simulated by a 20km-mesh global atmospheric model. *SOLA 1*:93-96.

Houghton, R. A. 2003. Revised estimates of the annual net flux of carbon to the atmosphere from changes in land use and land management 1850-2000. *Tellus Series B-Chemical and Physical Meteorology 55* (2):378-390.

Houghton, R. A. 2008. Carbon flux to the atmosphere from land-use changes: 1850-2005. In *TRENDS: A Compendium of Data on Global Change*. Carbon Dioxide Information Analysis Center, Oak Ridge National Laboratory, U.S. Department of Energy. Oak Ridge, TN.

Howat, I. M., I. Joughin, and T. A. Scambos. 2007. Rapid changes in ice discharge from Greenland outlet glaciers. *Science 315* (5818):1559-1561, **DOI:** 10.1126/science.1138478.

Hubbell, S. P., F. L. He, R. Condit, L. Borda-de-Agua, J. Kellner, and H. ter Steege. 2008. How many tree species and how many of them are there in the Amazon will go extinct? *Proceedings of the National Academy of Sciences of the United States of America 105* (SUPPL 1):11498-11504, **DOI:** 10.1073/pnas.0801915105.

Huber, M. 2008. A hotter greenhouse? *Science 321* (5887):353-354, **DOI:** 10.1126/science.1161170.

Huber, M. and L. C. Sloan. 2001. Heat transport, deep waters, and thermal gradients: coupled simulation of an Eocene Greenhouse Climate. *Geophysical Research Letters 28* (18):3481-3484.

Hutchins, D. A., M. R Mulholland, and F. Fu. 2009. Nutrient cycles and marine microbes in a CO_2-enriched ocean. *Oceanography 22* (4):128-145.

Huybrechts, P. 1993. Glaciological modeling of the late Cenozoic East Antarctic ice sheet: stability or dynamism? *Geografiska Annaler Series A—Physical Geography 75A*:221-238.

Huybrechts, P. and J. de Wolde. 1999. The dynamic response of the greenland and antarctic ice sheets to multiple-century climatic warming. *Journal of Climate 12* (8):2169-2188.

International Water Management Institute (IWMI). 2007. *Comprehensive Assessment of Water Management in Agriculture. 2007. Water for Food, Water for Life: A Comprehensive Assessment of Water Management in Agriculture.* London: Earthscan.

International, US and EU Climate Change Control Scenarios: Result from EMF 22 **Energy Economics,** Volume 31 (2009) Supplement 2.

IOC (Intergovernmental Oceanographic Commission) of UNESCO and the International CLIVAR Project Office. 2009. Ship-based repeat hydrography: a strategy for a sustained global programme, edited by M. Hood. IOC Technical Series, 89. IOCCP Reports, 17. ICPO Publication 142. UNESCO.

IPCC (Intergovernmental Panel on Climate Change). 2005. Safeguarding the ozone layer and the global climate system: issues related to hydrofluorocarbons and perfluorocarbons, IPCC/TEAP special report. New York: Cambridge University Press.

IPCC. 2007a. Contribution of Working Group I to the Fourth Assessment Report of the Intergovernmental Panel on Climate Change. In *Climate Change 2007: The Physical Science Basis*, edited by S. Solomon, D. Qin, M. Manning, Z. Chen, M. Marquis, K. B. Averyt, M. Tignor and H. L. Miller. New York: Cambridge University Press.

IPCC. 2007b. Contribution of Working Group II to the Fourth Assessment Report of the Intergovernmental Panel on Climate Change. In *Climate Change 2007: Impacts, Adaptation and Vulnerability*, edited by M. L. Parry, O. F. Canziani, J. P. Palutikof, P. J. van der Linden, and C. E. Hanson. New York: Cambridge University Press.

IPCC. 2007c. Contribution of Working Group I to the Fourth Assessment Report of the Intergovernmental Panel on Climate Change. In *Climate Change 2007: The Physical Science Basis*, Edited by S. Solomon, D. Qin, M. Manning, Z. Chen, M. Marquis, K. B. Averyt, M. Tignor, and H. L. Miller. New York: Cambridge University Press.

IPCC. 2007d. Contribution of Working Group II to the Fourth Assessment Report of the Intergovernmental Panel on Climate Change. In *Climate Change 2007: Impacts, Adaptation and Vulnerability*, edited by M. L. Parry, O. F. Canziani, J. P. Palutikof, P. J. van der Linden and C. E. Hanson. New York: Cambridge University Press.

IPCC. 2007e. Contribution of Working Groups I, II and III to the Fourth Assessment Report of the Intergovernmental Panel on Climate Change. In *Climate Change 2007: Synthesis Report*, edited by R. K. Pachauri and A. Reisinger. New York: Cambridge University Press.

Isaksen, K., R. E. Benestad, C. Harris, and J. L. Sollid. 2007. Recent extreme near-surface permafrost temperatures on Svalbard in relation to future climate scenarios. *Geophysical Research Letters 34* (17):L17502, **DOI:** 10.1029/2007GL031002.

Jansen, E., J. Overpeck, K. R. Briffa, J.-C. Duplessy, F. Joos, V. Masson-Delmotte, D. Olago, B. Otto-Bliesner, W. R. Peltier, S. Rahmstorf, R. Ramesh, D. Raynaud, D. Rind, O. Solomina, R. Villalba and D. Zhang. 2007. Palaeoclimate. In *Climate Change 2007: The Physical Science Basis. Contribution of Working Group I to the Fourth Assessment Report of the Intergovernmental Panel on Climate Change*, edited by S. Solomon, D. Qin, M. Manning, Z. Chen, M. Marquis, K. B. Averyt, M. Tignor and H. L. Miller. New York: Cambridge University Press.

Johannessen, O. M., K. Khvorostovsky, M. W. Miles, and L. P. Bobylev. 2005. Recent ice-sheet growth in the interior of Greenland. *Science 310* (5750):1013-1016, **DOI:** 10.1126/science.1115356.

Jones, C. D., P. M. Cox, and C. Huntingford. 2006. Climate-carbon cycle feedbacks under stabilization: uncertainty and observational constraints. *Tellus Series B-Chemical and Physical Meteorology 58* (5):603-613, **DOI:** 10.1111/j.1600-0889.2006.00215.x.

Joos, F., I. C. Prentice, S. Sitch, R. Meyer, G. Hooss, G. K. Plattner, S. Gerber, and K. Hasselmann. 2001. Global warming feedbacks on terrestrial carbon uptake under the Intergovernmental Panel on Climate Change (IPCC) emission scenarios. *Global Biogeochemical Cycles 15* (4):891-907.

Jorgenson, M. T., C. H. Racine, J. C. Walters, and T. E. Osterkamp. 2001. Permafrost degradation and ecological changes associated with a warming climate in central Alaska. *Climatic Change 48* (4):551-579.

Joshi, M. M., M. J. Webb, A. C. Maycock, and M. Collins. 2010. Stratospheric water vapour and high climate sensitivity in a version of the HadSM3 climate model. *Atmospheric Chemistry and Physics Discussions* 10:6241-6255.

Kaiser, R., A. Le Tertre, J. Schwartz, C. A. Gotway, W. R. Daley, and C. H. Rubin. 2007. The effect of the 1995 heat wave in Chicago on all-cause and cause-specific mortality. *American Journal of Public Health 97* (SUPPL 1).

Kalkstein, L. S., J. S. Greene, D. M. Mills, A. D. Perrin, J. P. Samenow, and J. C. Cohen. 2008. Analog European heat waves for US cities to analyze impacts on heat-related mortality. *Bulletin of the American Meteorological Society 89* (1):75, **DOI:** 10.1175/BAMS-89-1-75.

Karl, T. M. J. M. P. T. C. 2009. *Global Climate Change Impacts in the United States: A State of Knowledge Report from the U.S. Global Change Research Program.* New York: Cambridge University Press.

Kaser, G., J. G. Cogley, M. B. Dyurgerov, M. F. Meier, and A. Ohmura. 2006. Mass balance of glaciers and ice caps: consensus estimates for 1961-2004. *Geophysical Research Letters 33* (19):L19501, **DOI:** 10.1029/2006GL027511.

Kates, R. W., C. E. Colten, S. Laska, and S. P. Leatherman. 2006. Reconstruction of New Orleans after Hurricane Katrina: a research perspective. *Proceedings of the National Academy of Sciences of the United States of America 103* (40):14653-14660, **DOI:** 10.1073/pnas.0605726103.

Keeling, R. F., A. Körtzinger, and N. Gruber. 2010. Ocean deoxygenation in a warming world. *Annual Review of Marine Science 2* (1):199-229.

Kharin, V. V., F. W. Zwiers, X. B. Zhang, and G. C. Hegerl. 2007. Changes in temperature and precipitation extremes in the IPCC ensemble of global coupled model simulations. *Journal of Climate 20* (8):1419-1444, **DOI:** 10.1175/JCLI4066.1.

Khatiwala, S., F. Primeau, and T. Hall. 2009. Reconstruction of the history of anthropogenic CO_2 concentrations in the ocean. *Nature* 462 (7271):346-U110, **DOI:** 10.1038/nature08526.

Kirshen, P., K. Knee, and M. Ruth. 2008a. Climate change and coastal flooding in Metro Boston: impacts and adaptation strategies. *Climatic Change 90* (4):453-473, **DOI:** 10.1007/s10584-008-9398-9.

Kirshen, P., C. Watson, E. Douglas, A. Gontz, J. Lee, and Y. Tian. 2008b. Coastal flooding in the Northeastern United States due to climate change. *Mitigation and Adaptation Strategies for Global Change 13* (5-6):437-451.

Kitabata, H., K. Nishizawa, Y. Yoshida, and K. Maruyama. 2006. Permafrost thawing in circum-Arctic and highlands under climatic change scenario projected by Community Climate System Model (CCSM3). *SOLA 2*:53-56.

Kitoh, A. and T. Mukano. 2009. Changes in daily and monthly surface air temperature variability by multi-model global warming experiments. *Journal of the Meteorological Society of Japan 87* (3):513-524, **DOI:** 10.2151/jmsj.87.513.

Kleypas, J. A. and K. K. Yates. 2009. Coral reefs and ocean acidification. *Oceanography* 22 (4):108-117.

Knight, J. R. 2009. The Atlantic multidecadal oscillation inferred from the forced climate response in coupled general circulation models. *Journal of Climate 22* (7):1610-1625, **DOI:** 10.1175/2008JCLI2628.1.

Knowlton, K., C. Hogrefe, B. Lynn, C. Rosenzweig, J. Rosenthal, and P. L. Kinney. 2008. Impacts of heat and ozone on mortality risk in the New York City Metropolitan Region under a changing climate (pp. 143-160).

Knowlton, K., M. Rotkin-Ellman, G. King, H. G. Margolis, D. Smith, G. Solomon, R. Trent, and P. English. 2009. The 2006 California heat wave: impacts on hospitalizations and emergency department visits. *Environmental Health Perspectives 117* (1):61-67, **DOI:** 10.1289/ehp.11594.

Knutson, T. K., J. L. McBride, J. Chan, K. Emanuel, G. Holland, C. Landsea, I. M. Held, J. P. Kossin, A. K. Srivastava, and M. Sugi. 2010. Tropical cyclones and climate change. *Nature Geoscience 3* (3):157-163, **DOI:** 10.1038/NGEO779.

Knutson, T. R., J. J. Sirutis, S. T. Garner, G.A .Vecchi, and I. M. Held. 2008. Simulated reduction in Atlantic hurricane frequency under twenty-first-century warming conditions. *Nature Geoscience 1* (6):359-364, **DOI:** 10.1038/ngeo202.

Knutson, T. R. and R. E. Tuleya. 2004. Impact of CO_2-induced warming on simulated hurricane intensity and precipitation: sensitivity to the choice of climate model and convective parameterization. *Journal of Climate 17* (18):3477-3495.

Knutti, R. and G. C. Hegerl. 2008. The equilibrium sensitivity of the Earth's temperature to radiation changes. *Nature Geosciences 1* (11):735-743, **DOI:** 10.1038/ngeo337.

Knutti, R. and L. Tomassini. 2008. Constraints on the transient climate response from observed global temperature and ocean heat uptake. *Geophysical Research Letters 35* (9):L09701, **DOI:** 10.1029/2007GL032904.

Kraas, F. 2008. Megacities as global risk areas. In *Urban Ecology: An International Perspective on the Interaction between Humans and Nature*, edited by E. S. J. M. Marzluff, W. Endlicheret. 583-596. New York: Springer.

Krawchuk, M. A., M. A. Moritz, M.-A. Parisien, J. Van Dorn, and K. Hayhoe. 2009. Global pyrogeography: the current and future distribution of wildfire. *PLoS ONE 4* (4):e5102, **DOI:** 10.1371/journal.pone.0005102.

Krishnamurthy, A., J. K. Moore, N. Mahowald, C. Luo, S. C. Doney, K. Lindsay, and C. S. Zender. 2009. Impacts of increasing anthropogenic soluble iron and nitrogen deposition on ocean biogeochemistry. *Global Biogeochemical Cycles 23*:GB3016, **DOI:** 10.1029/2008GB003440.

Kwok, R. and D. A. Rothrock. 2009. Decline in Arctic sea ice thickness from submarine and ICESat records: 1958-2008. *Geophysical Research Letters 36*:L15501, **DOI:** 10.1029/2009GL039035.

Lachenbruch, A. H. and B. V. Marshall. 1986. Changing climate—geothermal evidence from permafrost in the Alaskan Arctic. *Science 234* (4777):689-696.

Lafferty, K. D. 2009. The ecology of climate change and infectious diseases. *Ecology 90* (4):888-900.

Lal, M., K. K. Singh, L. S. Rathore, G. Srinivasan, and S. A. Saseendran. 1998. Vulnerability of rice and wheat yields in NW India to future changes in climate. *Agricultural and Forest Meteorology 89* (2):101-114.

Lamarque, J. F., J. T. Kiehl, G. P. Brasseur, T. Butler, P. Cameron-Smith, W. D. Collins, W. J. Collins, C. Granier, D. Hauglustaine, P. G. Hess, E. A. Holland, L. Horowitz, M. G. Lawrence, D. McKenna, P. Merilees, M. J. Prather, P. J. Rasch, D. Rotman, D. Shindell, and P. Thornton. 2005. Assessing future nitrogen deposition and carbon cycle feedback using a multimodel approach: analysis of nitrogen deposition. *Journal of Geophysical Research-Atmospheres 110* (D19):D19303, **DOI:** 10.1029/2005JD005825.

Landsea, C. W. 2007. Counting Atlantic tropical cyclones back to 1900. *Eos 88* (18).

Landsea, C., G. A. Vecchi, L. Bengtsson, and T. R. Knutson. 2010. Impact of duration thresholds on Atlantic tropical cyclone counts. *Journal of Climate*, in press.

Langdon, C. and M. J. Atkinson. 2005. Effect of elevated pCO(2) on photosynthesis and calcification of corals and interactions with seasonal change in temperature/irradiance and nutrient enrichment. *Journal of Geophysical Research-Oceans 110* (C9):C09S07, **DOI:** 10.1029/2004JC002576.

Larow, T. E., Y. K. Lim, D. W. Shin, E. P. Chassignet, and S. Cocke. 2008. Atlantic basin seasonal hurricane simulations. *Journal of Climate 21* (13):3191-3206, **DOI:** 10.1175/2007JCLI2036.1.

Latif, M., W. Park, H. Ding, and N. S. Keenlyside. 2009. Internal and external North Atlantic Sector variability in the Kiel Climate Model. *Meteorologische Zeitschrift 18:*433-443.

Laurance, W. F. and D. C. Useche. 2009. Environmental synergisms and extinctions of tropical species. *Conservation Biology 23* (6):1427-1437, **DOI:** 10.1111/j.1523-1739.2009.01336.x.

Lawrence, D. M. and A. G. Slater. 2005. A projection of severe near-surface permafrost degradation during the 21st century. *Geophysical Research Letters 32* (24):L24401, **DOI:** 10.1029/2005GL025080.

Lawrence, D. M., A. G. Slater, V. E. Romanovsky, and D. J. Nicolsky. 2008a. Sensitivity of a model projection of near-surface permafrost degradation to soil column depth and representation of soil organic matter. *Journal of Geophysical Research-Earth Surface 113* (F2): F02011, **DOI:** 10.1029/2007JF000883.

Lawrence, D. M., A. G. Slater, R. A. Tomas, M. M. Holland, and C. Deser. 2008b. Accelerated Arctic land warming and permafrost degradation during rapid sea ice loss. *Geophysical Research Letters 35* (11):L11506, **DOI:** 10.1029/2008GL033985.

Lawton, J. H. and R. M. May. 1995. *Extinction Rates*. Oxford: Oxford University Press.

Le Quere, C., C. Rodenbeck, E. T. Buitenhuis, T. J. Conway, R. Langenfelds, A. Gomez, C. Labuschagne, M. Ramonet, T. Nakazawa, N. Metzl, N. P. Gillett, and M. Meimann. 2008. Saturation of the Southern Ocean CO_2 sink due to recent climate change. *Science 319* (5863):1735-1738, **DOI:** 10.1126/science.1136188.

Le Quere, C., M. R. Raupach, J. G. Canadell, G. Marland, L. Bopp, P. Ciais, T. J. Conway, S. C. Doney, R. A. Feely, P. Foster, P. Friedlingstein, K. Gurney, R. A. Houghton, J. I. House, C. Huntingford, P. E. Levy, M. R. Lomas, J. Majkut, N. Metzl, J. P. Ometto, G. P. Peters, I. C. Prentice, J. T. Randerson, S. W. Running, J. L. Sarmiento, U. Schuster, S. Sitch, T. Takahashi, N. Viovy, G. R. van der Werf, and F. I. Woodward. 2009. Trends in the sources and sinks of carbon dioxide. *Nature Geoscience 2* (12):831-836, **DOI:** 10.1038/ngeo689.

Leakey, A. D. B. 2009. Rising atmospheric carbon dioxide concentration and the future of C-4 crops for food and fuel. *Proceedings of the Royal Society B-Biological Sciences 276* (1666):2333-2343, **DOI:** 10.1098/rspb.2008.1517.

Lean, J. L. and D. H. Rind. 2009. How will Earth's surface temperature change in future decades? *Geophysical Research Letters 36:*L15708, **DOI:** 10.1029/2009GL038932.

Lemke, P., J. Ren, R. B. Alley, I. Allison, J. Carrasco, G. Flato, Y. Fujii, G. Kaser, P. Mote, R. H. Thomas and T. Zhang. 2007. Observations: changes in snow, ice and frozen ground. In *Climate Change 2007: The Physical Science Basis. Contribution of Working Group I to the Fourth Assessment Report of the Intergovernmental Panel on Climate Change*, edited by S. Solomon, D. Qin, M. Manning, Z. Chen, M. Marquis, K. B. Averyt, M. Tignor, and H.L. Miller. Cambridge: Cambridge University Press.

Lenderink, G. and E. Van Meijgaard. 2008. Increase in hourly precipitation extremes beyond expectations from temperature changes. *Nature Geoscience 1* (8):511-514, **DOI:** 10.1038/ngeo262.

Lenton, A., F. Codron, L. Bopp, N. Metzl, P. Cadule, A. Tagliabue, and J. Le Sommer. 2009. Stratospheric ozone depletion reduces ocean carbon uptake and enhances ocean acidification. *Geophysical Research Letters 36*:L12606, **DOI:** 10.1029/2009GL038227.

Levermann, A., J. Mignot, S. Nawrath, and S. Rahmstorf. 2007. The role of northern sea ice cover for the weakening of the thermohaline circulation under global warming. *Journal of Climate 20* (16):4160-4171, **DOI:** 10.1175/JCLI4232.1.

Levin, L. A., W. Ekau, A. J. Gooday, F. Jorissen, J. J. Middelburg, S. W. A. Naqvi, C. Neira, N. N. Rabalais, and J. Zhang. 2009. Effects of natural and human-induced hypoxia on coastal benthos. *Biogeosciences* 6 (10):2063-2098.

Liao, W. and D. H. Reed. 2009. Inbreeding-environment interactions increase extinction risk. *Animal Conservation 12* (1):54-61, **DOI:** 10.1111/j.1469-1795.2008.00220.x.

Lin, J. T., K. O. Patten, K. Hayhoe, X. Z. Liang, and D. J. Wuebbles. 2008. Effects of future climate and biogenic emissions changes on surface ozone over the United States and China. *Journal of Applied Meteorology and Climatology 47* (7):1888-1909, **DOI:** 10.1175/2007JAMC1681.1.

Lins, H. F. and J. R. Slack. 1999. Streamflow trends in the United States. *Geophysical Research Letters 26* (2):227-230.

Lins, H. F. and J. R. Slack. 2005. Seasonal and regional characteristics of US streamflow trends in the United States from 1940 to 1999. *Physical Geography 26* (6):489-501.

Littell, J. S., D. McKenzie, D. L. Peterson, and A. L. Westerling. 2009. Climate and wildfire area burned in western U. S. ecoprovinces, 1916-2003. *Ecological Applications 19* (4):1003-1021.

Liu, J. P., J. A. Curry, and D. G. Martinson. 2004. Interpretation of recent Antarctic sea ice variability. *Geophysical Research Letters 31* (2):L02205, **DOI:** 10.1029/2003GL018732.

Liu, J., Z. Zhang, R. M. Horton, C. Wang and X. Ren. 2007. Variability in North Pacific sea ice and East Asia-North Pacific winter climate. *Journal of Climate 20* (10):1991-2001, **DOI:** 10.1175/JCLI4105.1.

Loarie, S. R., P. B. Duffy, H. Hamilton, G. P. Asner, C. B. Field, and D. D. Ackerly. 2009. The velocity of climate change. *Nature 462* (7276):1052-U111, **DOI:** 10.1038/nature08649.

Lobell, D. B. and C. B. Field. 2007. Global scale climate—crop yield relationships and the impacts of recent warming. *Environmental Research Letters 2* (1):014002, **DOI:** 10.1088/1748-9326/2/1/014002.

Lobell, D. B. and M. B. Burke. 2008. Why are agricultural impacts of climate change so uncertain? The importance of temperature relative to precipitation. *Environmental Research Letters 3* (3):034007, **DOI:** 10.1088/1748-9326/3/3/034007.

Long, S. P., E. A. Ainsworth, A. D. B. Leakey, J. Nosberger, and D. R. Ort. 2006. Food for thought: lower-than-expected crop yield stimulation with rising CO_2 concentrations. *Science 312* (5782):1918-1921, **DOI:** 10.1126/science.1114722.

Loucks, C., S. Barber-Meyer, M. A. A. Hossain, A. Barlow, and R. M. Chowdhury. 2010. Sea level rise and tigers: predicted impacts to Bangladesh's Sundarbans mangroves. *Climatic Change 98* (1-2):291-298, **DOI:** 10.1007/s10584-009-9761-5.

Lovenduski, N. S., N. Gruber, S. C. Doney, and I. D. Lima. 2007. Enhanced CO_2 outgassing in the Southern Ocean from a positive phase of the Southern Annular Mode. *Global Biogeochemical Cycles 21* (2):GB2026, **DOI:** 10.1029/2006GB002900.

Lovenduski, N. S., N. Gruber, and S. C. Doney. 2008. Toward a mechanistic understanding of the decadal trends in the Southern Ocean carbon sink. *Global Biogeochemical Cycles 22* (3):GB3016, **DOI:** 10.1029/2007GB003139.

Lunt, D. J., A. M. Haywood, G. A. Schmidt, U. Salzmann, P. J. Valdes, and H. J. Dowsett. 2010. Earth system sensitivity inferred from Pliocene modelling and data. *Nature Geoscience 3* (1):60-64, **DOI:** 10.1038/NGEO706.

Lüthi, D., M. Le Floch, B. Bereiter1, T. Blunier, J. Barnola, U. Siegenthaler, D. Raynaud, J. Jouzel, H. Fischer, K. Kawamura, and T. F. Stocker. 2008. High-resolution carbon dioxide concentration record 650,000-800,000 years before present. *Nature 453* (7193):379-382, **DOI:** 10.1038/nature06949.

Macdonald, G. J. 1990. Role of methane clathrates in past and future climates. *Climatic Change* 16:247-281.

MacMynowski, D. P. and T. L. Root. 2007. Climate and the complexity of migratory phenology: sexes, migratory distance, and arrival distributions. *International Journal of Biometeorology 51* (5):361-373, **DOI:** 10.1007/s00484-006-0084-1.

Madsen, H. and F. Jakobsen. 2004. Cyclone induced storm surge and flood forecasting in the northern Bay of Bengal. *Coastal Engineering 51* (4):277-296, **DOI:** 10.1016/j.coastaleng.2004.03.001.

Malhi, Y., J. T. Roberts, R. A. Betts, T. J. Killeen, W. H. Li, and C. A. Nobre. 2008. Climate change, deforestation, and the fate of the Amazon. *Science 319* (5860):169-172, **DOI:** 10.1126/science.1146961.

Malhi, Y., L. Aragao, D. Galbraith, C. Huntingford, R. Fisher, P. Zelazowski, S. Sitch, C. McSweeney, and P. Meir. 2009. Exploring the likelihood and mechanism of a climate-change-induced dieback of the Amazon rainforest. *Proceedings of the National Academy of Sciences of the United States of America 106* (49):20610-20615, **DOI:** 10.1073/pnas.0804619106.

Manabe, S. and R. T. Wetherald. 1980. Distribution of climate change resulting from an increase in CO_2 content of the atmosphere. *Journal of the Atmospheric Sciences 37* (1):99-118.

Marčelja, S. 2009. The timescale and extent of thermal expansion of the oceans due to climate change. *Ocean Science 6* (3):2975-2992.

Markoff, M. S. and A. C. Cullen. 2008. Impact of climate change on Pacific Northwest hydropower. *Climatic Change 87* (3-4):451-469, **DOI:** 10.1007/s10584-007-9306-8.

Martens, W. J. M. 1998. Climate change, thermal stress and mortality changes. *Social Science and Medicine* 46:331-334.

Martin, P. H. and M. G. Lefebvre. 1995. Malaria and climate—sensitivity of malaria potential transmission to climate. *Ambio 24* (4):200-207.

Maslanik, J. A., C. Fowler, J. Stroeve, S. Drobot, J. Zwally, D. Yi, and W. Emery. 2007. A younger, thinner Arctic ice cover: increased potential for rapid, extensive sea-ice loss. *Geophysical Research Letters 34* (24):L24501, **DOI:** 10.1029/2007GL032043.

Matsumoto, K., J. L. Sarmiento, R. M. Key, O. Aumont, J. L. Bullister, K. Caldeira, J. M. Campin, S. C. Doney, H. Drange, J. C. Dutay, M. Follows, Y. Gao, A. Gnanadesikan, N. Gruber, A. Ishida, F. Joos, K. Lindsay, E. Maier-Reimer, J. C. Marshall, R. J. Matear, P. Monfray, A. Mouchet, R. Najjar, G. K. Plattner, R. Schlitzer, R. Slater, P. S. Swathi, I. J. Totterdell, M. F. Weirig, Y. Yamanaka, A. Yool, and J. C. Orr. 2004. Evaluation of ocean carbon cycle models with data-based metrics. *Geophysical Research Letters 31* (7):L07303, **DOI:** 10.1029/2003GL018970.

Matthews, H. D. 2006. Emissions targets for CO_2 stabilization as modified by carbon cycle feedbacks. *Tellus Series B-Chemical and Physical Meteorology 58* (5):591-602, **DOI:** 10.1111/j.1600-0889.2006.00200.x.

Matthews, H. D. and K. Caldeira. 2008. Stabilizing climate requires near-zero emissions. *Geophysical Research Letters 35* (4):L04705, **DOI:** 10.1029/2007GL032388.

Matthews, H. D., N. P. Gillett, P. A. Stott, and K. Zickfeld. 2009. The proportionality of global warming to cumulative carbon emissions. *Nature 459* (7248):829-U823, **DOI:** 10.1038/nature08047.

Matthews, R. B. 1995. *Modeling the Impact Of Climate Change On Rice Production in Asia.* Wallingford: CAB International in association with the International Rice Research Institute.

Maurer, E. P., I. T. Stewart, C. Bonfils, P. B. Duffy, and D. Cayan. 2007. Detection, attribution, and sensitivity of trends toward earlier streamflow in the Sierra Nevada. *Journal of Geophysical Research-Atmospheres, 112* (D11):D11118, **DOI:** 10.1029/2006JD008088.

May, R. M. 2010. Ecological science and tomorrow's world. *Philosophical Transactions of the Royal Society B-Biological Sciences 365* (1537):41-47, **DOI:** 10.1098/rstb.2009.0164.

McCabe, G. J. and D. M. Wolock. 2010. Long-term variability in Northern Hemisphere snow cover and associations with warmer winters. *Climatic Change 99* (1):141-153, **DOI:** 10.1007/s10584-009-9675-2.

McConnell, J. R., R. Edwards, G. L. Kok, M. G. Flanner, C. S. Zender, E. S. Saltzman, J. R. Banta, D. R. Pasteris, M. M. Carter, and J. D. W. Kahl. 2007. 20th-century industrial black carbon emissions altered Arctic climate forcing. *Science 317* (5843):1381-1384, **DOI:** 10.1126/science.1144856.

McGeehin, M. A. and M. Mirabelli. 2001. The potential impacts of climate variability and change on temperature-related morbidity and mortality in the United States. *Environmental Health Perspectives 109* (SUPPL. 2):185-189.

Mecking, S., C. Langdon, R. A. Feely, C. L. Sabine, C. A. Deutsch, and D. H. Min. 2008. Climate variability in the North Pacific thermocline diagnosed from oxygen measurements: an update based on the US CLIVAR/CO_2 Repeat Hydrography cruises. *Global Biogeochemical Cycles 22* (3):GB3015, **DOI:** 10.1029/2007GB003101.

Medina-Ramon, M. and J. Schwartz. 2007. Temperature, temperature extremes, and mortality: a study of acclimatisation and effect modification in 50 US cities. *Occupational and Environmental Medicine 64* (12):827-833, **DOI:** 10.1136/oem.2007.033175.

Meehl, G. A., T. F. Stocker, W. D. Collins, P. Friedlingstein, A. T. Gaye, J. M. Gregory, A. Kitoh, R. Knutti, J. M. Murphy, A. Noda, S. C. B. Raper, and I. G. Watterson, A. J. W. a. Z.-C. Z. 2007. Global Climate Projections. In *Climate Change 2007: The Physical Science Basis. Contribution of Working Group I to the Fourth Assessment Report of the Intergovernmental Panel on Climate Change*, edited by S. Solomon, M. M. D. Qin, Z. Chen, M. Marquis, K. B. Averyt, M. Tignor, and H .L. Miller. Cambridge: Cambridge University Press.

Meinshausen, M., N. Meinshausen, W. Hare, S. C. B. Raper, K. Frieler, R. Knutti, D. J. Frame, and M. R. Allen. 2009. Greenhouse-gas emission targets for limiting global warming to 2 degrees C. *Nature 458* (7242):1158-U1196, **DOI:** 10.1038/nature08017.

Meleshko, V. P., G. S. Golitsyn, V. A. Govorkova, P. F. Demchenko, A. V. Eliseev, V. M. Kattsov, et al. 2005. Anthropogenic climate change in Russia in the twenty-first century: an ensemble of climate model projections. *Russian Meteorology and Hydrology* (4):22-30.

Mellor, P. S. 2000. Replication of arboviruses in insect vectors. *Journal of Comparative Pathology 123* (4):231-247, **DOI:** 10.1053/jcpa.2000.0434.

Menzel, A., T. H. Sparks, N. Estrella, E. Koch, A. Aasa, R. Ahas, K. Alm-Kubler, P. Bissolli, O. Braslavska, A. Briede, F. M. Chmielewski, Z. Crepinsek, Y. Curnel, A. Dahl, C. Defila, A. Donnelly, Y. Filella, K. Jatcza, F. Mage, A. Mestre, O. Nordli, J. Penuelas, P. Pirinen, V. Remisova, H. Scheifinger, M. Striz, A. Susnik, A. J. H. Van Vliet, F. E. Wielgolaski, S. Zach, and A. Zust. 2006. European phenological response to climate change matches the warming pattern. *Global Change Biology 12* (10):1969-1976.

Mickley, L. J., D. J. Jacob, B. D. Field, and D. Rind. 2004. Effects of future climate change on regional air pollution episodes in the United States. *Geophysical Research Letters 31* (24):1-4 L24103, **DOI:** 10.1029/2004GL021216.

Miguel, E., S. Satyanath, and E. Sergenti. 2004. Economic shocks and civil conflict: an instrumental variables approach. *Journal of Political Economy 112* (4):725-753.

Miholcsa, T., A. Toth, and T. Csorgo. 2009. Change of timing of autumn migration in acrocephalus and locustella genus. *Acta Zoologica Academiae Scientiarum Hungaricae 55* (2):175-185.

Milkov, A. V. 2004. Global estimates of hydrate-bound gas in marine sediments: how much is really out there? *Earth-Science Reviews 66* (3-4):183-197, **DOI:** 10.1016/j.earscirev.2003.11.002.

MEA (Millennium Ecosystem Assessment). 2005. *Ecosystems and Human Well-being: Synthesis*. Washington DC: Island Press.

Miller, A. W., A. C. Reynolds, C. Sobrino, and G. F. Riedel. 2009. Shellfish face uncertain future in high CO_2 world: influence of acidification on oyster larvae calcification and growth in estuaries. *PLoS ONE 4* (5):e5661, **DOI:** 10.1371/journal.pone.0005661.

Miller, N. L., K. Hayhoe, J. Jin, and M. Auffhammer. 2008. Climate, extreme heat, and electricity demand in California. *Journal of Applied Meteorology and Climatology 47* (6):1834-1844, **DOI:** 10.1175/2007JAMC1480.1.

Miller, R. L., G. A. Schmidt, and D. T. Shindell. 2006. Forced annular variations in the 20th century intergovernmental panel on climate change fourth assessment report models. *Journal of Geophysical Research-Atmospheres 111* (D18):D18101, **DOI:** 10.1029/2005JD006323.

Milly, P. C. D., K. A. Dunne, and A. V. Vecchia. 2005. Global pattern of trends in streamflow and water availability in a changing climate. *Nature 438* (7066):347-350, **DOI:** 10.1038/nature04312.

Mimura, N. and H. Yokoki. 2004. Sea level changes and vulnerability of the coastal region of East Asia in response to global warming. In *Changes in the Human-Monsoon System of East Asia in the Context of Global Change*, edited by C. E. Fu, J. R. E. Freney, and J. W. B. E. Stewart. Hackensack, NJ:World Scientific.

Min, S. K., X. B. Zhang, and F. Zwiers. 2008. Human-induced arctic moistening. *Science 320* (5875):518-520, **DOI:** 10.1126/science.1153468.

Mitchell, J. F. B., T. C. Johns, M. Eagles, W. J. Ingram, and R. A. Davis. 1999. Towards the construction of climate change scenarios. *Climatic Change 41* (3-4):547-581.

Mitchell, T. D. 2003. Pattern scaling—An examination of the accuracy of the technique for describing future climates. *Climatic Change 60* (3):217-242.

Mitrovica, J. X. and G. A. Milne. 2002. On the origin of late Holocene sea-level highstands within equatorial ocean basins. *Quaternary Science Reviews 21* (20-22), 2179-2190, ISSN 0277-3791, **DOI:** 10.1016/S0277-3791(02)00080-X.

Mohan, J. E., L. H. Ziska, R. B. Thomas, R. C. Sicher, K. George, J. S. Clark, and W. H. Schlesinger. 2008. Biomass and toxicity responses of poison ivy (Toxicodendron radicans) to elevated atmospheric CO_2: reply. *Ecology 89* (2):585-587.

Montes-Hugo, M., S. C. Doney, H. W. Ducklow, W. Fraser, D. Martinson, S. E. Stammerjohn, and O. Schofield. 2009. Recent changes in phytoplankton communities associated with rapid regional climate change along the Western Antarctic Peninsula. *Science 323* (5920):1470-1473, **DOI:** 10.1126/science.1164533.

Moritz, C., J. L. Patton, C. J. Conroy, J. L. Parra, G. C. White, and S. R. Beissinger. 2008. Impact of a century of climate change on small-mammal communities in Yosemite National Park, USA. *Science 322* (5899):261-264, **DOI:** 10.1126/science.1163428.

Moss, R. H., J. A. Edmonds, K. A. Hibbard, M. R. Manning, S. K. Rose, D. P. Van Vuuren, T. R. Carter, S. Emori, M. Kainuma, T. Kram, G. A. Meehl, J. F. B. Mitchell, N. Nakicenovic, K. Riahi, S. J. Smith, R. J. Stouffer, A. M. Thomson, J. P. Weyant, and T. J. Wilbanks. 2010. The next generation of scenarios for climate change research and assessment. *Nature 463* (7282):747-756, **DOI:** 10.1038/nature08823.

Mote, P. W. 2003. Trends in snow water equivalent in the Pacific Northwest and their climatic causes. Geophysical Research Letter 30 (12):1601, **DOI:** 10.1029/2003GL017258.

Mote, P. W., A. F. Hamlet, M. P. Clark, and D. P. Lettenmaier. 2005. Declining mountain snowpack in western North America. *Bulletin of the American Meteorological Society 86* (1):39. **DOI:** 10.1175/BAMS-86-1-39.

Murphy, D. M., S. Solomon, R. W. Portmann, K. H. Rosenlof, P. M. Forster, and T. Wong. 2009. An observationally based energy balance for the Earth since 1950. *Journal of Geophysical Research-Atmospheres 114*:D17107, **DOI:** 10.1029/2009JD012105.

Murphy, J. M., B. B. B. Booth, M. Collins, G. R. Harris, D. M. H. Sexton, and M. J. Webb. 2007. A methodology for probabilistic predictions of regional climate change from perturbed physics ensembles. *Philosophical Transactions of the Royal Society a-Mathematical Physical and Engineering Sciences 365* (1857):1993-2028, **DOI:** 10.1098/rsta.2007.2077.

Myneni, R. B., C. J. Tucker, G. Asrar, and C. D. Keeling. 1998. Interannual variations in satellite-sensed vegetation index data from 1981 to 1991 (vol 103, pg 6145, 1998). *Journal of Geophysical Research-Atmospheres 103* (D16):19839-19839.

Nelson, F. E. 2003. (Un)frozen in time. *Science 299* (5613):1673-1675.

Nelson, F. E., N. I. Shiklomanov, G. R. Mueller, K. M. Hinkel, D. A. Walker and J. G. Bockheim. 1997. Estimating active-layer thickness over a large region: Kuparuk River Basin, Alaska, U.S.A. *Arctic and Alpine Research*, 29 (4):367-378.

NPCC (New York Panel on Climate Change). 2009. Climate change and New York City: creating flexible adaptation pathways. *Annals of the New York Academy of Sciences*. New York.

NYCOEM (New York City Office of Emergency Management). 2009. Planning for emergencies: 2009 hazard management plan. Office of the Mayor, New York City. Available online at: *http://www.nyc.gov/html/oem/html/about/planning_hazard_mitigation.*

Nicholls, R. J. 2004. Coastal flooding and wetland loss in the 21st century: changes under the SRES climate and socio-economic scenarios. *Global Environmental Change 14* (1):69-86, **DOI:** 10.1016/j.gloenvcha.2003.10.007.

Nicholls, R. J. and R. S. J. Tol. 2006. Impacts and responses to sea-level rise: a global analysis of the SRES scenarios over the twenty-first century. *Philosophical Transactions of the Royal Society a-Mathematical Physical and Engineering Sciences 364* (1841):1073-1095, **DOI:** 10.1098/rsta-2006.1754.

Nicholls, R. J., P. P. Wong, V. R. Burkett, J. O. Codignotto, J. E. Hay, R. F. McLean, S. Ragoonaden and C. D. Woodroffe. 2007. Coastal systems and low-lying areas. In *Climate Change 2007: Impacts, Adaptation and Vulnerability. Contribution of Working Group II to the Fourth Assessment Report of the Intergovernmental Panel on Climate Change, Climate Change 2007: Impacts, Adaptation and Vulnerability. Contribution of Working Group II to the Fourth Assessment Report of the Intergovernmental Panel on Climate Change*, edited by O. F. C. M. L. Parry, J. P. Palutikof, P .J. van der Linden, and C. E. Hanson. 315-356. Cambridge: Cambride University Press.

Nordas, R. and N. P. Gleditsch. 2007. Climate change and conflict. *Political Geography 26* (6):627-638, **DOI:** 10.1016/j.polgeo.2007.06.003.

NRC (National Research Council). 2008. *Potential Impacts of Climate Change on U.S. Transportation.* Transportation Research Board Special Report 290. Transportation Research Board, Washington, DC: National Academies Press.

NRC. 2009. *Ecological Impacts Of Climate Change.* Washington, DC: National Academies Press

NRC. 2010. *Development of an Integrated Science Strategy for Ocean Acidification Monitoring, Research, and Impacts Assessment.* Ocean Studies Board, Washington, DC: National Academies Press.

Nye, J. A., J. S. Link, J. A. Hare, and W. J. Overholtz. 2009. Changing spatial distribution of fish stocks in relation to climate and population size on the Northeast United States continental shelf. *Marine Ecology-Progress Series 393*:111-129, **DOI:** 10.3354/meps08220.

Oaks, S. C. 1991. *Malaria: Obstacles and Opportunities: a Report of the Committee for the Study on Malaria Prevention and Control: Status Review and Alternative Strategies,* Division of International Health, Institute of Medicine. Washington, DC: National Academy Press.

Ogden, N. H., M. Bigras-Poulin, K. Hanincova, A. Maarouf, C. J. O'Callaghan, and K. Kurtenbach. 2008. Projected effects of climate change on tick phenology and fitness of pathogens transmitted by the North American tick Ixodes scapularis. *Journal of Theoretical Biology 254* (3):621-632, **DOI:** 10.1016/j.jtbi.2008.06.020.

O'Gorman, P. A. and T. Schneider. 2009. The physical basis for increases in precipitation extremes in simulations of 21st-century climate change. *Proceedings of the National Academy of Sciences of the United States of America 106* (35):14773-14777, **DOI:** 10.1073/pnas.0907610106.

Oouchi, K., J. Yoshimura, H. Yoshimura, R. Mizuta, S. Kusunoki, and A. Noda. 2006. Tropical cyclone climatology in a global-warming climate as simulated in a 20 km-mesh global atmospheric model: frequency and wind intensity analyses. *Journal of the Meteorological Society of Japan 84* (2):259-276.

Orr, J. C., V. J. Fabry, O. Aumont, L. Bopp, S. C. Doney, R. A. Feely, A. Gnanadesikan, N. Gruber, A. Ishida, F. Joos, R. M. Key, K. Lindsay, E. Maier-Reimer, R. Matear, P. Monfray, A. Mouchet, R. G. Najjar, G. K. Plattner, K. B. Rodgers, C. L. Sabine, J. L. Sarmiento, R. Schlitzer, R. D. Slater, I. J. Totterdell, M. F. Weirig, Y. Yamanaka, and A. Yool. 2005. Anthropogenic ocean acidification over the twenty-first century and its impact on calcifying organisms. *Nature 437* (7059):681-686, **DOI:** 10.1038/nature04095.

Ostro, B. D., L. A. Roth, R. S. Green, and R. Basu. 2009. Estimating the mortality effect of the July 2006 California heat wave. *Environmental Research 109* (5):614-619, **DOI:** 10.1016/j.envres.2009.03.010.

Overpeck, J. T., R. S. Webb, and T. Webb. 1992. Mapping eastern North-American vegetation change of the past 18 ka—no-analogs and the future. *Geology 20* (12):1071-1074.

Overpeck, J. T., B. L. Otto-Bliesner, G. H. Miller, D. R. Muhs, R. B. Alley, and J. T. Kiehl. 2006. Paleoclimatic evidence for future ice-sheet instability and rapid sea-level rise. *Science 311* (5768):1747-1750, **DOI:** 10.1126/science.1115159.

Pacala, S. W., G. C. Hurtt, D. Baker, P. Peylin, R. A. Houghton, R. A. Birdsey, L. Heath, E. T. Sundquist, R. F. Stallard, P. Ciais, P. Moorcroft, J. P. Caspersen, E. Shevliakova, B. Moore, G. Kohlmaier, E. Holland, M. Gloor, M. E. Harmon, S. M. Fan, J. L. Sarmiento, C. L. Goodale, D. Schimel, and C. B. Field. 2001. Consistent land- and atmosphere-based US carbon sink estimates. *Science 292* (5525):2316-2320.

Pagani, M., K. Caldeira, D. Archer, and J. C. Zachos. 2006. An ancient carbon mystery. *Science 314* (5805):1556-1557, **DOI:** 10.1126/science.1136110.

Pall, P., M. R. Allen, and D. A. Stone. 2007. Testing the Clausius-Clapeyron constraint on changes in extreme precipitation under CO_2 warming. *Climate Dynamics 28* (4):351-363, **DOI:** 10.1007/s00382-006-0180-2.

Parmesan, C. 2006. Ecological and evolutionary responses to recent climate change. *Annual Review of Ecology Evolution and Systematics 37*:637-669, **DOI:** 10.1146/annurev.ecolsys.37.091305.110100.

Parmesan, C. 2007. Influences of species, latitudes and methodologies on estimates of phenological response to global warming. *Global Change Biology 13* (9):1860-1872, **DOI:** 10.1111/j.1365-2486.2007.01404.x.

Parmesan, C. and G. Yohe. 2003. A globally coherent fingerprint of climate change impacts across natural systems. *Nature 421* (6918):37-42, **DOI:** 10.1038/nature01286.

Partap, U. M. A., T. E. J. Partap, and H. E. Yonghua. 2001. Pollination failure in apple crop and farmers' management strategies in Hengduan Mountains, China. *Acta Horticulturae 561*:225-230.

Pascual, M., J. A. Ahumada, L. F. Chaves, X. Rodo, and M. Bouma. 2006. Malaria resurgence in the East African highlands: temperature trends revisited. *Proceedings of the National Academy of Sciences of the United States of America 103* (15):5829-5834, **DOI:** 10.1073/pnas.0508929103.

Patz, J. A., D. Campbell-Lendrum, T. Holloway, and J. A. Foley. 2005. Impact of regional climate change on human health. *Nature 438* (7066):310-317, **DOI:** 10.1038/nature04188.

Pavlov, A. V. and N. G. Moskalenko. 2002. The thermal regime of soils in the north of Western Siberia. *Permafrost and Periglacial Processes 13* (1):43-51, **DOI:** 10.1002/ppp.409.

Payne, J. T., A. W. Wood, A. F. Hamlet, R. N. Palmer, and D. P. Lettenmaier. 2004. Mitigating the effects of climate change on the water resources of the Columbia River Basin. *Climatic Change 62* (1-3):233-256.

Perry, A. L., P. J. Low, J. R. Ellis, and J. D. Reynolds. 2005. Climate change and distribution shifts in marine fishes. *Science 308* (5730):1912-1915, **DOI:** 10.1126/science.1111322.

Petrenko, V. V., A. M. Smith, E. J. Brook, D. Lowe, K. Riedel, G. Brailsford, Q. Hua, H. Schaefer, N. Reeh, R. F. Weiss, D. Etheridge, and J. P. Severinghaus. 2009. (CH4)-C-14 measurements in Greenland ice: investigating last glacial termination CH4 sources. *Science 324* (5926):506-508, **DOI:** 10.1126/science.1168909.

Peylin, P., P. Bousquet, C. Le Quere, S. Sitch, P. Friedlingstein, G. McKinley, N. Gruber, P. Rayner, and P. Ciais. 2005. Multiple constraints on regional CO_2 flux variations over land and oceans. *Global Biogeochemical Cycles 19* (1):GB1011, **DOI:** 10.1029/2003GB002214.

Pfeffer, W. T., J. T. Harper, and S. O'Neel. 2008. Kinematic constraints on glacier contributions to 21st-century sea-level rise. *Science 321* (5894):1340-1343, **DOI:** 10.1126/science.1159099.

Pierce, D. W., T. P. Barnett, H. G. Hidalgo, T. Das, C. Bonfils, B. D. Santer, G. Bala, M. D. Dettinger, D. R. Cayan, A. Mirin, A. W. Wood, and T. Nozawa. 2008. Attribution of declining Western U.S. snowpack to human effects. *Journal of Climate 21* (23):6425-6444, **DOI:** 10.1175/2008JCLI2405.1.

Pierrehumbert, R. T. 2002. The hydrologic cycle in deep-time climate problems. *Nature 419* (6903):191-198, **DOI:** 10.1038/nature01088.

Pierrehumbert, R. T. 2004. Warming the world. *Nature 432* (7018):677, **DOI:** 10.1038/432677a.

Pierrehumbert, R. T. 2010. *Principles of Planetary Climate*. Cambridge: Cambridge University Press (in press).

Pierrehumbert R. T., H. Brogniez, and R. Roca. 2007. On the relative humidity of the Earth's atmosphere. In *The Global Circulation of the Atmosphere*, edited by T. Schneider and A. Sobel. Princeton: Princeton University Press.

Plattner, G. K., R. Knutti, F. Joos, T. F. Stocker, W. von Bloh, V. Brovkin, D. Cameron, E. Driesschaert, S. Dutkiewicz, M. Eby, N. R. Edwards, T. Fichefet, J. C. Hargreaves, C. D. Jones, M. F. Loutre, H. D. Matthews, A. Mouchet, S. A. Muller, S. Nawrath, A. Price, A. Sokolov, K. M. Strassman, and A. J. Weaver. 2008. Long-term climate commitments projected with climate-carbon cycle models. *Journal of Climate 21* (12):2721-2751, **DOI:** 10.1175/2007JCLI1905.1.

Pollard, D. and R. M. DeConto. 2005. Hysteresis in Cenozoic Antarctic ice-sheet variations. *Global and Planetary Change 45* (1-3):9-21, **DOI:** 10.1016/j.gloplacha.2004.09.011.

Pollard, D. and R. M. DeConto. 2009. Modelling West Antarctic ice sheet growth and collapse through the past five million years. *Nature 458* (7236):329-U389, **DOI:** 10.1038/nature07809.

Polovina, J. J., E. A. Howell, and M. Abecassis. 2008. Ocean's least productive waters are expanding. *Geophysical Research Letters 35* (3):L03618, **DOI:** 10.1029/2007GL031745.

Polyakov, I. V., A. Beszczynska, E. C. Carmack, I. A. Dmitrenko, E. Fahrbach, I. E. Frolov, et al. 2005. One more step toward a warmer Arctic. *Geophysical Research Letters 32* (17): L17605, **DOI:** 10.1029/2005GL023740.

Polyakov, I. V., V. A. Alexeev, U. S. Bhatt, E. I. Polyakova, and X. Zhang. 2009. North Atlantic warming: patterns of long-term trend and multidecadal variability. *Climate Dynamics 34* (2-3):439-457, **DOI:** 10.1007/s00382-008-0522-3.

Portner, H. O. and A. P. Farrell. 2008. Ecology physiology and climate change. *Science 322* (5902):690-692, **DOI:** 10.1126/science.1163156.

Post, E., M. C. Forchhammer, M. S. Bret-Harte, T. V. Callaghan, T. R. Christensen, B. Elberling, A. D. Fox, O. Gilg, D. S. Hik, T. T. Høye, R. A. Ims, E. J., D. R. Klein, J. Madsen, A. D. McGuire, S. Rysgaard, D. E. Schindler, I. Stirling, M. P. Tamstorf, N. J. C. Tyler, R. van der Wal, J. Welker, P. A. Wookey, N. M. Schmidt, and P. Aastrup. 2009. Ecological dynamics across the Arctic associated with recent climate change. *Science 325* (5946):1355-1358, **DOI:** 10.1126/science.1173113.

Primack, R. B. and A. J. Miller-Rushing. 2009. The role of botanical gardens in climate change research. *New Phytologist 182* (2):303-313, **DOI:** 10.1111/j.1469-8137.2009.02800.x.

Primack, R. B., H. Higuchi, and A. J. Miller-Rushing. 2009. The impact of climate change on cherry trees and other species in Japan. *Biological Conservation 142* (9):1943-1949, **DOI:** 10.1016/j.biocon.2009.03.016.

Pritchard, H. D., R. J. Arthern, D. G. Vaughan, and L. A. Edwards. 2009. Extensive dynamic thinning on the margins of the Greenland and Antarctic ice sheets. *Nature 461* (7266):971-975, **DOI:** 10.1038/nature08471.

Quaas, J., O. Boucher, N. Bellouin, and S. Kinne. 2008. Satellite-based estimate of the direct and indirect aerosol climate forcing. *Journal of Geophysical Research-Atmospheres 113* (D5):D05204, **DOI:** 10.1029/2007JD008962.

Rabalais, N. N., R. J. Díaz, L. A. Levin, R. E. Turner, D. Gilbert, and J. Zhang. 2010. Dynamics and distribution of natural and human-caused hypoxia. *Biogeosciences 7* (2):585-619.

Rabinowitz, D., S. Cairns, T. Dillon. 1986. Seven forms of rarity and their frequency in the flora of the British Isles. In *Conservation Biology: The Science of Scarcity and Diversity*, edited by M. Soule. 182-204. Suderland, Massachussetts: Sinauer Associates.

Ramanathan, V., H. Akimoto, P. Bonasoni, M. Brauer, G. Carmichael, C. E. Chung, Y. Feng, S. Fuzzi, S. I. Hasnain, M. Iyngararasan, A. Jayaraman, M. G. Lawrence, T. Nakajima, T. S. Panwar, M. V. Ramana, M. Rupakheti, S. Weidemann, S.-C. Yoon, Y. Zhang and A. Zhu. 2008. *Atmospheric Brown Clouds and Regional Climate Change, Part I of Atmospheric Brown Clouds: Regional Assessment Report with Focus on Asia*. Nairobi, Kenya.

Ramaswamy, V., P. Forster, V. P. Artaxo, T. Berntsen, R. Betts, D.W. Fahey, J. Haywood, J. Lean, D.C. Lowe, G. Myhre, J. Nganga, R. Prinn, G. Raga, M. Schulz, and R. Van Dorland. 2007. Changes in Atmospheric Constituents and in Radiative Forcing. In: Climate Change 2007: The Physical Science Basis. Contribution of Working Group I to the Fourth Assessment Report of the Intergovernmental Panel on Climate Change [Solomon, S., D. Qin, M. Manning, Z. Chen, M. Marquis, K.B. Averyt, M.Tignor, and H.L. Miller (eds.)]. Cambridge, United Kingdom and New York: Cambridge University Press.

Randall, D. A., R. A. Wood, S. Bony, R. Colman, T. Fichefet, J. Fyfe, V. Kattsov, A. Pitman, J. Shukla, J. Srinivasan, R. J. Stouffer, A. Sumi and K. E. Taylor. 2007. Climate models and their evaluation. In *Climate Change 2007: The Physical Science Basis. Contribution of Working Group I to the Fourth Assessment Report of the Intergovernmental Panel on Climate Change*, edited by S. Solomon, D. Qin, M. Manning, Z. Chen, M. Marquis, K. B. Averyt, M. Tignor, and H. L. Miller. New York, Cambridge University Press.

Randerson, J. T., H. Liu, M. G. Flanner, S. D. Chambers, Y. Jin, P. G. Hess, G. Pfister, M. C. Mack, K. K. Treseder, L. R. Welp, F. S. Chapin, J. W. Harden, M. L. Goulden, E. Lyons, J. C. Neff, E. A. G. Schuur, and C. S. Zender. 2006. The impact of boreal forest fire on climate warming. *Science 314* (5802):1130-1132, **DOI:** 10.1126/science.1132075.

Randles, C. A. and Ramaswamy, V. 2008. Absorbing aerosols over Asia: a geophysical fluid dynamics laboratory general circulation model sensitivity study of model response to aerosol optical depth and aerosol absorption. *Journal of Geophysical Research-Atmospheres 113* (D21):D21203, **DOI:** 10.1029/2008JD010140.

Raper, S. C. B., J. M. Gregory, and R. J. Stouffer. 2002. The role of climate sensitivity and ocean heat uptake on AOGCM transient temperature response. *Journal of Climate 15* (1):124-130.

Raphael, M. N. 2007. The influence of atmospheric zonal wave three on Antarctic sea ice variability. *Journal of Geophysical Research-Atmospheres 112* (D12):D12112, **DOI:** 10.1029/2006JD007852.

Raupach, M. R., G. Marland, P. Ciais, C. Le Quere, J. G. Canadell, G. Klepper, and C. B. Field. 2007. Global and regional drivers of accelerating CO_2 emissions. *Proceedings of the National Academy of Sciences of the United States of America 104* (24):10288-10293, **DOI:** 10.1073/pnas.0700609104.

Raymo, M. E., D. Rind, and W. F. Ruddiman. 1990. Climatic effects of reduced Arctic sea ice limits in the GISS II general circulation model. *Paleooceanography* 5:367-382.

Rayner, P. J., I. G. Enting, R. J. Francey, and R. Langenfelds. 1999. Reconstructing the recent carbon cycle from atmospheric CO_2, delta C-13 and O-2/N-2 observations. *Tellus Series B-Chemical and Physical Meteorology 51* (2):213-232.

Regan, H. M., R. Lupia, A. N. Drinnan, and M. A. Burgman. 2001. The currency and tempo of extinction. *American Naturalist 157* (1):1-10.

Rennermalm, A. K, L. C. Smith, J. C. Stroeve, and V. W. Chu. 2009. Does sea ice influence Greenland ice sheet surface-melt? *Environmental Research Letters 4* (2). **DOI:** 10.1088/1748-9326/4/2/024011.

Ridley, J. K., P. Huybrechts, J. M. Gregory, and J. A. Lowe. 2005. Elimination of the Greenland ice sheet in a high CO_2 climate. *Journal of Climate 18* (17):3409-3427.

Ridley, J., J. Lowe, and D. Simonin. 2008. The demise of Arctic sea ice during stabilisation at high greenhouse gas concentrations. *Climate Dynamics 30* (4):333-341, **DOI:** 10.1007/s00382-007-0291-4.

Rigby, M., R. G. Prinn, P. J. Fraser, P. G. Simmonds, R. L. Langenfelds, J. Huang, D. M. Cunnold, L. P. Steele, P. B. Krummel, R. F. Weiss, S. O'Doherty, P. K. Salameh, H. J. Wang, C. M. Harth, J. Muhle, and L. W. Porter. 2008. Renewed growth of atmospheric methane. *Geophysical Research Letters 35* (22):L22805, **DOI:** 10.1029/2008GL036037.

Rignot, E. 2006. Changes in ice dynamics and mass balance of the Antarctic ice sheet. *Philosophical Transactions of the Royal Society a-Mathematical Physical and Engineering Sciences 364* (1844):1637-1655, 10.1098/rsta.2006.1793.

Rignot, E. and P. Kanagaratnam. 2006. Changes in the velocity structure of the Greenland ice sheet. *Science 311* (5763):986-990, **DOI:** 10.1126/science.1121381.

Rigor, I. G. and J. M. Wallace. 2004. Variations in the age of Arctic sea-ice and summer sea-ice extent. *Geophysical Research Letters 31* (9):L09401, **DOI:** 10.1029/2004GL019492.

Robine, J. M., S. L. K. Cheung, S. Le Roy, H. Van Oyen, C. Griffiths, J. P. Michel, and F. R. Herrmann. 2008. Death toll exceeded 70,000 in Europe during the summer of 2003. *Comptes Rendus—Biologies 331* (2):171-178, **DOI:** 10.1016/j.crvi.2007.12.001.

Rödenbeck, C., S. Houweling, M. Gloor, and M. Heimann. 2003. CO_2 flux history 1982-2001 inferred from atmospheric data using a global inversion of atmospheric transport. *Atmospheric Chemistry and Physics 3:*1919-1964.

Roe, G. H. and M. B. Baker. 2007. Why is climate sensitivity so unpredictable? *Science 318* (5850):629-632, **DOI:** 10.1126/science.1144735.

Root, T. 1988a. Energy constraints on avian distribution and abundances. *Ecology 69* (2):330-339.

Root, T. 1988b. Environmental-factors associated with avian distributional boundaries. *Journal of Biogeography 15* (3):489-505.

Root T. L. and S. H. Schneider. 2002. Overview and implications for wildlife. *Wildlife Responses to Climate Change: North American Case Studies*, edited by S. H. Schneider and T. L. Root. 1-56. Washington, DC: Island Press.

Root, T. L., J. T. Price, K. R. Hall, S. H. Schneider, C. Rosenzweig, and J. A. Pounds. 2003. Fingerprints of global warming on wild animals and plants. *Nature 421* (6918):57-60, **DOI:** 10.1038/nature01333.

Root, T. L., D. P. MacMynowski, M. D. Mastrandrea, and S. H. Schneider. 2005. Human-modified temperatures induce species changes: joint attribution. *Proceedings of the National Academy of Sciences of the United States of America 102* (21):7465-7469, **DOI:** 10.1073/pnas.0502286102.

Rose, J. B., S. Daeschner, D. R. Easterling, F. C. Curriero, S. Lele, and J. A. Patz. 2000. Climate and waterborne disease outbreaks. *Journal American Water Works Association 92* (9):77-87.

Rosenthal, D. H., H. K. Gruenspecht, and E. A. Moran. 1995. Effects of global warming on energy use for space heating and cooling in the United States. *Energy Journal 16* (2):77-96.

Rosenzweig, C. 1993. *Climate Change and World Food Supply.* Oxford: Environmental Change Unit, University of Oxford.

Rosenzweig, C. and M. L. Parry. 1994. Potential impact of climate-change on world food-supply. *Nature 367* (6459):133-138.

Royer, J. F., S. Planton, and M. Deque. 1990. A sensitivity experiment for the removal of Arctic sea ice with the French spectral general circulation model. *Climate Dynamics* 5:1-17.

Russell, J. L., K. W. Dixon, A. Gnanadesikan, R. J. Stouffer, and J. R. Toggweiler. 2006. The Southern Hemisphere westerlies in a warming world: propping open the door to the deep ocean. *Journal of Climate 19* (24):6382-6390.

Sabine, C. L., R. A. Feely, N. Gruber, R. M. Key, K. Lee, J. L. Bullister, R. Wanninkhof, C. S. Wong, D. W. R. Wallace, B. Tilbrook, F. J. Millero, T. H. Peng, A. Kozyr, T. Ono, and A. F. Rios. 2004. The oceanic sink for anthropogenic CO_2. *Science 305* (5682):367-371.

Saino, N., D. Rubolini, E. Lehikoinen, L. V. Sokolov, A. Bonisoli-Alquati, R. Ambrosini, G. Boncoraglio, and A. P. Moller. 2009. Climate change effects on migration phenology may mismatch brood parasitic cuckoos and their hosts. *Biology Letters 5* (4):539-541, **DOI:** 10.1098/rsbl.2009.0312.

Salisbury, J., M. Green, C. Hunt, and J. Campbell. 2008. Coastal acidification by rivers: a threat to shellfish? *Eos 89* (50):513.

Sanderson, B. M., C. Piani, W. J. Ingram, D. A. Stone, and M. R. Allen. 2008. Towards constraining climate sensitivity by linear analysis of feedback patterns in thousands of perturbed-physics GCM simulations. *Climate Dynamics 30* (2-3):175-190, **DOI:** 10.1007/s00382-007-0280-7.

Santer, B. D., T. M. L. Wigley, M. E. Schlesinger, and J. F. B. Mitchell. 1990. *Developing Climate Scenarios from Equilibrium GCM Results.* Hamburg: Max Planck Institute fur Meteorologie.

Santer, B. D., C. Mears, F. J. Wentz, K. E. Taylor, P. J. Gleckler, T. M. L. Wigley, T. P. Barnett, J. S. Boyle, W. Bruggemann, N. P. Gillett, S. A. Klein, G. A. Meehl, T. Nozawa, D. W. Pierce, P. A. Stott, W. M. Washington, and M. F. Wehner. 2007. Identification of human-induced changes in atmospheric moisture content. *Proceedings of the National Academy of Sciences of the United States of America 104* (39):15248-15253, **DOI:** 10.1073/pnas.0702872104.

Sarmiento, J. L., T. M. C. Hughes, R. J. Stouffer, and S. Manabe. 1998. Simulated response of the ocean carbon cycle to anthropogenic climate warming. *Nature 393* (6682):245-249.

Sarmiento, J. L., R. Slater, R. Barber, L. Bopp, S. C. Doney, A. C. Hirst, J. Kleypas, R. Matear, U. Mikolajewicz, P. Monfray, V. Soldatov, S. A. Spall, and R. Stouffer. 2004. Response of ocean ecosystems to climate warming. *Global Biogeochemical Cycles 18* (3):GB3003, **DOI:** 10.1029/2003GB002134.

Sazonova, T. S., V. E. Romanovsky, J. E. Walsh, and D. O. Sergueev. 2004. Permafrost dynamics in the 20th and 21st centuries along the East Siberian transect. *Journal of Geophysical Research-Atmospheres 109* (D1):D01108, **DOI:** 10.1029/2003JD003680.

Schar, C., P. L. Vidale, D. Luthi, C. Frei, C. Haberli, M. A. Liniger, and C. Appenzeller. 2004. The role of increasing temperature variability in European summer heatwaves. *Nature 427* (6972):332-336.

Scherrer, S. C., C. Appenzeller, M. A. Liniger, and C. Schar. 2005. European temperature distribution changes in observations and climate change scenarios. *Geophysical Research Letters 32* (19):L19705, **DOI:** 10.1029/2005GL024108.

Schimel, D. S., J. I. House, K. A. Hibbard, P. Bousquet, P. Ciais, P. Peylin, B. H. Braswell, M. J. Apps, D. Baker, A. Bondeau, J. Canadell, G. Churkina, W. Cramer, A. S. Denning, C. B. Field, P. Friedlingstein, C. Goodale, M. Heimann, R. A. Houghton, J. M. Melillo, B. Moore, D. Murdiyarso, I. Noble, S. W. Pacala, I. C. Prentice, M. R. Raupach, P. J. Rayner, R. J. Scholes, W. L. Steffen, and C. Wirth. 2001. Recent patterns and mechanisms of carbon exchange by terrestrial ecosystems. *Nature 414* (6860):169-172.

Schlenker, W. and M. J. Roberts. 2009. Nonlinear temperature effects indicate severe damages to US crop yields under climate change. *Proceedings of the National Academy of Sciences of the United States of America 106* (37):15594-15598, **DOI:** 10.1073/pnas.0906865106.

Schlenker, W. and D. B. Lobell. 2010. Robust negative impacts of climate change on African agriculture. *Environmental Research Letters* (1):014010, **DOI:** 10.1088/1748-9326/5/1/014010.

Schlesinger, M. 1986. Equilibrium and transient climatic warming induced by increased atmospheric CO_2. *Climate Dynamics 1*:35-51.

Scholze, M., W. Knorr, N. W. Arnell, and I. C. Prentice. 2006. A climate-change risk analysis for world ecosystems. *Proceedings of the National Academy of Sciences of the United States of America, 103* (35):13116-13120, **DOI:** 10.1073/pnas.0601816103.

Schummer, M. L., R. M. Kaminski, A. H. Raedeke, and D. A. Graber. 2010. Weather-related indices of autumn-winter dabbling duck abundance in middle North America. In press.

Schuur, E. A. G., J. Bockheim, J. G. Canadell, E. Euskirchen, C. B. Field, S. V. Goryachkin, et al. 2008. Vulnerability of permafrost carbon to climate change: implications for the global carbon cycle. *Bioscience 58* (8):701-714.

Seager, R., M. Ting, I. Held, Y. Kushnir, J. Lu, G. Vecchi, H. P. Huang, N. Harnik, A. Leetmaa, N. C. Lau, C. H. Li, J. Velez, and N. Naik. 2007. Model projections of an imminent transition to a more arid climate in southwestern North America. *Science 316* (5828):1181-1184, **DOI:** 10.1126/science.1139601.

Sen Gupta, A., A. Santoso, A. S. Taschetto, C. C. Ummenhofer, J. Trevena, and M. H. England. 2009. Projected changes to the southern hemisphere ocean and sea ice in the IPCC AR4 climate models. *Journal of Climate 22* (11):3047-3078, **DOI:** 10.1175/2008JCLI2827.1.

Serreze, M. C., D. H. Bromwich, M. P. Clark, A. J. Etringer, T. J. Zhang, and R. Lammers. 2002. Large-scale hydro-climatology of the terrestrial Arctic drainage system. *Journal of Geophysical Research-Atmospheres 108* (D2):8160, **DOI:** 10.1029/2001JD000919.

Serreze, M. C., M. M. Holland, and J. Stroeve. 2007. Perspectives on the Arctic's shrinking sea-ice cover. *Science 315* (5818):1533-1536, **DOI:** 10.1126/science.1139426.

Shea, K., R. Truckner, R. Weber, and D. Peden. 2008. Climate change and allergic disease. *Journal of Allergy and Clinical Immunology 122* (3):443-453, **DOI:** 10.1016/j.jaci.2008.06.032.

Sheffield, J. and E. F. Wood. 2008. Projected changes in drought occurrence under future global warming from multi-model, multi-scenario, IPCC AR4 simulations. *Climate Dynamics 31* (1):79-105, **DOI:** 10.1007/s00382-007-0340-z.

Sheridan, S. C., A. J. Kalkstein, and L. S. Kalkstein. 2009. Trends in heat-related mortality in the United States, 1975-2004. *Natural Hazards 50* (1):145-160, **DOI:** 10.1007/s11069-008-9327-2.

Sherry, R. A., X. H. Zhou, S. L. Gu, J. A. Arnone, D. S. Schimel, P. S. Verburg, et al. 2007. Divergence of reproductive phenology under climate warming. *Proceedings of the National Academy of Sciences of the United States of America 104* (1):198-202, **DOI:** 10.1073/pnas.0605642104.

Sherwood, S. C. and M. Huber. 2010. An adaptability limit to climate change due to heat stress. *Proceedings of the National Academy of Sciences of the United States of America 107* (21):9552-9555, **DOI:** 10.1073/pnas.0913352107.

Shevliakova, E., S. W. Pacala, S. Malyshev, G. C. Hurtt, P. C. D. Milly, J. P. Caspersen, L. T. Sentman, J. P. Fisk, C. Wirth, and C. Crevoisier. 2009. Carbon cycling under 300 years of land use change: Importance of the secondary vegetation sink. *Global Biogeochemical Cycles* 23:GB2022, **DOI:** 10.1029/2007GB003176.

Shiklomanov, I. A. 1999. World freshwater resources (world water resources and their use). 7 Place de Fontenoy 75352 Paris: UNESCO International Hydrology Program CD Roms. Also available online at *http://webworld.unesco.org/ihp_db/publications*.

Shimada, K., T. Kamoshida, M. Itoh, S. Nishino, E. Carmack, F. McLaughlin, S. Zimmermann, and A. Proshutinsky. 2006. Pacific Ocean inflow: influence on catastrophic reduction of sea ice cover in the Arctic Ocean. *Geophysical Research Letters 33* (8):L08605, **DOI:** 10.1029/2005GL025624.

Shustack, D. P., A. D. Rodewald, and T. A. Waite. 2009. Springtime in the city: exotic shrubs promote earlier greenup in urban forests. *Biological Invasions 11* (6):1357-1371, **DOI:** 10.1007/s10530-008-9343-x.

Silverman, J., B. Lazar, and J. Erez. 2007. Effect of aragonite saturation, temperature, and nutrients on the community calcification rate of a coral reef. *Journal of Geophysical Research-Oceans 112* (C5):C05004, **DOI:** 10.1029/2006JC003770.

Silverman, J., B. Lazar, L. Cao, K. Caldeira, and J. Erez. 2009. Coral reefs may start dissolving when atmospheric CO_2 doubles. *Geophysical Research Letters 36:*L05606, **DOI:** 10.1029/2008GL036282.

Sinervo, B., F. Méndez-de-la-Cruz, D. B. Miles, B. Heulin, E. Bastiaans, M. Villagrán-Santa Cruz, R. Lara-Resendiz, N. Martínez-Méndez, M. L. Calderón-Espinosa, R. N. Meza-Lázaro, H. Gadsden, L. J. Avila, M. Morando, I. J. De la Riva, P. V. Sepulveda, C. F. D. Rocha, N. Ibargüengoytía, C. A. Puntriano, M. Massot, V. Lepetz, T. A. Oksanen, D. G. Chapple, A. M. Bauer, W. R. Branch, J. Clobert, and J. W. Sites, Jr. 2010. Erosion of lizard diversity by climate change and altered thermal niches. *Science 328* (5980):894-899, **DOI:** 10.1126/science.1184695.

Sitch, S., C. Huntingford, N. Gedney, P. E. Levy, M. Lomas, S. L. Piao, R. Betts, P. Ciais, P. Cox, P. Friedlingstein, C. D. Jones, I. C. Prentice, and F. I. Woodward. 2008. Evaluation of the terrestrial carbon cycle, future plant geography and climate-carbon cycle feedbacks using five Dynamic Global Vegetation Models (DGVMs). *Global Change Biology 14* (9):2015-2039, **DOI:** 10.1111/j.1365-2486.2008.01626.x.

Smith, F. A. and J. L. Betancourt. 1998. Response of bushy-tailed woodrats (Neotoma cinerea) to late Quaternary climatic change in the Colorado Plateau. *Quaternary Research 50* (1):1-11.

Smith, S. L., M. M. Burgess, and A. E. Taylor. 2003. High Arctic permafrost observatory at Alert, Nunavut—analysis of a 23-year dataset. *Proceedings of the 8th International Conference on Permafrost*: 1073-1078.

Solomon, S., G. K. Plattner, R. Knutti, and P. Friedlingstein. 2009. Irreversible climate change due to carbon dioxide emissions. *Proceedings of the National Academy of Sciences of the United States of America 106* (6):1704-1709, **DOI:** 10.1073/pnas.0812721106.

Solomon, S., K. Rosenlof, R. Portmann, J. Daniel, S. Davis, T. Sanford, and G. K. Plattner. 2010. Contributions of stratospheric water vapor to decadal changes in the rate of global warming. *Science 327* (5970):1219-1223, **DOI:** 10.1126/science.1182488.

Spracklen, D. V., L. J. Mickley, J. A. Logan, R. C. Hudman, R. Yevich, M. D. Flannigan, and A. L. Westerling. 2009. Impact of climate change from 2000 to 2050 on wildfire activity and carbonaceous aerosol concentrations in the western United States. *Journal of Geophysical Research 114*, **DOI:** 10.1029/2008JD010966.

Stainforth, D. A., T. Aina, C. Christensen, M. Collins, N. Faull, D. J. Frame, J. A. Kettleborough, S. Knight, A. Martin, J. M. Murphy, C. Piani, D. Sexton, L. A. Smith, R. A. Spicer, A. J. Thorpe, and M. R. Allen. 2005. Uncertainty in predictions of the climate response to rising levels of greenhouse gases. *Nature 433* (7024):403-406, **DOI:** 10.1038/nature03301.

Steinacher, M., F. Joos, T. L. Frölicher, L. Bopp, P. Cadule, S. C. Doney, et al. 2009. Projected 21st century decrease in marine productivity: a multi-model analysis. *Biogeosciences Discussion 6* (4):7933-7981.

Steinacher, M., F. Joos, T. L. Frölicher, L. Bopp, P. Cadule, V. Cocco, S. C. Doney, M. Gehlen, K. Lindsay, J. K. Moore, B. Schneider, and J. Segschneider. 2010. Projected 21st century decrease in marine productivity: a multi-model analysis. *Biogeosciences 7* (3):979-1005.

Stendel, M. and J. H. Christensen. 2002. Impact of global warming on permafrost conditions in a coupled GCM. *Geophysical Research Letters 29* (13):1632, **DOI:** 10.1029/2001GL014345.

Stott, P. A., J. F. B. Mitchell, M. R. Allen, T. L. Delworth, J. M. Gregory, G. A. Meehl, and B. D. Santer. 2006. Observational constraints on past attributable warming and predictions of future global warming. *Journal of Climate 19* (13):3055-3069.

Stott, P. A., R. T. Sutton, and D. M. Smith. 2008. Detection and attribution of Atlantic salinity changes. *Geophysical Research Letters 35* (21):L21702.

Stouffer, R. J. 2004. Time scales of climate response. *Journal of Climate 17* (1):209-217.

Stroeve, J., M. M. Holland, W. Meier, T. Scambos, and M. Serreze. 2007. Arctic sea ice decline: faster than forecast. *Geophysical Research Letters 34* (9):L09501, **DOI:** 10.1029/2007GL029703.

Stroeve J., M. Serreze, S. Drobot, S. Gearhead, M. Holland, J. Maslanik, W. Meier, T. Scambos. 2008. Arctic sea ice plummets in 2007. *Eos 89* (2):13-14, **DOI:** 10.1029/2008EO020001

Sturm, M., C. Racine, and K. Tape. 2001. Climate change—increasing shrub abundance in the Arctic. *Nature 411* (6837):546-547.

Sugi, M., A. Noda, and N. Sato. 2002. Influence of the global warming on tropical cyclone climatology: an experiment with the JMA global model. *Journal of the Meteorological Society of Japan 80* (2):249-272.

Sugiyama, M., H. Shiogama, and S. Emori. 2010. Precipitation extreme changes exceeding moisture content increases in MIROC and IPCC climate models. *Proceedings of the National Academy of Sciences 107*:571-575.

Swetnam, T. W. and J. Betancourt. 1998. Mesoscale disturbance and ecological response to decadal climate variability in the American Southwest. *Journal of Climate 11* (12):3128-3147.

Tagaris, E., K. J. Liao, A. J. Delucia, L. Deck, P. Amar, and A. G. Russell. 2009. Potential impact of climate change on air pollution-related human health effects. *Environmental Science & Technology 43* (13):4979-4988, **DOI:** 10.1021/es803650w.

Tao, Z., A. Williams, H. C. Huang, M. Caughey, and X. Z. Liang. 2007. Sensitivity of U.S. surface ozone to future emissions and climate changes. *Geophysical Research Letters 34* (8):L08811, **DOI:** 10.1029/2007GL029455.

Tape, K., M. Sturm, and C. Racine. 2006. The evidence for shrub expansion in Northern Alaska and the Pan-Arctic. *Global Change Biology 12* (4):686-702, **DOI:** 10.1111/j.1365-2486.2006.01128.x.

Tebaldi, C., K. Hayhoe, J. M. Arblaster, and G. A. Meehl. 2006. Going to the extremes: an intercomparison of model-simulated historical and future changes in extreme events. *Climatic Change 79* (3-4):185-211.

Thacker, M. T. F., R. Lee, R. I. Sabogal, and A. Henderson. 2008. Overview of deaths associated with natural events, United States, 1979-2004. *Disasters 32* (2):303-315, **DOI:** 10.1111/j.0361-3666.2008.01041.X.

Thompson, D. W. J. and J. M. Wallace. 2000. Annular modes in the extratropical circulation. Part I: Month-month variability. *Journal of Climate*, 13:1000-1016.

Thompson, D. W. J., J. M. Wallace, and G. C. Hegerl. 2000. Annular modes in the extratropical circulation. part II: trends. *Journal of Climate 13* (5):1018-1036.

Thompson, D. W. J., J. M. Wallace, P. D. Jones, and J. J. Kennedy. 2009. Identifying signatures of natural climate variability in time series of global-mean surface temperature: methodology and insights. *Journal of Climate 22* (22):6120-6141, **DOI:** 10.1175/2009JCLI3089.1.

Thornton, P. E., S. C. Doney, K. Lindsay, J. K. Moore, N. Mahowald, J. T. Randerson, I. Fung, J. F. Lamarque, J. J. Feddema, and Y. H. Lee. 2009a. Carbon-nitrogen interactions regulate climate-carbon cycle feedbacks: results from an atmosphere-ocean general circulation model. *Biogeosciences 6* (10):2099-2120.

Thornton, P. K., J. van de Steeg, A. Notenbaert, and M. Herrero. 2009b. The impacts of climate change on livestock and livestock systems in developing countries: a review of what we know and what we need to know. *Agricultural Systems 101* (3):113-127, **DOI:** 10.1016/j.agsy.2009.05.002.

Timmermann, A., J. Oberhuber, A. Bacher, M. Esch, M. Latif, and E. Roeckner. 1999. Increased El Niño frequency in a climate model forced by future greenhouse warming. *Nature 398* (6729):694-697.

Tingley, M. W., W. B. Monahan, S. R. Beissinger, and C. Moritz. 2009. Birds track their Grinnellian niche through a century of climate change. *Proceedings of the National Academy of Sciences of the United States of America 106:*19637-19643, **DOI:** 10.1073/pnas.0901562106.

Tol, R. S. J. 2007. The double trade-off between adaptation and mitigation for sea level rise: an application of FUND. *Mitigation and Adaptation Strategies for Global Change 12* (5):741-753.

Trenberth, K. E. 1999. Conceptual framework for changes of extremes of the hydrological cycle with climate change. *Climatic Change 42* (1):327-339.

Trenberth, K. E., J. Fasullo, and L. Smith. 2005. Trends and variability in column-integrated atmospheric water vapor. *Climate Dynamics 24* (7-8):741-758, **DOI:** 10.1007/s00382-005-0017-4.

Trumper, K., M. Bertzky, B. Dickson, G. van der Heijden, M. Jenkins, and P. Manning. 2009. The Natural Fix? The role of ecosystems in climate mitigation. A UNEP rapid response assessment. Cambridge, UK: United Nations Environment Programme, UNEP-WCMC.

Turner, J., J. C. Comiso, G. J. Marshall, T. A. Lachlan-Cope, T. Bracegirdle, T. Maksym, M. P. Meredith, Z. Wang, and A. Orr. 2009. Non-annular atmospheric circulation change induced by stratospheric ozone depletion and its role in the recent increase of Antarctic sea ice extent. *Geophysical Research Letters*, **DOI:** 10.1029/2009GL037524.

UN (United Nations). 2008. World urbanization prospects: the 2007 revision population database. United Nations Population Division. Available online at: *http://esa.un.org/unup/*.

USCCSP (United States Climate Change Science Program). 2008a. *Analyses of the Effects of Global Change on Human Health and Welfare and Human Systems*. SAP 4.6.

USCCSP. 2008b. *The Effects of Climate Change on Agriculture, Land Resources, Water Resources, and Biodiversity in the United States*. SAP 4.3.

USCCSP. 2008c. *Weather and Climate Extremes in a Changing Climate: Regions of focus—North America, Hawaii, Caribbean, and U.S. Pacific Islands*. SAP 4.3.

USCCSP. 2009. *Coastal Sensitivity to Sea-level Rise: A focus on the Mid-Atlantic region*. SAP 4.1.

U.S. EPA (United States Environmental Protection Agency). 2008. *Inventory of US Greenhouse Gas Emissions and Sink: 1990-2006*.

USGCRP (United States Global Change Research Program). 2009. *Global Climate Change Impacts in the United States*. Cambridge: Cambridge University Press.

Van der Werf, G. R., J. T. Randerson, L. Giglio, G. J. Collatz, P. S. Kasibhatla, and A. F. Arellano. 2006. Interannual variability in global biomass burning emissions from 1997 to 2004. *Atmospheric Chemistry and Physics 6*:3423-3441.

Vecchi, G. A. and T. R. Knutson. 2008. On estimates of historical north Atlantic tropical cyclone activity. *Journal of Climate 21* (14):3580-3600, **DOI:** 10.1175/2008JCLI2178.1.

Vecchi, G. A., K. L. Swanson, and B. J. Soden. 2008. Climate change: whither hurricane activity? *Science 322* (5902):687-689, **DOI:** 10.1126/science.1164396.

Velders, G. J. M., D. W. Fahey, J. S. Daniel, M. McFarland, and S. O. Andersen. 2009. The large contribution of projected HFC emissions to future climate forcing. *Proceedings of the National Academy of Sciences of the United States of America 106* (27):10949-10954, **DOI:** 10.1073/pnas.0902817106.

Vermeer, M. and S. Rahmstorf. 2009. Global sea level linked to global temperature. *Proceedings of the National Academy of Sciences of the United States of America 106*:21527-21532, **DOI:** 10.1073/pnas.0907765106.

Veron, J. E. N., O. Hoegh-Guldberg, T. M. Lenton, J. M. Lough, D. O. Obura, P. Pearce-Kelly, C. R. C. Sheppard, M. Spalding, M. G. Stafford-Smith, and A. D. Rogers. 2009. The coral reef crisis: the critical importance of < 350 ppm CO_2. *Marine Pollution Bulletin 58* (10):1428-1436, **DOI:** 10.1016/j.marpolbul.2009.09.009.

Vicuña, S., R. Leonardson, M. W. Hanemann, L. L. Dale, and J. A. Dracup. 2008. Climate change impacts on high elevation hydropower generation in California's Sierra Nevada: a case study in the Upper American River. *Climatic Change 87*:S123-S137, **DOI:** 10.1007/s10584-007-9365-x.

Vinnikov, K. Y., A. Robock, R. J. Stouffer, J. E. Walsh, C. L. Parkinson, D. J. Cavalieri, J. F. B. Mitchell, D. Garrett, and V. F. Zakharov. 1999. Global warming and Northern Hemisphere sea ice extent. *Science 286* (5446):1934-1937.

Vizcaino, M., U. Mikolajewicz, M. Gröger, E. Maier-Reimer, G. Schurgers and A. M. E. Winguth. 2008. Long-term ice sheet-climate interactions under anthropogenic greenhouse forcing simulated with a complex Earth System Model. *Climate Dynamics* 31:665-690, **DOI:** 10.1007/s00382-008-0369-7.

Vörösmarty, C. J., P. Green, J. Salisbury, and R. B. Lammers. 2000. Global water resources: vulnerability from climate change and population growth. *Science 289* (5477):284-288.

Vrac, M., K. Hayhoe, and M. Stein. 2007. Identification and intermodel comparison of seasonal circulation patterns over North America. *International Journal of Climatology 27* (5):603-620, **DOI:** 10.1002/joc.1422.

Walker, M. D., C. H. Wahren, R. D. Hollister, G. H. R. Henry, L. E. Ahlquist, J. M. Alatalo, M. S. Bret-Harte, M. P. Calef, T. V. Callaghan, A. B. Carroll, H. E. Epstein, I. S. Jonsdottir, J. A. Klein, B. Magnusson, U. Molau, S. F. Oberbauer, S. P. Rewa, C. H. Robinson, G. R. Shaver, K. N. Suding, C. C. Thompson, A. Tolvanen, O. Totland, P. L. Turner, C. E. Tweedie, P. J. Webber, and P. A. Wookey. 2006. Plant community responses to experimental warming across the tundra biome. *Proceedings of the National Academy of Sciences of the United States of America 103* (5):1342-1346, **DOI:** 10.1073/pnas.0503198103.

Walsh, M. B. P. J. A. C. S. D. S. 2008. *The Effects of Climate Change on Agriculture, Land Resources, Water Resources, and Biodiversity in the United States*. Washington, DC: U.S. Climate Change Science Program.

Walter, K. M., S. A. Zimov, J. P. Chanton, D. Verbyla, and F. S. Chapin. 2006. Methane bubbling from Siberian thaw lakes as a positive feedback to climate warming. *Nature 443* (7107):71-75, **DOI:** 10.1038/nature05040.

Wang, G. L. 2005. Agricultural drought in a future climate: results from 15 global climate models participating in the IPCC 4th assessment. *Climate Dynamics 25* (7-8):739-753, **DOI:** 10.1007/s00382-005-0057-9.

Wang, M. Y. and J. E. Overland. 2009. A sea ice free summer Arctic within 30 years? *Geophysical Research Letters 36:*L07502, **DOI:** 10.1029/2009GL037820.

Wang, X. L. L., V. R. Swail, F. W. Zwiers, X. B. Zhang, and Y. Feng. 2009. Detection of external influence on trends of atmospheric storminess and northern oceans wave heights. *Climate Dynamics 32* (2-3):189-203, **DOI:** 10.1007/s00382-008-0442-2.

Wara, M. W. 2005. Permanent El Niño-like conditions during the Pliocene warm period (29 July, pg 758, 2005). *Science 313* (5794):1739.

Washington, W. M., R. Knutti, G. A. Meehl, H. Y. Teng, C. Tebaldi, D. Lawrence, L. Buja, and W. G. Strand. 2009. How much climate change can be avoided by mitigation? *Geophysical Research Letters 36:*L08703, **DOI:** 10.1029/2008GL037074.

Wassmann, R., S. V. K. Jagadish, S. Heuer, A. Ismail, E. Redona, R. Serraj, R. K. Singh, G. Howell, H. Pathak, and K. Sumfleth. 2009. Climate change affecting rice production: the physiological and agronomic basis for possible adaptation strategies. *Advances in Agronomy 101*:59-122, **DOI:** 10.1016/S0065-2113(08)00802-X.

Watson, A. J., U. Schuster, D. C. E. Bakker, N. R. Bates, A. Corbiere, M. Gonzalez-Davila, T. Friedrich, J. Hauck, C. Heinze, T. Johannessen, A. Kortzinger, N. Metzl, J. Olafsson, A. Olsen, A. Oschlies, X. A. Padin, B. Pfeil, J. M. Santana-Casiano, T. Steinhoff, M. Telszewski, A. F. Rios, D. W. R. Wallace, and R. Wanninkhof. 2009. Tracking the variable North Atlantic sink for atmosphere CO_2. *Science* **326** (5958):1391-1393, **DOI:** 10.1126/science.1177394.

Watterson, I. G. 1996. Non-dimensional measures of climate model performance. *International Journal of Climatology 16* (4):379-391.

Watterson, I. G. 2008. Calculation of probability density functions for temperature and precipitation change under global warming. *Journal of Geophysical Research-Atmospheres 113* (D12): D12106, **DOI:** 10.1029/2007JD009254.

Weaver, A. J., M. Eby, E. C. Wiebe, C. M. Bitz, P. B. Duffy, T. L. Ewen, A. F. Fanning, M. M. Holland, A. MacFadyen, H. D. Matthews, K. J. Meissner, O. Saenko, A. Schmittner, H. X. Wang, and M. Yoshimori. 2001. The UVic Earth System Climate Model: model description, climatology, and applications to past, present and future climates. *Atmos. Ocean 39* (4):361-428.

Weaver, A. J., K. Zickfeld, A. Montenegro, and M. Eby. 2007. Long term climate implications of 2050 emission reduction targets. *Geophysical Research Letters 34* (19):L19703, **DOI:** 10.1029/2007GL031018.

Westerling, A. L. and B. P. Bryant. 2008. Climate change and wildfire in California. *Climatic Change 87* (1):1-19, **DOI:** 10.1007/s10584-007-9363-z.

White, M. A., K. M. de Beurs, K. Didan, D. W. Inouye, A. D. Richardson, O. P. Jensen, J. O'Keefe, G. Zhang, R. R. Nemani, W. J. D. van Leeuwen, J. F. Brown, A. de Wit, M. Schaepman, X. M. Lin, M. Dettinger, A. S. Bailey, J. Kimball, M. D. Schwartz, D. D. Baldocchi, J. T. Lee, and W. K. Lauenroth. 2009. Intercomparison, interpretation, and assessment of spring phenology in North America estimated from remote sensing for 1982-2006. *Global Change Biology 15* (10):2335-2359, **DOI:** 10.1111/j.1365-2486.2009.01910.x.

Wigley, T. M. L., R. Richels, and J. A. Edmonds. 1996. Economic and environmental choices in the stabilization of atmospheric CO_2 concentrations. *Nature 379* (6562):240-243.

Wilbanks, T. J., P. Romero Lankao, M. Bao, F. Berkhout, S. Cairncross, J.-P. Ceron, M. Kapshe, R. Muir-Wood and R. Zapata-Marti. 2007. Industry, settlement and society. In *Climate Change 2007: Impacts, Adaptation and Vulnerability. Contribution of Working Group II to the Fourth Assessment Report of the Intergovernmental Panel on Climate Change*, edited by O. F. C. M. L. Parry, J. P. Palutikof, P. J. van der Linden, and C. E. Hanson. 357-390. Cambridge: Cambridge University Press.

Willett, K. M., N. P. Gillett, P. D. Jones, and P. W. Thorne. 2007. Attribution of observed surface humidity changes to human influence. *Nature 449* (7163):710-712, **DOI:** 10.1038/nature06207.

Williams, I. N., R. T. Pierrehumbert, and M. Huber. 2009. Global warming, convective threshold and false thermostats. *Geophysical Research Letters 36*:L21805, **DOI:** 10.1029/2009GL039849.

Williams, K. D., W. J. Ingram, and J. M. Gregory. 2008. Time variation of effective climate sensitivity in GCMs. *Journal of Climate 21* (19):5076-5090, **DOI:** 10.1175/2008JCLI2371.1.

Winton, M. 2006a. Amplified Arctic climate change: what does surface albedo feedback have to do with it? *Geophysical Research Letters 33* (3):L03701, **DOI:** 10.1029/2005GL025244.

Winton, M. 2006b. Does the Arctic sea ice have a tipping point? *Geophysical Research Letters 33* (23):L23504, **DOI:** 10.1029/2006GL028017.

Winton, M., K. Takahashi, and I. M. Held. 2010. Importance of ocean heat uptake efficacy to transient climate change. *Journal of Climate 23* (9):2333-2344, **DOI:** 10.1175/2009JCLI3139.1.

Wolf, M., M. Friggens, and J. Salazar-Bravo. 2009. Does weather shape rodents? climate related changes in morphology of two heteromyid species. *Naturwissenschaften 96* (1):93-101.

Wood, A. W., L. R. Leung, V. Sridhar, and D. P. Lettenmaier. 2004. Hydrologic implications of dynamical and statistical approaches to downscaling climate model outputs. *Climatic Change 62* (1-3):189-216.

Wootton, J. T., C. A. Pfister, and J. D. Forester. 2008. Dynamic patterns and ecological impacts of declining ocean pH in a high-resolution multi-year dataset. *Proceedings of the National Academy of Sciences of the United States of America 105* (48):18848-18853, **DOI:** 10.1073/pnas.0810079105.

WDR (World Development Report). 2010. *Development and Climate Change. 2010.* Washington, DC: World Bank.

Wright, S. J., H. C. Muller-Landau, and J. Schipper. 2009. The future of tropical species on a warmer planet. *Conservation Biology 23* (6):1418-1426, **DOI:** 10.1111/j.1523-1739.2009.01337.x.

WWAP (World Water Assessment Program). 2008. Changes in the global water cycle. In *Water in a Changing World: 3rd UN World Water Development Report.* Available at: www.unesco.org/water/wwap/wwdr/wwdr3.

Yamaguchi, K., A. Noda, and A. Kitoh. 2005. The changes in permafrost induced by greenhouse warming: a numerical study applying multiple-layer ground model. *Journal of the Meteorological Society of Japan 83* (5):799-815.

Yohe, G., E. Malone, A. Brenkert, M. Schlesinger, H. Meij and X. Xing. 2006. Global distributions of vulnerability to climate change, *Integrated Assessment Journal 6*:35-44.

Yohe, G. W., R. D. Lasco, Q. K. Ahmad, N. W. Arnell, S. J. Cohen, C. Hope, A. C. Janetos and R. T. Perez. 2007. Perspectives on climate change and sustainability. In *Climate Change 2007: Impacts, Adaptation and Vulnerability. Contribution of Working Group II to the Fourth Assessment Report of the Intergovernmental Panel on Climate Change*, edited by O. F. C. M. L. Parry, J. P. Palutikof, P. J. van der Linden, and C. E. Hanson. 811-841. Cambridge: Cambridge University Press.

Yokohata, T., S. Emori, N. Nozawa, Y. Tsushima, T. Ogura, and M. Kimoto. 2005. Climate response to volcanic forcing: validation of climate sensitivity of a coupled atmosphere-ocean general circulation model. *Geophysical Research Letters 32* (21):L21710, **DOI:** 10.1029/2005GL023542.

Zeebe, R. E., J. C. Zachos, and G. R. Dickens. 2009. Carbon dioxide forcing alone insufficient to explain Palaeocene-Eocene Thermal Maximum warming. *Nature Geoscience 2* (8):576-580.

Zhang, D. D., P. Brecke, H. F. Lee, Y. Q. He, and J. Zhang. 2007a. Global climate change, war, and population decline in recent human history. *Proceedings of the National Academy of Sciences of the United States of America 104* (49):19214-19219, **DOI:** 10.1073/pnas.0703073104.

Zhang, R. 2008. Coherent surface-subsurface fingerprint of the Atlantic meridional overturning circulation. *Geophysical Research Letters 35* (20):L20705, **DOI:** 10.1029/2008GL035463.

Zhang, T. et al. 2005. Spatial and temporal variability in active layer thickness over the Russian Arctic drainage basin. *Journal of Geophys. Research 110*,D16101, **DOI:** 10.1029/2004JD005642.

Zhang, T., R. G. Barry, K. Knowles, J. A. Heginbottom, and J. Brown. 1999. Statistics and characteristics of permafrost and ground-ice distribution in the Northern Hemisphere. *Polar Geography 23*(2):132-154.

Zhang, T. J., F. E. Nelson, and S. Gruber. 2007b. Introduction to special section: permafrost and seasonally frozen ground under a changing climate. *Journal of Geophysical Research-Earth Surface 112* (F2):F02S01, **DOI:** 10.1029/2007JF000821.

Zhang, T., R. G. Barry, K. Knowles, J. A. Heginbottom, and J. Brown. 2008. Statistics and characteristics of permafrost and ground-ice distribution in the Northern Hemisphere. *Polar Geography 31* (1-2):47-68.

Zhang, X. B., F. W. Zwiers, G. C. Hegerl, F. H. Lambert, N. P. Gillett, S. Solomon, P. A. Stott, and T. Nozawa. 2007c. Detection of human influence on twentieth-century precipitation trends. *Nature 448* (7152):461-U464, **DOI:** 10.1038/nature06025.

Zhao, M., I. M. Held, S. J. Lin, and G. A. Vecchi. 2009. Simulations of global hurricane climatology, interannual variability, and response to global warming using a 50-km resolution GCM. *Journal of Climate 22* (24):6653-6678, **DOI:** 10.1175/2009JCLI3049.1.

Zickfeld, K., M. Eby, H. D. Matthews, and A. J. Weaver. 2009. Setting cumulative emissions targets to reduce the risk of dangerous climate change. *Proceedings of the National Academy of Sciences of the United States of America 106* (38):16129-16134, **DOI:** 10.1073/pnas.0805800106.

Ziska, L. H., D. M. Blumenthal, G. B. Runion, E. R. Hunt, and H. Diaz-Soltero. 2010. Invasive species and climate change: an agronomic perspective. *Climate Change*. in press.

Ziska, L. H., P. R. Epstein, and W. H. Schlesinger. 2009. Rising CO_2, climate change, and public health: exploring the links to plant biology. *Environmental Health Perspectives 117* (2):155-158.

Appendix A

Statement of Task

The stabilization of atmospheric greenhouse gas concentrations and the avoidance of serious or irreversible impacts on earth's climate system are a matter of critical concern in both scientific and policy arenas. Using the most current science available, this study will evaluate the implications of different atmospheric concentration target levels and explain the uncertainties inherent in the analyses to assist policy makers as they make decisions about stabilization target levels for atmospheric greenhouse gas concentrations.

This study will:

- Evaluate a range of greenhouse gas stabilization targets and describe the types and scales of impacts likely associated with different ranges, including discussion of the associated uncertainties, timescale of impacts, and potential serious or irreversible impacts.[1]

This study will focus on evaluating the implications of a range of GHG stabilization targets, but it will not involve the committee's assessment of what stabilization targets are technically feasible nor their normative judgment on what targets are most appropriate.

[1]This study will consider cumulative emissions of the "basket" of gases considered in the Kyoto protocol (CO_2, CH4, N2O, HFCs, PFCs, SF6). Any particular overall stabilization target could include varying combinations of emissions targets for the different gases.

Appendix B

Committee Membership

Susan Solomon (NAS) is a senior scientist at the National Oceanic and Atmospheric Administration (NOAA), where she has been a researcher since receiving her Ph.D. degree in chemistry from the University of California at Berkeley in 1981. She made some of the first measurements in the Antarctic that showed that chlorofluorocarbons were responsible for the stratospheric ozone hole, and she pioneered the theoretical understanding of the surface chemistry that causes it. In March 2000, she received the National Medal of Science, the United States' highest scientific honor, for "key insights in explaining the cause of the Antarctic ozone hole." Her current research focuses on chemistry-climate coupling, and she served as co-chair of the science panel for the Intergovernmental Panel on Climate Change (2007) report. Dr. Solomon was elected to the National Academy of Sciences in 1992 and is a foreign member of the Academie des Sciences in France, the European Academy, and the Royal Society.

David S. Battisti is The Tamaki Endowed Chair of Atmospheric Sciences at the University of Washington. Dr. Battisti's research is focused on understanding the natural variability of the climate system. He is especially interested in understanding how the interactions between the ocean, atmosphere, land, and sea ice lead to variability in climate on time scales from seasonal to decades. He is also working on the impacts of climate variability and climate change on food production in Mexico, Indonesia, and China. Dr. Battisti received a Ph.D. in Atmospheric Sciences (1988) from the University of Washington.

Scott Doney is a senior scientist in the Department of Marine Chemistry and Geochemistry at the Woods Hole Oceanographic Institution (WHOI). He graduated with a Ph.D. from the Massachusetts Institute of Technology/ Woods Hole Oceanographic Institution Joint Program in Oceanography in

1991 and was a postdoctoral fellow and later a scientist at the National Center for Atmospheric Research, before returning to Woods Hole in 2002. He was awarded the James B. Macelwane Medal from the American Geophysical Union in 2000, a Aldo Leopold Leadership Fellow in 2004, and the WHOI W. Van Alan Clark Sr. Chair in 2007. His scientific interests span oceanography, climate, and biogeochemistry. Much of his research focuses on how the global carbon cycle and ocean ecology respond to natural and human-driven climate change, which may act to either dampen or accelerate climate trends. A current focus is on ocean acidification due to the invasion into the ocean of carbon dioxide and other chemicals from fossil-fuel burning. He is currently the chair of the U.S. Ocean Carbon and Biogeochemistry Program and the U.S. Ocean Carbon and Climate Change Program.

Katharine Hayhoe is an atmospheric scientist and research associate professor in the Department of Geosciences at Texas Tech University. Her research focuses on quantifying the potential impacts of human activities at the regional scale, including evaluating the ability of coupled atmosphere-ocean general circulation models to simulate real-world phenomena and developing new techniques to generate scientifically robust, high-resolution projections. She is the author of more than 40 peer-reviewed publications, several book chapters, and numerous reports, including the U.S. Global Change Research Program's 2009 report, *Global Climate Change Impacts in the United States.*

Isaac Held (NAS) majored in physics at the University of Minnesota, continued on in physics to obtain a master's degree from the State University of New York at Stony Brook, and then started his career of research into climate dynamics at Princeton University, where he received his Ph.D. in Atmospheric and Oceanic Sciences in 1976. He has spent most of his career at NOAA's Geophysical Fluid Dynamics Laboratory, where he is currently a Senior Research Scientist and conducts studies on climate dynamics and climate modeling. He is also a lecturer with rank of professor at Princeton University, in its Atmospheric and Oceanic Sciences Program, and is an Associate Faculty member in Princeton's Applied and Computational Mathematics Program and in the Princeton Environmental Institute. Dr. Held is a fellow of the American Meteorological Society (1991) and the American Geophysical Union (1995) and a member of the National Academy of Sciences (2003). Governmental awards include a Department of Commerce Gold Medal (1999) for "world leadership in studies of climate dynamics" and a NOAA Presidential Rank Award (2005). He recently received the AMS Carl Gustav Rossby Gold Medal (2008) for "fundamental insights into the dynamics of the Earth's climate through studies of idealized models and comprehensive climate simulations."

Dennis Lettenmaier is the Robert and Irene Sylvester Professor of Civil Engineering at the University of Washington. Dr. Lettenmaier's interests include hydroclimatology, surface water hydrology, and hydrologic aspects of remote sensing. He was a recipient of ASCE's Huber Research Prize in 1990 and the American Geophysical Union Hydrology Section Award in 2000, and he is a fellow of the American Geophysical Union, the American Meteorological Society, and the American Association for the Advancement of Science. He is the author of more than 200 journal articles. He is the past Chief Editor of the American Meteorological Society's *Journal of Hydrometeorology*, and he is President-Elect of the American Geophysical Union Hydrology Section. He was elected to the National Academy of Engineering in 2010. Dr. Lettenmaier is a member of the NRC Committee on Hydrologic Science, and has served on other NRC committees and panels including the Committee on the National Ecological Observatory Network (2003-2004), the Committee for Earth Science and Applications from Space: A Community Assessment and Strategy for the Future (2005-2007), and the Committee on Scientific Bases of Colorado River Basin Water Management (2006-2007). Dr. Lettenmaier received his Ph.D. in Civil Engineering (1975) from the University of Washington.

David Lobell is an assistant professor at Stanford University in environmental earth system science, and a center fellow in Stanford's Program on Food Security and the Environment. His research focuses on identifying opportunities to raise crop yields in major agricultural regions, with a particular emphasis on adaptation to climate change. His current projects span Africa, South Asia, Mexico, and the United States and involve a range of tools including remote sensing, GIS, and crop and climate models. Prior to his current appointment, Dr. Lobell was a senior research scholar at FSE from 2008-2009 and a Lawrence Post-doctoral Fellow at Lawrence Livermore National Laboratory from 2005-2007. He received a Ph.D. in Geological and Environmental Sciences from Stanford University in 2005, and a Sc.B. in Applied Mathematics, magna cum laude from Brown University in 2000.

H. Damon Matthews is assistant professor and university research fellow in the Department of Geography Planning and Environment at Concordia

University. He obtained a B.Sc. in Environmental Science from Simon Fraser University in 1999 and a Ph.D. in Earth and Ocean Sciences from the University of Victoria in 2004. Prior to joining Concordia University in January 2007, he held a post-doctoral fellowship at the University of Calgary and worked as a post-doctoral researcher at the Carnegie Institution at Stanford. Dr. Matthews currently teaches courses on the climate system, climate change, and environmental modeling at Concordia University. His research is aimed at better understanding the many possible interactions between human activities, natural ecosystems, and future climate change, and contributing to the scientific knowledge base required to promote the development of sound national and international climate policy. Dr. Matthews holds several current research grants for projects to investigate the uncertainties associated with current terrestrial carbon sinks in the context of expected future climate changes. He has published a number of research papers in the area of global climate modeling, with particular emphasis on the role of the global carbon cycle in the climate system, estimating allowable emissions for climate stabilization, and understanding our commitment to long-term climate warming.

Raymond T. Pierrehumbert is the Louis Block Professor in Geophysical Sciences at The College at the University of Chicago, having earlier served on the atmospheric science faculties of Massachusetts Institute of Technology and Princeton. His Climate Systems Center project has worked to bring modern software design techniques to the problem of climate simulation, and his research on climate dynamics has covered phenomena ranging from global warming to deep-time paleoclimate to climate of other planets.

He has also collaborated with David Archer on the University of Chicago's global warming curriculum. He was a lead author of the IPCC Third Assessment Report and a co-author of the National Research Council study on abrupt climate change. He is a fellow of the American Geophysical Union, has been a John Simon Guggenheim Fellow, and, in recognition of his work on climate, he has been named Chevalier de l'Ordre des Palmes Academiques by the Republic of France. Dr. Pierrehumbert's book on comparative planetary climate, *Principles of Planetary Climate,* will be published in December 2010 by Cambridge University Press. Another book, *The Warming Papers,* written in collaboration with David Archer, will be appearing from Wiley Blackwell. He received his Ph.D. from the Massachusetts Institute of Technology.

Marilyn Raphael is a professor in the Department of Geography at the University of California, Los Angeles. Her research interests are in climate variability and change particularly in the Southern Hemisphere. Her research focuses on understanding the interaction between Antarctic sea-ice variability and the large-scale atmosphere and includes global climate modeling with an emphasis on improving the simulation of sea ice and the atmosphere in the Southern Hemisphere. Dr. Raphael also does work on the Santa Ana Winds of California. Dr. Raphael received her Ph.D. in geography from The Ohio State University.

Richard Richels is senior technical executive for global climate change research at the Electric Power Research Institute (EPRI) in Washington, DC. His current research focus is the economics of mitigating greenhouse gas emissions, development and application of integrated assessment models for informing climate change policy making, and the incorporation of uncertainty into climate-related decision making. Dr. Richels has served on a number of national and international advisory panels, including committees of the Department of Energy, the Environmental Protection Agency, and the National Research Council. Dr. Richels has served as a lead author for the Intergovernmental Panel on Climate Change's (IPCC) Second, Third, and Fourth Scientific Assessments, contributing to chapters on mitigation, adaptation, and integrated assessment. He also served on the Synthesis Team for the U.S. National Assessment of Climate Change Impacts on the United States, was a lead author for the U.S. Climate Change Science Program Study on Future Emissions and Atmospheric Concentrations, and served on the Scientific Steering Committee for the U.S. Carbon Cycle Program. He currently serves on the National Research Council's Climate Research Committee; the Advisory Committee for Carnegie-Mellon University's Center for Integrated Study of the Human Dimensions of Global Change; and the U.S. government's Climate Change Science Program Product Development Advisory Committee.

Terry L. Root is a senior fellow at the Center for Environmental Science and Policy in the Institute for International Studies at Stanford University. Dr. Root's work focuses on large-scale ecological questions investigating factors shaping the ranges and abundances of animals, primarily birds. Her small-scale studies have focused on possible mechanisms, such as physiological constraints, that may be helping to generate the observed large-scale patterns. Her work demonstrated that climate and/or vegetation are important factors shaping the ranges and abundances of birds and may help forecast the possible consequences of global warming on animal communities. In 1990, she received the Presidential Young Investigator Award from the National Science Foundation and in 1992 was selected as a Pew Scholar in Conservation and the Environment and Aldo Leopold Leadership Fellow in 1999. She received her bachelor's degree in Mathematics and Statistics at the University of New Mexico, her master's degree in Biology at the University of Colorado in 1982, and her Ph.D. in biology from Princeton University in 1987. She has served on the National Research Council Committee on Environmental Indicators.

Konrad Steffen is a professor at the Cooperative Institute for Environmental Research/University of Colorado at Boulder, teaching climatology and remote sensing since 1990. His research involves the study of processes related to climate variability and change, cryospheric interaction in polar regions, and sea level rise based on in-situ measurements, satellite observations, and model approximations. He has lead field expeditions to the Greenland ice sheet and other Arctic regions for 33 consecutive years to measure the dynamic response of the ice masses under a warming climate. He is also the director of the University of Colorado's Cooperative Institute for Environmental Research (CIRES), the largest research unit on the University of Colorado, Boulder campus. He earned his Ph.D. from the Swiss Federal Institute of Technology in Zurich in 1983.

Claudia Tebaldi is a research scientist at Climate Central, a research-media organization dedicated to the communication of the science of climate change and a part-time adjunct faculty member in the Department of Statistics at University of British Columbia, Vancouver. She has a Ph.D. in Statistics from Duke University. Her work focuses on applications of statistical modeling to various aspects of climate change research: observed changes, future projections and their uncertainties, changes in climate extremes, and climate change impacts, especially in the hydrological and agricultural sectors. She is a contributing author to the Fourth Assessment Report of the IPCC.

Gary Yohe is the Woodhouse/Sysco Professor of Economics at Wesleyan University; he has been on the faculty at Wesleyan for more than 30 years. He was educated at the University of Pennsylvania and received his Ph.D. in Economics from Yale University in 1975. Most of his work has focused attention on the mitigation and adaptation/impacts sides of the climate issue from a risk-management perspective. He is a senior member of the Intergovernmental Panel on Climate Change. Involved with the IPCC since the mid 1990s, he served as a lead author for four different chapters in the Third Assessment Report that was published in 2001 and as convening lead author for the last chapter of the contribution of Working Group II to the Fourth Assessment Report. In that Assessment, he also worked with the core writing team to prepare the overall Synthesis Report. Dr. Yohe serves as a member of the New York City Panel on Climate Change and the standing Committee on the Human Dimensions of Global Change of the National Academy of Sciences. He has testified before the Senate Foreign Relations Committee on the "Hidden (climate change) Cost of Oil" on March 30, 2006, the Senate Energy Committee on the Stern Review on February 14, 2007, and the Senate Banking Committee on "Material Risk from Climate Change and Climate Policy" on October 31, 2007. In addition to accepting an invitation to join the Adaptation Subcommittee of the Governor's Steering Committee on Climate Change (CT), he is served on the Adaptation Panel of the National Academy of Sciences' initiative on America's Climate Choices.

Appendix C

Methods

2.1 MODELS

Models used in Section 2.1.

The UVIC model used here is ESCM version 2.8, which includes a 19-layer ocean general circulation model. The ocean model is coupled to a dynamic-thermodynamic sea-ice model and an energy-moisture balance model of the atmosphere. The land surface and terrestrial vegetation are represented by a simplified version of the Hadley Center's MOSES land-surface scheme coupled to the dynamic vegetation model TRIFFID. Ocean carbon is simulated by means of an OCMIP-type inorganic carbon-cycle model (J. Orr, R. Najjar, C. Sabine, and F. Joos, Abiotic how-to document, 2000, available at *http://www.ipsl.jussieu.fr/OCMIP*) and a marine ecosystem model solving prognostic equations for nutrients, phyto-plankton, zooplankton, and detritus. The model has participated in a number of model intercomparison projects including the C4MIP, the Paleoclimate Modeling Intercomparison Project (PMIP), and the coordinated thermohaline circulation experiments. See Zickfeld et al. (2009) and references therein. The Bern model used in this study is the Bern2.5CC EMIC described in Plattner et al. (2008) and Joos et al. (2001); it is compared to other models in Plattner et al. It is a coupled climate-carbon cycle model of intermediate complexity that consists of a zonally averaged dynamic ocean model, a one-layer atmospheric energy-moisture balance model, and interactive representations of the marine and terrestrial carbon cycles.

CHAPTER 3.2

Table 3.2 and associated discussion in text:

This section refers to two theoretical estimates of climate sensitivity carried out using realistic CO_2 and water vapor radiative transfer based on the NCAR Community Climate Model radiation code, but employing an idealized vertical profile of temperature and humidity. The general approach is the same as that outlined in Chapter 4 of Pierrehumbert (2010). The temperature profile consists of a moist adiabat patched on to an isothermal stratosphere. The stratospheric water vapor mixing ratio was assumed vertically uniform, at a value equal to the mixing ratio at the tropopause. For the case of fixed water vapor content (no water vapor feedback) the tropospheric water vapor mixing ratio was held fixed at a value corresponding to 50% relative humidity *computed for the unperturbed temperature.* For the case including water vapor feedback, the mixing ratio was allowed to increase with temperature so as to hold the tropospheric relative humidity fixed at 50%. In both cases, the radiation calculation was done for clear-sky conditions.

4.5 TEMPERATURE EXTREMES

The steps of the analysis, which we applied to the grid point scale, are:

1. From the 22 CMIP3 models' runs available for 20C3M we extract annual values of average TAS in June-July-August and December-January-February.
2. We then form anomalies from the 1971-2000 mean and compute their distribution (i.e., a set of quantiles).
3. We choose a high quantile (95%, 100%) as benchmark against which to evaluate the change in likelihood of exceedances in a warmer climate.
4. We then superimpose spatial patterns of change in seasonal average temperature derived through pattern scaling for a series of representative changes in global average temperature. (Pattern scaling gives us a robust geographical pattern of seasonal temperature changes that scales linearly with values of global average temperature.) For each choice of global average temperature change, this will shift uniformly the quantiles of the distribution to the right. In our example below we choose additive patterns corresponding to 1°C, 2°C, and 3°C global average warming.
5. We finally compute what fraction of the newly derived distribution lays to the right of the chosen threshold/benchmark.

5.1 FOOD PRODUCTION, PRICES, AND HUNGER

Methods summary for food figure

Left-hand panel:

For broad regions, yield losses per °C of local warming were taken from Figure 5.2 in the Working Group 2 reports of the Fourth Assessment Report of the IPCC (Easterling et al., 2007). These estimates include estimates of CO_2 effects but without explicit modeling of adaptation. The mean and one standard error for each level of warming were approximated from the figure. Local temperature changes were converted to global temperature levels using a value of 1.5°C local per global °C for mid-to-high latitudes and 1.2°C local per global °C for low latitudes.

Note that since several of these studies are based on experiments where climate is allowed to equilibrate with doubled CO_2 levels, while others were taken from transient simulations (e.g., based on SRES scenarios), the CO_2 levels for different amounts of warming likely varied by study, with the equilibrium studies likely underestimating CO_2 levels for a given warming amount.

Right-hand panel:

Yield losses per °C of local warming were taken from the following studies: U.S. maize and soybean (Schlenker and Roberts, 2009); Asia rice (Matthews et al., 1995); India wheat (Lal et al., 1998); Africa maize (Schlenker and Lobell, 2010). For each region, global temperatures were converted to local temperature change based on the patterns in Section 4.2. The yield effects of higher CO_2 were estimated based on a recent meta-analysis (Ainsworth et al., 2008). CO_2 levels for each temperature value were based on the values reported in Section 3.2, and assuming a ratio of CO_2 to CO_2-equivalent equal to the average of the SRES scenarios (ratio is 1.05 for 1°C, 0.93 for 2°C, and 0.8 for 3°C and warmer).

Standard errors were estimated by propagating estimates of standard errors for (1) local temperature change for a given global temperature; (2) crop yield response to temperature; (3) CO_2 levels for a given global temperature; and (4) yield response at a CO_2 level. Propagation was done with the standard equations:

$$\mathrm{Var}(a \times b) = E[a]^2 \times \mathrm{Var}(b) + \mathrm{Var}(a) \times E[b]^2 + \mathrm{Var}(a) \times \mathrm{Var}(b) \quad [1]$$

$$\mathrm{Var}(f(X)) \approx (f'(E[X]))^2 \times \mathrm{Var}(X) \quad [2]$$

The shaded region in the figure corresponds to the likely range, which is defined as the 67% confidence interval or ± one standard error.

CHAPTER 6

Figure 6.1 and associated discussion in text:

The very long-term warming in Figure 6.1 was computed on the basis of the CO_2 concentrations at 1,000, 5,000 and 10,000 years in the LTMIP ensemble of carbon-cycle models (Archer et al., 2009). The minimum, median, and maximum climate sensitivity from Table 3.1 was applied to each member of the ensemble in order to produce the range of estimated warming. Only ensemble members that included sediment dissolution feedback were used in the calculation, but the ensemble includes simulations with and without climate feedback on carbon uptake. Because the climate feedback invariably increases the long-term CO_2 value, the no-feedback case defines the lower end of the estimated warming, corresponding to a case in which the climate feedback on uptake is negligible. The upper end of the warming is underestimated in this calculation, because the climate feedback should increase when a climate sensitivity higher than that used in the carbon-cycle model is applied, but this effect was not taken into account because there was no reliable methodology for doing so within the ensemble of published results. The LTMIP ensemble states results for 1,000 GtC and 5,000 GtC cumulative emissions. Temperatures for intermediate values of cumulative emissions were obtained by interpolating $\log(CO_2)$ linearly in the cumulative emissions, based on the geochemical principles laid out in Caldeira and Kasting (1993).

Appendix D

Acronym List

A1B	a "medium" emissions scenario; part of the SRES in the IPCC
AAO	Antarctic Oscillation
AAR	Accumulation Area Ratios
ACC	America's Climate Choices
ACIA	Arctic Climate Impact Assessment
AOGCM	Atmosphere-ocean General Circulation Model
AR5	Intergovernmental Panel on Climate Change, Fifth Assessment Report
ASO	August, September, October
BECS	bioenergy with carbon capture and sequestration
C4MIP	Coupled Carbon Cycle Climate Model Intercomparison Project
Ca^{2+}	calcium ions
$CaCO_3$	calcium carbonate
CALFED	Calfed Bay-Delta Program
CCSP SAP	Climate Change Science Program Synthesis and Assessment Product
CH_4	methane
CMIP3	Coupled Model Intercomparison Project phase 3
CO_2	carbon dioxide
CO_3^{2}	carbonate ions
CRU	Climate Research Unit (of the University of East Anglia)
D&A	detection and attribution
DIC	dissolved inorganic carbon
DJF	December, January, February

ECMWF	European Centre for Medium-Range Weather Forecasts
EMICs	Earth models of intermediate complexity
ENSO	El Niño/Southern Oscillation
EPA	Environmental Protection Agency
ESCM	Earth System Climate Model
FEMA	Federal Emergency Management Agency
FPN	fraction of positive minus negative estimates
GCM	general circulation model
GDP	gross domestic product
GFDL	Geophysical Fluid Dynamics Laboratory
GHG	greenhouse gas
GISS	Goddard Institute for Space Studies
GISS AGCM	GISS's Atmospheric General Circulation Model
GT	gigaton
GtC	gigaton of carbon (gigaton = 1×10^9 tons)
GWP	global warming potential
H^+	hydrogen ion
HadISST	Hadley Centre Global Sea Ice and Sea Surface Temperature
HFCs	hydrofluorocarbons
IMAGE	Integrated Model to Assess the Global Environment
IPCC AR3	Intergovernmental Panel on Climate Change, Third Assessment Report
IPCC AR4	Intergovernmental Panel on Climate Change, Fourth Assessment Report
IS92a	global scenario of the IPCC
ITCZ	Intertropical Convergence zone
JAS	July, August, September
JFM	January, February, March
JGOFS	Joint Global Ocean Flux Study
JJA	June, July, August
LIG	interglacial time period
LMSL	local mean sea level
LTMIP	Long Term Model Intercomparison Project

m	meters
MiniCAM	Mini-Climate Change Assessment Model
MOC	meridional overturning circulation
MSL	mean sea level
N_2O	nitrous oxide
NASA	National Aeronautics and Space Administration
NCDC	National Climatic Data Center
NH	Northern Hemisphere
NOAA	National Oceanic and Atmospheric Administration
NPCC	New York City Panel on Climate Change
NRC	National Research Council
NSIDC	National Snow and Ice Data Center
NYCOEM	New York City Office of Emergency Management
PCM	Parallel Climate Model
pCO_2	pressure of carbon dioxide
PDO	Pacific Decadal Oscillation
PDSI	Palmer Drought Severity Index
PETM	Paleocene-Eocene Thermal Maximum
PgC	petagram of carbon (petagram = 1×10^{15} grams)
pH	potentiometric hydrogen ion concentration, $pH = -\log[H^+]$
ppm	parts per million
ppmv	parts per million by volume
RCM	Regional Climate Model
RCP	representative carbon pathways
RF	radiative forcing
SAM	Southern (Hemisphere) Annular Mode
SAT	surface air temperature
SCD	snow cover duration
SLE	sea level equivalent
SLR	sea level rise
SPM	Summary for Policymakers (of IPCC)
SRES	Special Report on Emissions Scenarios
SST	sea surface temperature
SWE	snow water equivalent

TA total alkalinity
TCR transient climate response

U.S. United States
USCCSP United States Climate Change Science Program
USCCSP SAP Climate Change Science Program Synthesis and Assessment Product
USGCRP United States Global Change Research Program
UVIC University of Victoria (British Columbia, Canada) model

VIC Variable Infiltration Capacity

WG1, WG2 Working Group 1 or 2 (of IPCC)
WMGG well-mixed greenhouse gases
DWMO World Meteorological Organization
WOCE World Ocean Circulation Experiment
WWAP World Water Assessment Programme